THEORY OF DISPERSED MULTIPHASE FLOW

Publication No. 49
of the Mathematics Research Center
The University of Wisconsin–Madison

Academic Press Rapid Manuscript Reproduction

THEORY OF DISPERSED MULTIPHASE FLOW

Edited by

RICHARD E. MEYER
Mathematics Research Center
The University of Wisconsin
Madison, Wisconsin

Proceedings of an Advanced Seminar
Conducted by the Mathematics Research Center
The University of Wisconsin–Madison
May 26–28, 1982

1983

ACADEMIC PRESS
A Subsidiary of Harcourt Brace Jovanovich, Publishers

New York London
Paris San Diego San Francisco São Paulo Sydney Tokyo Toronto

ACADEMIC PRESS, INC.
111 Fifth Avenue, New York, New York 10003

United Kingdom Edition published by
ACADEMIC PRESS, INC. (LONDON) LTD.
24/28 Oval Road, London NW1 7DX

Library of Congress Cataloging in Publication Data
Main entry under title:

Theory of dispersed multiphase flow.

(Publication no. 49 of the Mathematics Research Center, the University of Wisconsin--Madison)
Includes index.
1. Multiphase flow--Congresses. 2. Dispersion--Congresses. I. Meyer, Richard E., Date
II. University of Wisconsin--Madison. Mathematics Research Center. III. Series: Publication of the Mathematics Research Center, the University of Wisconsin--Madison ; ho. 49.
QA3.U45 no. 49 [TA357] 510s [532'.051] 82-24404
ISBN 0-12-493120-0

PRINTED IN THE UNITED STATES OF AMERICA

83 84 85 86 9 8 7 6 5 4 3 2 1

Contents

Senior Contributors

Numbers in parentheses indicate the pages on which the authors' contributions begin.

Andreas Acrivos *(81), Department of Chemical Engineering, Stanford University, Stanford, California 94305*

George D. Ashton *(271), Cold Regions Research and Engineering Laboratory, Corps of Engineers, Department of the Army, Hanover, New Hampshire 03755*

James R. Brock *(135), Department of Chemical Engineering, The University of Texas at Austin, Austin, Texas 78712*

Barton Dahneke *(97), Eastman Kodak Company, and Department of Chemical Engineering, University of Rochester, Rochester, New York 14642*

Masao Doi *(35), Department of Physics, Tokyo Metropolitan University, Setagaya-ku, Tokyo, Japan*

Donald A. Drew *(173), Mathematics Department, Rensselaer Polytechnic Institute, Troy, New York 12181*

George M. Homsy *(57), Department of Chemical Engineering, Stanford University, Stanford, California 94305*

Roy Jackson *(291), Department of Chemical Engineering, University of Houston, Houston, Texas 77004*

David F. McTigue *(227), Sandia National Laboratories, Albuquerque, New Mexico 87185*

Jace W. Nunziato *(191), Sandia National Laboratories, Albuquerque, New Mexico 87185*

George Papanicolaou *(73), Courant Institute of Mathematical Sciences, New York University, New York, New York 10012*

Ronald E. Rosensweig *(359), Exxon Research and Engineering Company, Linden, New Jersey 07036*

William B. Russel *(1), Department of Chemical Engineering, Princeton University, Princeton, New Jersey 08544*

Stuart B. Savage *(339), Department of Civil Engineering and Applied Mechanics, McGill University, Montreal, Canada H3A 2K6*

Leen van Wijngaarden *(251), Technische Hogeschool Twente, Enschede, The Netherlands*

Preface

An advanced seminar on the motion of multiphase fluids was held in Madison in May 1982 by the Mathematics Research Center of the University of Wisconsin. This volume collects the addresses of the fifteen international experts who discussed the recent advances in this field, to which they have been leading contributors, and the state to which our knowledge on the subject has now been brought. Their topics range widely from solutions of long chain polymers in liquids to magnetic control of particle suspensions in fluid streams, from aerosols to dense granular flows and to ice crystals or vapor bubbles dispersed in river waters. Collectively, they give a remarkably comprehensive and balanced picture of the active field of scientific enquiry that has arisen from concern with a wide variety of issues affecting our daily lives.

The Mathematics Research Center is most of all indebted to the authors for the excellence of their contributions. It also records its debt to the United States Army, which sponsored the conference under its Contract No. DAAG29-80-C-0041, to the National Science Foundation, which supported it by Grants CPE-8203292 and MCS-7927062(1) under the direction of its Chemical and Process Engineering and Mathematical Sciences Divisions, respectively, and to the Exxon Research Foundation, which added to the support. The editor also wishes to thank Gladys Moran, for the handling of the conference details, and Judith Siesen, for assembling the volume and index.

Richard E. Meyer

Effects of Interactions between Particles on the Rheology of Dispersions

W. B. Russel

1. INTRODUCTION

Any theoretical treatment of multiphase flow which proposes to predict the details of a macroscopic process, e.g. the mixture theories described elsewhere in this volume, requires constitutive equations relating fluxes of momentum, mass, and energy to gradients in the local state variables. For concentrated suspensions of submicron particles in Newtonian fluids an ample body of literature demonstrates the profound effects of interparticle forces on the coefficients in such relations, i.e. viscosities, diffusivities, and sedimentation coefficients. Nonideal behavior frequently arises, in the sense that these parameters depend strongly on concentration while the relationship between stress and rate of strain often becomes nonlinear and history dependent. Unfortunately, in most experiments the complexity of the chemistry and the transport processes or incomplete characterization of either obscures the link between specific colloidal forces and the macroscopic effect.

A preceding review [1] of the rheology of colloidal suspensions assimilated experimental results from well-characterized model systems together with idealized theories which treat explicitly the colloidal and hydrodynamic forces. The comparison demonstrates the quantitative validity of the theories for some limiting cases and the qualitatively correct trends suggested under other conditions. This paper focusses

ISBN 0-12-493120-0

more narrowly on the rheology of stable suspensions, elaborating further on the relationship between the dilute, or pair interaction, regime and higher, or semi-dilute, concentrations. The earlier review and several others in the literature [2-6] should be consulted for more extensive references and a broader view of the subject.

2. QUALITATIVE INTERPRETATIONS

A dimensional analysis of the relevant forces (Table I) provides a qualitative appreciation for the behavior of colloidal systems. The presence or absence of non-Newtonian effects depends primarily on the rate of diffusion compared to the relative convection of spheres, or the rotation of rods, by an imposed shear flow. The ratio comprises a Peclét number which must become O(1) for the flow to disrupt significantly the suspension microstructure and produce a nonlinear and time-dependent rheology. For normal liquids or gases shear rates of 10^{10}-10^{11} s^{-1} are required [7]; hence these invariably behave as Newtonian fluids. In low viscosity media such as water, macromolecules or particles must attain dimensions of ~10^{-1} μm or greater for viscous effects to compete with Brownian motion at moderate shear rates and dilute concentrations. Conversely, hydrodynamic forces dominate the motion of particles larger than ~10 μm in water, making relaxation processes slow and equilibrium difficult to achieve. In unstable systems the aggregation of small particles into extensive networks or large flocs produces similar effects. Here we concentrate on stable suspensions in the intermediate, or colloidal, regime where both the near equilibrium state, at low shear rates or small strains, and the non-Newtonian phenomena at higher shear rates are observable.

Within this size range the character of the rheology varies with both the particle geometry and the relative strength of the non-hydrodynamic interactions. Rodlike particles, of course, generate non-Newtonian effects at infinite dilution since shear affects their orientation [8]. With rigid spheres, however, only relative positions are significant, so non-Newtonian stresses must emerge from interactions at finite concentrations. Clearly the colloidal stability of the suspension becomes critical; even among stable systems the nature of the repulsive force strongly influences the rheology. To illustrate this we consider three situations:

(a) neutral stability with $A/kT << 1$

(b) steric stability with $(1/2 - \chi)a\Delta^2/V_s >> 1$, and

(c) electrostatic stability with $\varepsilon\psi_0{}^2a/kT >> 1$.

The range of the repulsive potential distinguishes these three cases. Hard sphere repulsions with zero range, in principle, provide neutral stabilty. Steric interactions first appear at separations Δ comparable to the dimensions of the adsorbed macromolecules which are normally small relative to the particle radius [9]. The Debye length κ^{-1}, characterizing the range of electrostatic repulsions, varies substantially with ionic strength, however, from ~0.3 nm at 1.0M NaCl to ~1.0 μm in distilled water. Superposition of the London-van der Waals attraction with range on the order of the particle radius generates complete interaction potentials of the form shown in Figure 1.

Note that in good solvents the steric interaction overcomes the ubiquitous van der Waals attractions near contact to provide true thermodynamic stability; at separations beyond twice the layer thickness it decays quite rapidly to zero. With electrostatic interactions a deep primary minimum may persist, but a maximum much larger than kT imparts kinetic stability on time scales of practical interest. At separations on the order of the Debye length the potential decays exponentially as counterions comprising the electrical double layers effectively screen the surface charges.

The dimensional analysis indicates that for hard sphere repulsions the reduced shear viscosity μ/μ_0 should depend only on the volume fraction ϕ and the Peclêt number. Indeed the data of Krieger [10] for polystyrene latices, carefully adjusted to approximate hard sphere behavior, conclusively demonstrate this point. At $\phi = 0.50$ reduced viscosities for latices of several different particle sizes in several fluids collapse onto a single curve when plotted against the Peclêt number (Figure 2). Even at this high volume fraction the viscosity and shear-thinning effects remain modest, ranging from the low shear value of ~24 to the hydrodynamically dominated limit of ~12. Below $\phi \sim 0.20$ no shear thinning was detected.

These effects of Brownian motion and viscous forces alone provide the basis for gauging the effects of other interparticle forces.

TABLE I

Time Scales		Dimensionless Group
convection	γ^{-1}	-
diffusion	$\frac{6\pi\mu_0 a^3}{kT}$	$\frac{6\pi\mu_0 a^3}{kT}\gamma$
Interaction Energies		
thermal	kT	-
dispersion	A	$\frac{A}{kT}$
electrostatic	$\varepsilon\psi_0^2 a$	$\frac{\varepsilon\psi_0^2 a}{kT}$
steric	$(\frac{1}{2} - \chi)\frac{a\Delta^2}{V_s}kT$	$(\frac{1}{2} - \chi)\frac{a\Delta^2}{V_s}$
Length Scales		
radius	a	-
mean separation	$2a(\phi^{-1/3}-1)$	$2(\phi^{-1/3}-1)$
attraction	δ	δ/a
repulsion	κ^{-1}, Δ	$a\kappa, \Delta/a$

Note list of nomenclature on pp. 32-34.

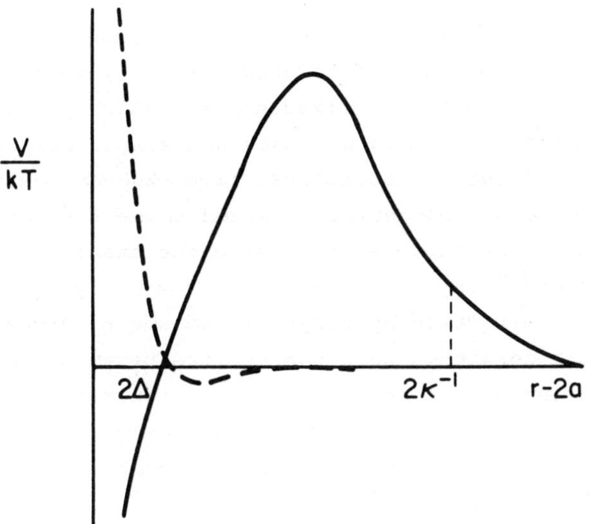

Figure 1. Form of total interaction potential for electrostatic (——) and steric (----) stability.

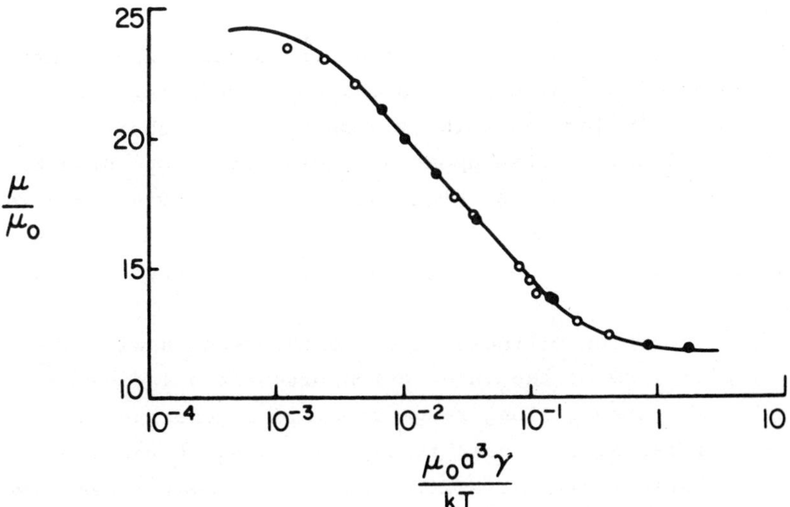

Figure 2. Relative viscosity vs. Peclét number for monodisperse polystyrene latices at ϕ = 0.50 in different fluids [10]: — water, o benzyl alcohol, ● m-cresol.

At low to moderate concentrations steric stabilization produces qualitatively similar effects. Willey and Macosko [11] noted the superposition of reduced viscosities of polyvinyl chloride plastisols at a fixed concentration provided $\chi < 1/2$, corresponding to good solvents and stable suspensions (Figure 3). The measured viscosities, from ~200 at low shear to ~30 at high shear, substantially exceed those for hard spheres, however. The authors attributed the stability to solvated chains extending into the fluid from a crystalline core, causing the effective hydrodynamic volume to exceed that based on the bulk density. Hard sphere-like behavior seems reasonable in light of the steepness of the interaction potential suggested by Figure 1, but the appearance of discontinuous dilatancy at considerably higher concentrations [11,12,13] limits the applicability of this analogy.

Aqueous suspensions stabilized electrostatically manifest more dramatic effects as a result of the long and variable range of the interparticle potential. The data of Krieger and Eguiluz [14] in Figure 4 reveals both quantitative and qualitative differences from the rheology produced by Brownian motion in concert with hard sphere-like repulsions. At high ionic strength the viscosities are low and vary only slightly with shear rate in accord with the hard sphere experiments. Reducing the ionic strength and thereby increasing the Debye length markedly increases the viscosity at low shear rates, eventually leading to an apparent yield stress or infinite zero-shear viscosity. At high shear rates the viscosity remains comparable to that for hard spheres, a tremendous shear thinning effect which no longer scales simply with the Peclét number.

Juxtaposition of these experimental results with the qualitative form of the interaction potentials indicates clearly that strong, long range repulsions generate much higher viscosities at low shear rates than do short range or hard sphere interactions. At sufficiently high shear rates, however, viscous forces overwhelm the colloidal forces thereby reducing the viscosity to a common level primarily dependent on the volume fraction. The following sections quantify these observations by first retracing analyses of pair interactions

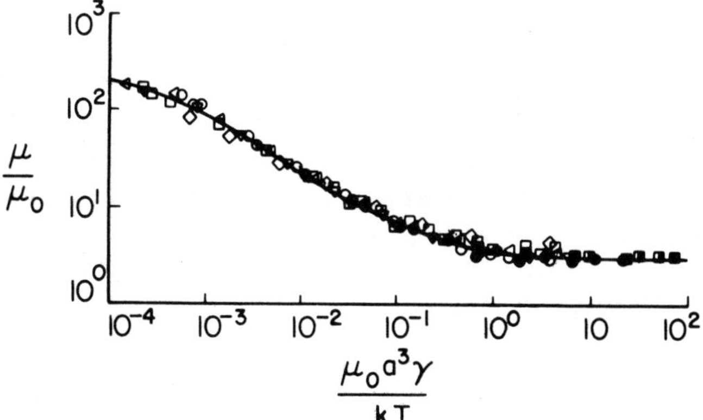

Figure 3. Relative viscosity vs. Peclét number for sterically stabilized monodisperse polyvinyl chloride spheres in several solvents at ϕ = 0.20 [11]: ◇ diethyl o-phthalate, χ = 0.38; □ di-n-butyl o-phthalate, χ = -0.02; ◁ di-n-hexyl o-phthalate, χ = -0.04; o di-2-ethylhexyl o-phthalate, χ = 0.06; ▽ di-2-ethylhexyl isophthalate. Shadings indicate different particle sizes.

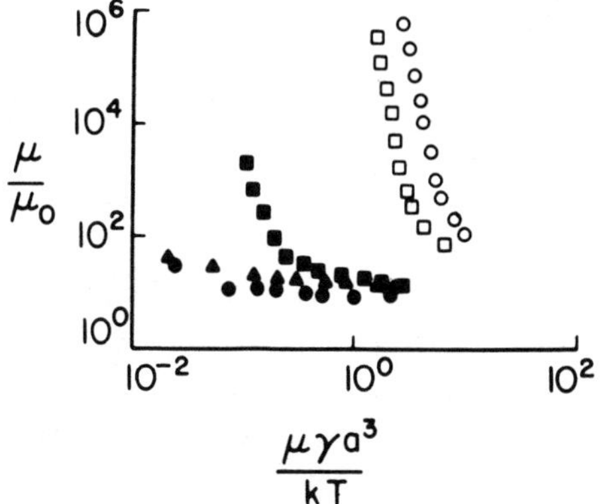

Figure 4. Effect of added electrolyte on the relative viscosity of polystyrene latex suspension with a = 0.11 μm at ϕ = 0.40 [14]: o deionized; □ 1.876 × 10^{-4} M HCl; ■ 1.876 × 10^{-3} M HCl; ● 1.876 × 10^{-2} M HCl; ▲ 9.378 × 10^{-2} M HCl.

which for dilute suspensions rigorously and accurately relate the effects of Brownian motion and electrostatic forces to the bulk rheology. Then we briefly examine the existing theories for concentrated suspensions of spheres and semi-dilute suspensions of rigid rods to illustrate the smooth transition from the dilute regime. With electrostatically stabilized suspensions, however, the qualitative change in behavior at low ionic strengths and finite concentrations necessitates a fundamentally different theory described in the final section.

3. PAIR INTERACTION THEORIES

General Formulation. Interactions between two particles at dilute concentrations comprise the simplest, meaningful problem in the context of colloidal phenomena. This level of complexity includes the nonhydrodynamic forces responsible for the non-idealities reviewed in the previous section yet preserves a configuration amenable to rigorous treatment of the hydrodynamics, colloidal forces, and statistical mechanics. Despite the superficial similarity to the calculation of second virial coefficients, this analysis poses greater difficulty and holds more interest for several reasons. First, non-equilibrium effects are essential, accounting for a substantial fraction of the zero-shear viscosity and all the non-Newtonian behavior at higher Peclét numbers. Secondly, all the colloidal forces contribute directly to the stress, in addition to altering the microstructure and thereby modulating the hydrodynamic portion. Finally, straightforward attempts to average the pair contributions invariably lead to non-convergent integrals which require proper renormalization. The resulting $O(\phi^2)$ term in the bulk stress varies with particle shape and interaction forces in a manner similar to the second virial coefficient, but also reveals much about the rheology pertinent to more concentrated suspensions.

The treatment requires that the length scale of the shearing motion be large compared to the average spacing between particles. Then the suspension behaves macroscopically as an effective continuum. Bulk properties are obtained by averaging over a statistically homogeneous representative volume V, which is small on the macroscale but large enough to contain many particles. Therein the shear flow is homogeneous

[15]. For dilute monodisperse suspensions affected only by pair interactions the microstructure within this volume is characterized fully by a pair density function $P(\underline{r}, \underline{q}_1, \underline{q}_2)$, specifying the probability of finding a second particle with orientation $\underline{q}_2$ at position $\underline{r}$ relative to a test particle with orientation $\underline{q}_1$. Normalization insures that

$$\frac{1}{V} \int_V P \, d\underline{r} d\underline{q}_1 d\underline{q}_2 = n \tag{1}$$

with n the average number density.

The pair density function must satisfy the conservation equation [8,16-19]

$$\frac{\partial P}{\partial t} + \nabla_{\underline{r}} \cdot \dot{\underline{r}} P + \sum_{i=1}^{2} \nabla_{\underline{q}_i} \cdot \dot{\underline{q}}_i P = 0 \tag{2}$$

with

$$\dot{\underline{r}} = \underline{U} - \underline{D}_t \cdot \nabla_{\underline{r}}(\ln P + V/kT)$$

$$\dot{\underline{q}}_i = \underline{\Omega}_i - \underline{D}_r \cdot \nabla_{\underline{q}_i}(\ln P + V/kT).$$

The boundary conditions

$$P \to n \quad \text{as} \quad |\underline{r}| \to \infty$$

$$\text{and} \quad P \, \dot{\underline{r}} \cdot \underline{n} = 0 \quad \text{at} \quad |\underline{r}| = 2a \qquad \text{(spheres)} \tag{3}$$

$$\text{or} \quad P(\dot{\underline{r}} + s_1\dot{\underline{q}}_1 + s_2\dot{\underline{q}}_2) \cdot \frac{\underline{q}_1 \times \underline{q}_2}{(1-(\underline{q}_1 \cdot \underline{q}_2)^2)^{1/2}} = 0 \quad \text{at} \ \rho = 2a \quad \text{(rods)}$$

specify a microstructure without long range order and zero flux at contact. This formulation applies to axisymmetric particles in general but the second boundary condition is specialized for the limits of spheres and slender rods. For the latter contact can occur at any distances s_1 and s_2 from the centers of mass where the perpendicular separation between the axes ρ equals the rod diameter.

The relative velocity of the centers of mass $\underline{U}$ and the rates of rotation $\underline{\Omega}_i$ due to the applied shear as well as the translational and rotary diffusion tensors, $\underline{D}_t$ and $\underline{D}_r$, depend on the configuration $(\underline{r}, \underline{q}_1, \underline{q}_2)$ because of hydrodynamic interactions. Solutions to the Stokes equations, e.g. in bispherical coordinates for spheres [16,20] and via slender body theory for rods [19], supply either exact or asymptotic representations. The interaction potential V encompasses long range electrostatic repulsions for which acceptable approximations are available [21]. Finally Brownian motion enters through the effective thermodynamic force $-kT\ \nabla_{\underline{r}} \ln P$ and torques $-kT\ \nabla_{\underline{q}_i} \ln P$ [8,16].

A general solution for P would lead to the full constitutive equation governing the suspension rheology. Unfortunately, the variable coefficients have precluded other than asymptotic solutions such as the weak flow expansion

$$P = P_0 + Pe \cdot P_1 + Pe^2 \cdot P_2 + \ldots \qquad (4)$$

for $Pe \ll 1$. The first term, the Boltzmann distribution $n \exp(-V/kT)$, represents the equilibrium between Brownian motion and the interparticle potential. It, together with the $O(Pe)$ term, determines the zero-shear viscosity and the linear viscoelastic response. The succeeding terms encompass nonlinear effects such as normal stresses and shear thinning in simple shear and the strain rate dependence of the elongational viscosity. The approach also can yield predictions of transient phenomena such as stress overshoot during startup and relaxation or recovery upon the cessation of flow. With colloidal systems, however, elastic effects generally are less pronounced than for polymeric fluids, but flocculated suspensions in particular are notorious for their time dependence [6].

Suitable averaging of this information on the microstructure determines the bulk stress in the suspension. The development below parallels that due to Batchelor and coworkers [15,16,20,22,23] and addresses only the deviatoric stresses of rheological import. Osmotic pressures are excluded. Thermodynamic stresses $\underline{\tau}$, due to Brownian motion and the potential interactions, remain important, nonetheless, and must be added to the mechanical stresses $\underline{\sigma}$ arising from viscous forces.

The latter satisfy the local momentum equation

$$\nabla \cdot \underline{\sigma} = -\sum_{i=1}^{N} \begin{cases} \int_{-\ell}^{\ell} \underline{f}_i(s)\,\delta(\underline{x}-\underline{x}_i-s\underline{q}_i)\,ds & \text{(rods)} \\ \\ \underline{F}_i\,\delta(\underline{x}-\underline{x}_i) & \text{(spheres).} \end{cases} \qquad (5)$$

$\underline{f}_i$ and $\underline{F}_i$ denote the total nonhydrodynamic forces acting on the ith particle, assigned to the center of a sphere but distributed along the axis of a rod. The corresponding bulk stress follows from volume averaging as

$$\langle\underline{\sigma}\rangle = \frac{1}{V}\int \underline{\sigma}\, dV = 2\mu_0\underline{E} + \frac{1}{V}\sum_{i=1}^{N}\underline{S}_i \tag{6}$$

with
$$\underline{S}_i = \int_{A_i} \underline{\sigma}\cdot\underline{n}(\underline{x}-\underline{x}_i)da - \underline{q}_i(\underline{q}_i\times\underline{T}_i)$$
$$\underline{T}_i = \int_{-\ell}^{\ell} s\underline{q}_i\times\underline{f}_i ds. \tag{7}$$

Further decomposition of the stresslet for pair interactions as

$$\underline{S}_i = \underline{S}_{i0} + \sum_{j\neq i}(\underline{S}_{ij}' + \underline{C}_{ij}^r\cdot\underline{T}_i + \underline{C}_{ij}^t\cdot\underline{F}_i) \tag{8}$$

identifies explicitly the dipoles induced in force- and torque-free particles by the applied shear, $\underline{S}_{i0}$ for an isolated particle plus $\underline{S}_{ij}'$ due to hydrodynamic interactions. The coupling tensors relating the stresslet to the applied force and torque decay to zero with increasing separation as

$$\underline{C}_{ij}^r \sim 1/r^3$$
$$\underline{C}_{ij}^t \sim 1/r^2,$$

indicating that non-hydrodynamic forces and torques contribute to the mechanical stress only through hydrodynamic interactions.

With (8) and the pair density function the bulk stress becomes

$$\langle\underline{\sigma}\rangle = n\int \underline{S}_{10} f(\underline{q}_1)d\underline{q}_1 + n\int (\underline{S}_{12}' + \underline{C}_{12}^r\cdot\underline{T}_1 + \underline{C}_{12}^t\cdot\underline{F}_1) P d\underline{r}d\underline{q}_1 d\underline{q}_2 \tag{9}$$

where
$$nf(\underline{q}_1) = \int P d\underline{r} d\underline{q}_2 \tag{10}$$

includes the effect of interactions on the orientation of the test particle.

As $r \to \infty$

$$\underline{S}'_{12}+\underline{C}^{r}_{12}\cdot\underline{T}_1 \sim \frac{4\pi}{3}\mu_0 \begin{cases} \ell^3 \varepsilon \underline{q}_1\underline{q}_1\underline{q}_1\cdot\underline{e}'(\underline{r},\underline{q}_2)\cdot\underline{q}_1 & \text{(rods)} \\ 5a^3\underline{e}'(\underline{r}) & \text{(spheres)} \end{cases} \tag{11}$$

with $\underline{e}'$ the disturbance rate of strain caused by an isolated particle. Unfortunately, $\underline{e}' \sim r^{-3}$, rendering the integral nonconvergent. Renormalization is straightforward, however, since the local and macroscopic rates of strain are related by

$$\begin{aligned} \underline{E} &= \frac{1}{V}\int \left(\underline{E} + \sum_{i=1}^{N} \underline{e}'(\underline{r}_i,\underline{q}_i)\right)dV \\ &= \underline{E} + n\int \underline{e}'(\underline{r},\underline{q}_2)f(\underline{q}_2)d\underline{r}d\underline{q}_2; \end{aligned} \tag{12}$$

hence, the volume average of $\underline{e}'$ must be zero. Subtraction of the appropriate multiple of zero from (9) then converges the second integral [20]. The integral of $\underline{C}_{12}{}^{t} \cdot \underline{F}_1$ always converges since $\underline{F}_1$ decays no slower than r^{-3}.

The perturbation to P caused by interactions decays as r^{-3}, making straightforward evaluation of f from (10) impossible as well. For spheres the normalization $\int P d\underline{r} = 1$ eliminates the problem, but for rods (1) does not suffice for evaluating the average orientation

$$\langle \underline{q}_1\underline{q}_1 \rangle = \int \underline{q}_1\underline{q}_1 P d\underline{r}d\underline{q}_2 d\underline{q}_1 \tag{13}$$

which appears in the first term of (9). Renormalization seems possible, but not yet completely certain for far-field interactions between slender rods [19]. The convergent expression obtained for $\langle \underline{q}_1\underline{q}_1 \rangle$ consists of an integral over the finite surface A_{12} enclosing the region in which $P \equiv 0$ due to the hard rod interaction potential, i.e. the excluded volume.

The thermodynamic stress, obtained from the change in the Helmholtz free energy induced by an arbitrary homogeneous deformation, is [16,23,24]

$$\langle \underline{\tau} \rangle = -\frac{1}{V}\sum_{i=1}^{N} \left(\underline{x}_i\underline{F}_i + \underline{q}_i\,\underline{q}_i \times \underline{T}_i\right)$$

$$\text{where } \underline{F}_i = -\nabla_{\underline{x}_i}(kT\ln P_N + V) \tag{14}$$

$$\underline{q}_i \times \underline{T}_i = -\nabla_{\underline{q}_i}(kT\ln P_N + V).$$

P_N is the N particle distribution function. The r^{-3} far-field behavior of P again frustrates direct integration at the pair interaction level. Batchelor [16] evaluated the Brownian force contribution by applying the divergence theorem to obtain

$$n \int \underline{x}_1 \underline{F}_1 P_N dC_N = -n\underline{I} \int P_N dC_N + n \int \left(\int_A \underline{x}_1 \underline{n} P_N d\underline{x}_1 - N \int_{A_{12}} \underline{x}_1 \underline{n} P_N d\underline{x}_1 \right) d\underline{q}_1 dC_{N-1} \quad (15)$$

and then noting that for $\underline{x}_1$ on the surface A bounding the representative volume

$$\int P_N d\underline{q}_1 dC_{N-1} = \frac{1}{V} \quad . \quad (16)$$

Since $\int P_N dC_N = 1$ and $\int_A \underline{x}_1 \underline{n}_1 d\underline{x}_1 = V\underline{I}$ the first two terms cancel, leaving only the convergent integral over the surface A_{12} defined earlier. The Brownian torque contribution reduces to

$$\int \underline{q}_1 \nabla_{q_1} f d\underline{q}_1 = 3\langle \underline{q}_1 \underline{q}_1 \rangle - \underline{I} \quad (17)$$

through (10) and the use of Green's theorem. Hence the renormalization developed for the stresslet suffices.

Before continuing it is instructive to examine the magnitude of the stress dipoles arising from the various forces for three models. Table II summarizes scalings based on the form of the bulk stress together with knowledge of far-field hydrodynamic interactions and the electrostatic force law. The excluded volume for the pair interaction illustrates the different scalings possible for thermodynamic and hydrodynamic interactions. The three cases span the possible orderings among the various hydrodynamic and thermodynamic effects for stable systems.

For hard spheres both the hydrodynamic and excluded volumes compare with the physical volume, $\sim a^3$. Hydrodynamic interactions are strong and long range producing a dipole,

$$n \int \mu_0 a^3 \underline{E} \left(\frac{a}{r}\right)^3 d\underline{r} \sim \phi \langle \underline{S}_{10} \rangle \ ,$$

of the same order of magnitude as the Brownian forces.

With charged spheres the balance between Brownian motion and the electrostatic force,

TABLE II

	V_{excl}	$\langle \underline{S}_{10} \rangle$	$\langle \underline{S}_{12}' \rangle$	$\langle \underline{S}^{Br} \rangle$	$\langle \underline{r}\nabla V \rangle$
hard spheres	a^3	$\mu_0 a^3 \underline{E}$	$\mu_0 a^3 \underline{E}^3 n a^3$	$\mu_0 a^3 \underline{E} n a^3$	0
charged spheres	L^3	$\mu_0 a^3 \underline{E}$	$\mu_0 a^3 \underline{E} n a^3$	0	$\mu_0 a L^2 \underline{E} n L^3$
rigid rods	$a\ell^2$	$\mu_0 \ell^3 \varepsilon \underline{E}$	$\mu_0 \ell^3 \varepsilon^2 \underline{E} n \ell^3$	$\mu_0 \ell^3 \varepsilon E$	0

$$1 \sim \alpha\, e^{-\kappa L}/\kappa L$$

with $\alpha = 4\pi\varepsilon\psi_0^2 a^3 (a\kappa) e^{2a\kappa}/kT$,
determines the characteristic separation

$$L \sim \kappa^{-1} \ln(\alpha/\ln(\alpha/\ln\alpha))$$

which scales the excluded volume and the stress dipole [22]. At low ionic strengths $L >> a$ so that the thermodynamic effects act at much longer range than the hydrodynamic. Consequently, $\langle \underline{r}\nabla V \rangle \sim (L/a)^5 \phi \langle \underline{S}_{10} \rangle$ and Brownian stresses disappear in the absence of near-field hydrodynamic interactions.

Rigid rods provide the counter-example with the hydrodynamic volume, $\varepsilon \ell^3$, greatly exceeding both the excluded volume, $a\ell^2$, and the physical volume, $a^2\ell$. Hydrodynamic interactions are weak but long range so $\langle \underline{S}_{12}' \rangle \sim \varepsilon n \ell^3 \langle \underline{S}_{10} \rangle$ with $\varepsilon << 1$. Unlike spheres, the orientational degrees of freedom with rods lead to Brownian torques and stresslets without interactions, i.e. $\sim \langle \underline{S}_{10} \rangle$. Excluded volume effects only become important for interaction potentials having a range comparable to ℓ.

Detailed Results. For interactions between spheres orientations are irrelevent and the O(Pe) expansion becomes [16]

$$P(\underline{r}) = n \exp(-V/kT)\left(1 - \frac{6\pi\mu a^3}{kT}\frac{\underline{r}\cdot\underline{E}\cdot\underline{r}}{r^2}\, g(r)\right) \tag{18}$$

with $g(r)$ satisfying

$$\frac{d}{dr} r^2 G(r) \frac{dg}{dr} - r^2 \frac{d}{dr}\left(\frac{V}{kT}\right) G(r) \frac{dg}{dr} - 6H(r)g = \frac{1}{a^2}\frac{d}{dr} r^3 A(r)$$

$$- \frac{3r^2}{a^2} B(r) - (1-A(r))r^3 \frac{d}{dr}\left(\frac{V}{kT}\right) . \tag{19}$$

A, B, G, and H are known functions coming from $\underline{U}$ and $\underline{D}_t$. The boundary conditions reduce to

$$\lim_{r\to 2a} \left(r^2 G(r) \frac{dg}{dr} - (1-A)r\right) = 0 \tag{20}$$

and $g \to 0$ as $r \to \infty$.

The righthandside of (19) describes the action of flow on the spherically symmetric equilibrium distribution. Vorticity plays no role, but straining causes pairs to accumulate where $\underline{r} \cdot \underline{E} \cdot \underline{r} < 0$ and depletes the population where $\underline{r} \cdot \underline{E} \cdot \underline{r} > 0$.

For hard spheres with full hydrodynamic interactions Batchelor [16] reported numerical solutions for g and determined the zero shear viscosity to be

$$\frac{\mu}{\mu_0} = 1 + 2.5\phi + 6.2\phi^2 + O(\phi^3) \quad . \tag{21}$$

The bulk of the $O(\phi^2)$ coefficient comes from the viscous stresses associated with the uniform equilibrium state, leaving a smaller contribution (0.99) to the Brownian stresses generated by the non-equilibrium pair distribution function. As illustrated in Figure 5 this prediction agrees reasonably well with data for polystyrene latices [25], particularly in light of the difficulty in suppressing long range potentials.

As noted above a strong electrostatic potential of the form

$$V = \varepsilon\psi_0^2\, a^2 \exp[-\kappa(r-2a)]/r \tag{22}$$

excludes pairs from separations less than L when $\alpha \gg 1$. At larger separations Brownian motion restores $P_0 \sim n$. If $L/2a \gg 1$ near-field hydrodynamic interactions become negligible, leaving $G \sim H \sim 1$ and $A \sim B \sim 0$ in (19) and thereby removing most of the troublesome variable coefficients.

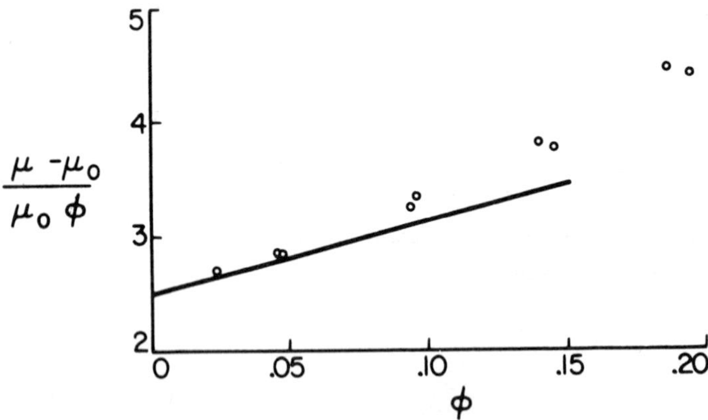

Figure 5. Reduced low shear viscosity of dilute polystyrene latex suspension with a = 0.42, 0.87 μm o [25] compared with theory of Batchelor [16] ——.

Under such conditions an asymptotic expansion for g has been constructed, with a thin transition layer at $r \sim L$ matched to the electrostatically dominated inner region and the diffusion dominated outer region [22]. The resulting Newtonian zero shear viscosity is

$$\frac{\mu}{\mu_0} = 1 + 2.5\beta\phi + 2.5(\beta\phi)^2 + \frac{3}{40}\left(\frac{L}{a}\right)^5 \phi^2 + O(\phi^3). \quad (23)$$

The coefficient $\beta > 1$, available as a function of $a\kappa$ and $e\psi_0/kT$ from Sherwood [26], accounts for the primary electroviscous effect associated with deformation of the electrical double layer about an isolated sphere. Residual far-field hydrodynamic interactions produce the first $O(\phi^2)$ term. But the dominant term is the non-equilibrium contribution from electrostatic interactions which conforms to the $(L/a)^5$ dependence expected from the scaling of the stress dipoles. Data for polystyrene latices [27] and bovine serum albumin [28] in Figure 6 confirm the large magnitude of the coefficient at low ionic strengths, $> 10^3$, and the accuracy of the theory over the full range.

The theory for rigid rods [19] is asymptotic in the slenderness ratio $\varepsilon \ll 1$ and requires two additional approximations to overcome the complexity of seven independent configurational variables. Hydrodynamic interactions are characterized to $O(\varepsilon)$ via Faxen's law, i.e. the first reflection, and reduced to their far-field form to provide tractable coefficients in (2). Decomposition of the pair density as

$$P = n\left(f_0(\underline{q}_1)f_0(\underline{q}_2) + \varepsilon g(\underline{r},\underline{q}_1,\underline{q}_2)\right) + O(\varepsilon^2) \quad (24)$$

and substitution into (2) generates on $O(1)$ homogeneous equation for f_0, the single particle orientation function, and an $O(\varepsilon)$ inhomogeneous equation for g, embodying the interactions. The forcing terms in the latter depend on f_0 and the hydrodynamic interactions. The known weak flow expansion for f_0 [29] thus permits the corresponding expansion for g to be determined analytically.

The result to $O(Pe^3)$ is a far-field solution with $g \sim O(\ell/r)^3$, including no effect of excluded volume. Evaluation

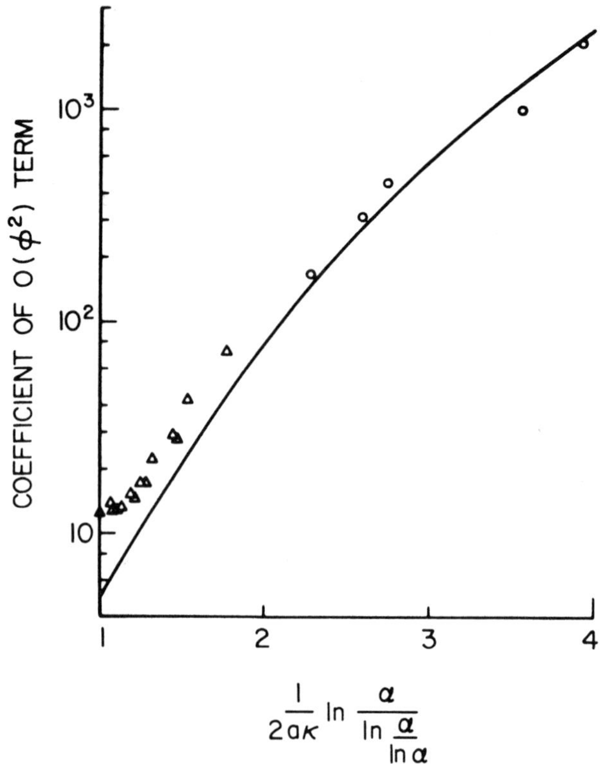

Figure 6. Coefficient of O(ϕ^2) term in low shear viscosity for polystyrene latices o [27] and bovine serum albumin Δ [28] compared with theory including electrostatic interactions [22].

of the renormalized bulk stress, however, requires integration over the surface A_{12} bounding the excluded volume. We assume the solution to remain valid to contact and approximate the awkward parallelopiped comprising the actual A_{12} with a sphere whose radius proves irrelevant. Despite these approximations and the uncertainty remaining about our renormalization of f the results described below appear to be a reasonable O(1) approximation to interactions between rigid rods.

For simple shear at rate γ the viscosity and normal stress coefficients are predicted as

$$\frac{\mu}{\mu_0} = 1 + [\eta]n + k[\eta]^2 n^2 + \ldots$$

$$\frac{N_1}{\mu_0\gamma} = \frac{1}{4} \, Pe[\eta]_0 n(1 + 0.73\,[\eta]_0 n + \ldots) \tag{25}$$

$$\frac{N_2}{\mu_0\gamma} = -\frac{1}{28} \, Pe[\eta]_0 n(1 - 1.63[\eta]_0 + \ldots)$$

with $[\eta] = [\eta]_0(1 - 0.020\, Pe^2 + \ldots)$

$$k = \frac{2}{5}\,(1 - 0.39\, Pe^2 + \ldots)$$

$$[\eta]_0 = \frac{16\pi\ell^3}{45}\,\varepsilon \tag{26}$$

$$Pe = \frac{3kT}{8\pi\mu_0\ell^3\varepsilon}\,(\underline{E}:\underline{E})^{1/2} \; .$$

Alignment of the isolated rods by the shear decreases the shear stress at $O(n)$ and, hence, the effective hydrodynamic volume $[\eta]$, but in the process generates non-zero normal stresses. The shear thinning at $O(n^2)$ reflects this, but also indicates a larger effect due to further alignment of the rods by hydrodynamic interactions.

For steady elongation the Trouton viscosity

$$\frac{\eta}{\mu_0} = 3(1 + [\eta]_T n + k_T[\eta]_T^2 n^2 + \ldots) \tag{27}$$

characterizes the fluid response with

$$[\eta]_T = [\eta]_0(1 + \frac{5}{56}\, Pe + \frac{1}{560}\, Pe^2 + \ldots)$$

$$k_T = \frac{2}{5}\,(1 + 0.282\, Pe + 0.038\, Pe^2 + \ldots) \; . \tag{28}$$

Here increased alignment with the flow produces a larger stress with the $O(n^2)$ term increasing somewhat faster than the $O(n)$.

Viscometric data on dilute suspensions of truly rigid rods without electroviscous effects are rare. Xanthan gum, with molecular weights of 10^3-10^4 kg/mole, at moderately high salt concentrations (~0.1 M NaCl) behaves much like an uncharged rigid rod [30] although some flexibility undeniably exists [31]. Figure 7 illustrates the interpretation of data from Chauvateau [32] via the rigid rod theory. Molecular

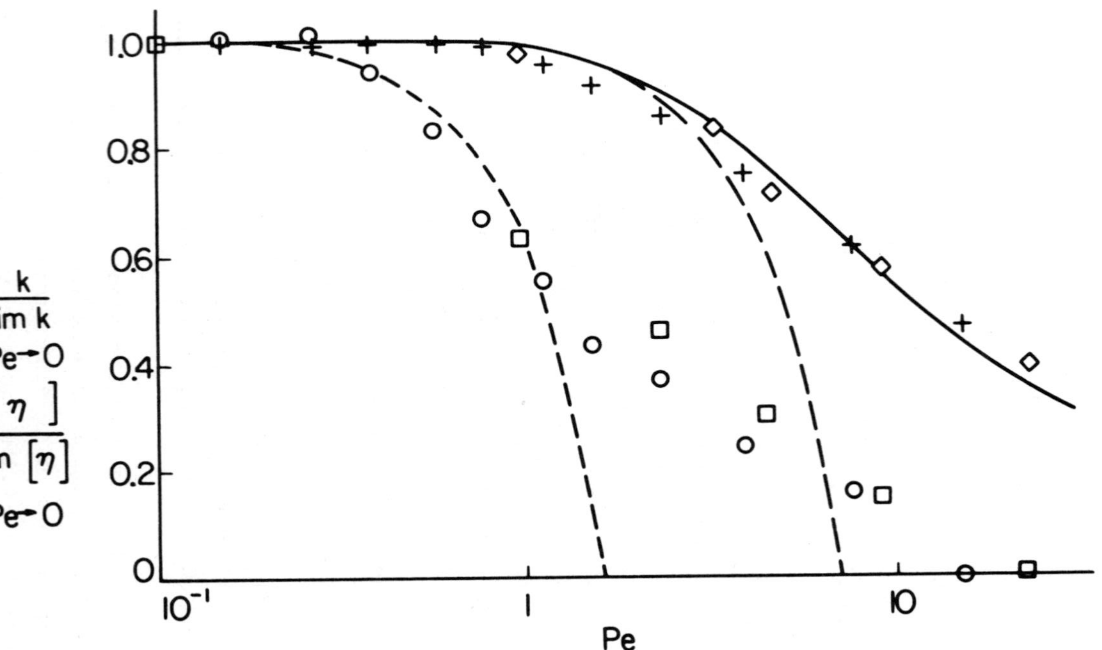

Figure 7. Data for dilute solutions of Xanthan gum in 0.1 M NaCl [32], [η] (+,◇) and k (o,□), compared with theories for rigid rods, exact (——) [33] and $O(Pe^2)$ expansions (---).

parameters extracted from the zero shear limit and shear rate dependence of [η] [19] differ from those measured by light scattering or deduced from the molecular structure, suggesting either flexibility or poly-dispersity. Nonetheless, the complete theory [33] tracks the data for [η] remarkably well over the entire range of shear rates. The a priori comparison between (26) and the measured Huggins coefficient reveals agreement within 10% in the low shear limit and a reasonable prediction of the shear-rate dependence up to Pe ~ O(1) where the weak flow expansion fails.

At this point one can safely conclude that pair interaction theories based on classical descriptions of hydrodynamic, Brownian, and colloidal forces can predict quantitatively the rheological properties of dilute suspensions. The story is by no means complete, however, since many interesting effects remain, particularly for strong flows and other types of forces such as London-van der Waals attractions and those resulting from polymer-particle interactions [34].

4. THE TRANSITION FROM DILUTE TO CONCENTRATED

One rationale for studying pair interactions is to gain insight into the behavior of concentrated suspensions. Hence one might inquire into the relationship between the dilute theories and either measurements of or theories for the viscosity of concentrated suspensions.

For hard sphere interactions the zero shear viscosities measured by Saunders [25] and Krieger [10] for polystyrene latices, combined in Figure 8, define the concentration dependence for $0.01 < \phi < 0.50$. The $O(\phi^2)$ theory remains reasonable up to $\phi \sim 0.20$-0.25 where $\mu/\mu_0 \sim 1.75$-2.00. No existing theory treats both the many body hydrodynamics and the microstructural effects satisfactorily beyond this point. The lubrication analysis of Frankel and Acrivos [35] constructs a reasonable picture of the hydrodynamics by focussing on the energy dissipated in the small gap between spheres near closest packing to determine

$$\frac{\mu}{\mu_0} = \frac{9}{8}\,\frac{(\phi/\phi_m)^{1/3}}{1-(\phi/\phi_m)^{1/3}} \,. \tag{29}$$

The packing fraction ϕ_m is left as an adjustable parameter. Unfortunately their treatment of the microstructure as a regular array and neglect of Brownian effects produces a concentration dependence inappropriate for colloidal systems as seen in Figure 8. The need for a more appropriate theory, recognizing the random microstructure at equilibrium and the perturbation from equilibrium due to shear, is clear.

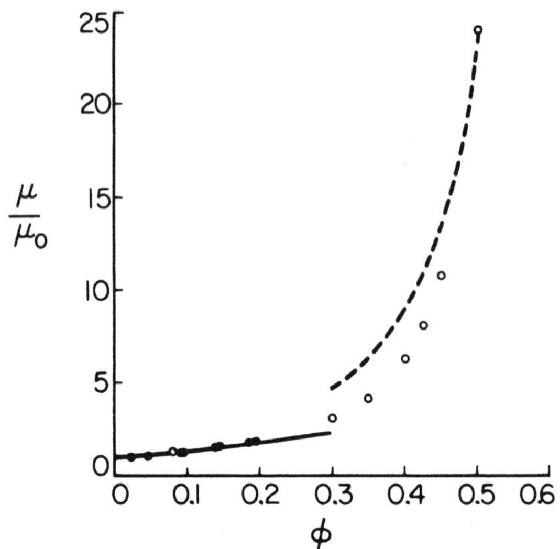

Figure 8. Zero shear viscosity for suspensions of hard spheres from pair interaction theory (——) [16] and lubrication theory (----) [35] compared with data for polystyrene latices from Krieger (o) [10] and Saunders (●) [25].

For rigid rods at semi-dilute concentrations, $\ell^{-3} \ll n \ll (a^2\ell)^{-1}$, Doi and Edwards [36] derived a physically appealing theory appropriate for the weak flows addressed here. At these concentrations an individual rod attempting to rotate encounters many neighbors at separations small compared with ℓ. By treating these interactions as contraints on the motion of the test rod they deduced the effective rotary diffusion coefficient to be $\beta \underline{D}_{r0}/(8n\ell^3)^2$. β is an unknown constant. Calculation of the orientational distribution function and the stresses then proceeds as at infinite dilution, although some controversy still remains with respect to the latter [37,38]. For Pe << 1 they obtained

$$\frac{\mu}{\mu_0} = \frac{36}{5\pi^2 \beta\varepsilon^2} \left(\frac{4\pi}{3} \varepsilon n \ell^3\right)^3 + O(Pe^2)$$
$$\frac{N_2}{\mu_0 \gamma} = -\frac{7}{2}\frac{N_2}{\mu_0\gamma} = \frac{144}{5\pi^4 (\beta\varepsilon)^2} \left(\frac{4\pi}{3} \varepsilon n \ell^3\right)^4 Pe. \qquad (30)$$

Figure 9 compares the pair interaction and semi-dilute theories with Chauvateau's data for Xanthan gum [32]. In this case the dilute theory suffices for $\mu/\mu_0 < 2-4$, while above $\mu/\mu_0 \sim 30$ the semi-dilute theory appears appropriate, if $\beta\varepsilon^2 \sim 210$. This implies $\beta \sim 10^4$, considerably larger than the O(1) value expected by Doi and Edwards [36] but reasonably consistent with recent dynamic light scattering measurements [39]. These results suggest a smooth transition from the upper limit of the dilute domain at $[\eta]_0 n \sim 1.5-1.75$ to the lower limit of the semi-dilute regime at $[\eta]_0 n \sim 5-6$.

The results for hard spheres and rigid rods together suggest that $O(\phi^2)$ theories for the Newtonian low shear viscosity remain accurate up to $k[\eta]_0 n = 6.2\phi^2/2.5\phi \sim 0.6$, followed by a smooth transition to a regime which is still Newtonian but has a much stronger dependence on concentration. For suspensions of charged spheres, however, the data of Krieger and Eguiluz [14] in Figure 4 and 10 reveals a distinctly different behavior at low ionic strengths or high volume fractions. In both cases there is little or no evidence of a Newtonian regime at low shear rates. Instead the viscosity diverges at a finite stress, indicating a solid-like equilibrium state.

This phenomenon does not contradict the pair interaction analysis and indeed can be explained qualitatively by the scaling arguments presented above [1]. By the criterion established for hard spheres and rods the range of validity of the pair interaction theory $\phi < 0.2(a/L)^5$ shrinks to zero as L/a becomes substantial at low ionic strengths. In fact, at finite ϕ the characteristic separation L can easily exceed the mean spacing in the suspension, $\sim 2a/\phi^{1/3}$. If $L/2a >> \phi^{1/3}$ multiparticle electrostatic interactions become sufficiently strong relative to Brownian motion to localize the particles on lattice sites corresponding to minima in the total potential energy [40,41]. The material then responds as a solid with an elastic modulus and yield stress dependent on the many-body electrostatic interactions. Prediction of these

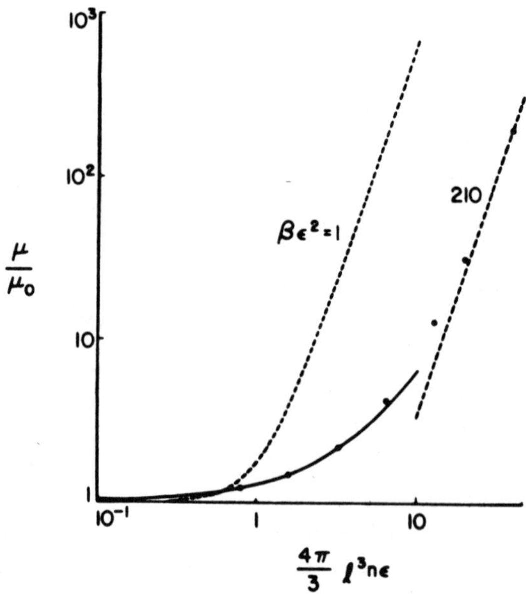

Figure 9. Zero shear limit of relative viscosity for suspensions of rigid rods from pair interaction theory [19] and semi-dilute theory [36] compared with data for Xanthan gum [32].

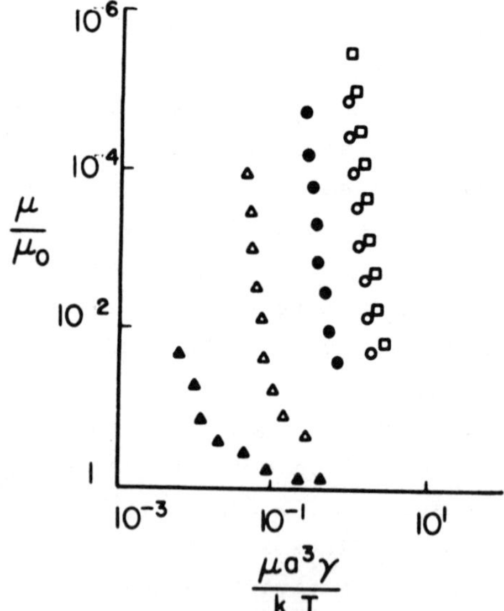

Figure 10. Effect of volume fraction on relative viscosity of deionized polystyrene latex suspensions with a = 0.213 μm [14]: ▲ ϕ = 0.05, △ ϕ = 0.010, ● ϕ = 0.20, o ϕ = 0.30, ■ ϕ = 0.40.

properties requires a substantially different approach outlined in the next section.

5. ELECTROSTATICALLY CONCENTRATED SUSPENSIONS

Several theoretical approaches have been adapted to calculate the thermodynamical and mechanical properties of electrostatically ordered systems. The quantities of interest are the order-disorder transition plus the osmotic pressure, shear modulus, dynamic viscosity, and yield stress of the ordered state. Methods borrowed from the statistical mechanics of dense liquids or ordered solids [42-45] generally assume pairwise additivity for the electrostatic potentials. The omission of counterions associated with the surface charges on the spheres in the computation of the Debye length used in this approach is a serious deficiency. The error is significant at the transition [46,47] and becomes larger with decreasing ionic strength or increasing concentration.

A self-consistent field theory [48] circumvents both problems through an ad hoc, but reasonable, mean-field treatment of the multiparticle interactions. A priori predictions follow for the equilibrium and viscoelastic properties of the ordered system in terms of the particle size, charge, and volume fraction and the amount of excess electrolyte. Agreement with data for polystyrene latices, while not perfect, is at least semi-quantitative [48]. Furthermore, the formalism identifies the correct way of adapting the pairwise additive approach to electrostatically concentrated suspensions.

The important dimensionless parameters which emerge from the theory are ϕ plus the surface charge $Q = eazq/\varepsilon kT$ and the concentration of excess electrolyte $N_0 = e^2a^2z^2n_0/\varepsilon kT$. Together with the dimensionless stress $e^2a^2z^2\underline{\sigma}/\varepsilon(kT)^2$ these highlight the strong particle size dependence of the phenomena [14,48]. For example, with a ~ 0.1 μm the characteristic ionic strength of 10^{-5}M should be associated with shear moduli of $\sim 10^2$ N/m^2 and osmotic pressures of ~10 mm H_2O. At 1.0 μm, however, the ionic strength falls below that of distilled water and the stresses would be too small to measure. The quantitative predictions and the experimental data confirm these expectations. The moduli and pressures attain these magnitudes and become independent of added electrolyte when $N_0 < Q\,\phi/(1-\phi)$, but decay to zero when $N_0 > Q\,\phi/(1-\phi)$, i.e. when the excess electrolyte exceeds the counterion concentration. The dynamic viscosity, on the other hand, remains comparable to that of the fluid and insensitive to the electrostatics.

The contrast with the steady shear experiments is dramatic. With small amplitude oscillations the large electrostatic energies are recoverable and primarily affect the shear modulus, leaving only the modest viscous stresses in the fluid for the dynamic viscosity. Steady deformation, however, forces the structure to yield, dissipating the electrostatic energy and producing enormous viscosities at low shear rates. Other studies [49] have found correlations between the elastic modulus and the yield stress characterizing the onset of flow, as one would expect from this interpretation.

As noted above the work also presents the possibility of developing a satisfactory pairwise additive treatment for concentrated systems in general. When solving for the local electrostatic potential ψ, we linearize about the mean potential in the fluid $\langle\psi\rangle$ so that

$$\psi = \langle\psi\rangle + \psi' \qquad (31)$$

$$\text{and} \quad n_k = c_k e^{-\frac{ez_k\langle\psi\rangle}{kT}}\left(1 - \frac{ez_k\psi'}{kT} + \ldots\right).$$

For a closed system containing a single z-z electrolyte added at concentration n_0 based on the total volume of suspension, the integration constants appearing in the Boltzmann distributions for the local ion densities follow directly from integration over the fluid volume together with the electroneutrality condition as

$$c_+ = \left(n_0 - \frac{3\phi q}{ae}\right) e^{\frac{ez\langle\psi\rangle}{kT}} / (1-\phi)$$

$$c_- = n_0 e^{-\frac{ez\langle\psi\rangle}{kT}} / (1-\phi). \qquad (32)$$

The spheres are taken as negatively charged, i.e. $q < 0$.

Substitution of the n_k into Poisson's equation then yields

$$\nabla^2\psi' = \kappa^2\psi' - \rho_0/\varepsilon$$

$$\text{with} \quad \kappa^2 = \frac{e^2z^2}{\varepsilon kT} \cdot \frac{2n_0 - \frac{3\phi q}{ae}}{1-\phi} \tag{33}$$

$$\rho_0 = -\frac{3\phi}{1-\phi}\frac{q}{a} .$$

Note that the mean potential $\langle\psi\rangle$ proves irrelevant and that no statement was needed about the microstructure. ρ_0 represents a uniform distribution of counterions throughout the fluid. The modified ionic strength appearing in κ^2 accounts for counterions [46,47] plus the reduction in the fluid volume due to the presence of solid particles.

Construction of a pairwise additive theory requires solution of (33) for two interacting spheres with either the potential or the charge density specified on their surfaces and $\psi' \sim \rho_0/\varepsilon\kappa^2$ in the bulk. The homogeneous solution to (33) is exactly that obtained for pair interactions at infinite dilution except that the modified κ^2 includes multiparticle interactions in a mean field sense. Since $\rho_0/\varepsilon\kappa^2$ does not vary with separation it has no affect on the potential energy of interaction. Hence only κ^2 need be modified in existing pairwise additive theories.

The implementation of this correction within a lattice model with face-centered-cubic symmetry is illustrated in Figure 11 [48]. The shear modulus is derived from the interaction potential via [42]

$$G' = 0.439\,\frac{\phi^{1/3}}{a}\,\frac{d^2V}{dx^2}(x_0) \tag{34}$$

$$\text{with} \quad x_0 = 2a\left(\left(\frac{0.74}{\phi}\right)^{1/3} - 1\right) \quad \text{and}$$

$$\frac{d^2V}{dx^2} = \frac{2\pi\varepsilon\psi_0^2}{a}\,\frac{e^{-\kappa x_0}}{1+x_0/2a}\left((a\kappa)^2 + \frac{a\kappa}{1+x_0/2a} + \frac{1}{2(1+x_0/2a)^2}\right). \tag{35}$$

This superposition approximation for V [21] errs by less than 10% for these conditions, i.e. $\kappa x_0 > 3$ [50]. As described previously [48] the surface potential relative to the bulk is extracted from electrophoresis measurements at infinite dilution but at the effective ionic strength corresponding to (33).

The results of the unmodified theory are clearly unsatisfactory, especially with respect to the effect of added electrolyte (Figure 11b). Recognition of the appropriate electrostatic conditions in the concentrated system, however, produces predictions in better accord with both the data and the self-consistent model. The values reported for the modified theory here differ from those in the original paper [48] due to a numerical error in the latter.

6. SUMMARY

This review attempts to illustrate the status of what might be called microstructural theories for the rheology of colloidal suspensions by examining three systems: hard spheres, charged spheres, and rigid rods. In each case predictions from rigorous pair interaction theories [16,19,22] agree nicely with data for dilute suspensions conforming to the assumptions of the theory. At higher concentrations the hard sphere and rigid rod suspensions enter a regime in which the zero shear viscosity remains Newtonian but increases much more rapidly with concentration. For rigid rods a satisfactory theory for these semi-dilute concentrations appears to exist [36]. Ultimately both systems undergo a disorder-to-order transition with profound rheological consequences discussed in part elsewhere in this volume [51].

For charged spheres at low ionic strengths long range electrostatic interactions produce ordered suspensions at quite low volume fractions. The yield stress characterizing the zero shear limit lies beyond the capability of existing theories, but a self-consistent field model [48] predicts the linear viscoelastic response satisfactorily. Furthermore it may form the basis for accurate pairwise additive theories for electrostatic effects in concentrated but disordered dispersions.

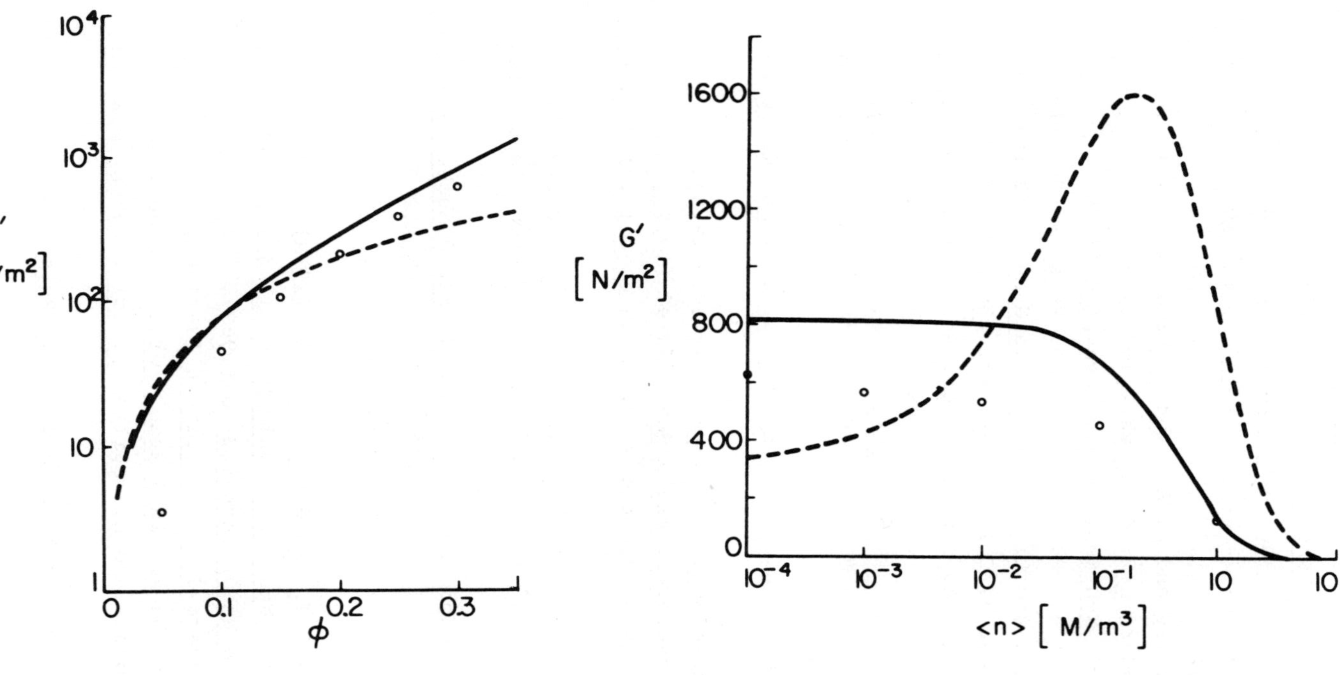

Figure 11a, 11b. Comparison of predictions from the pair-wise additive theory with measured values of shear modulus (o) for ordered polystyrene latex with a = 0.53 μm and $e\psi_o/kT$ = 2.20 at infinite dilution: (a) without added electrolyte, (b) effect of added electrolyte at ϕ = 0.30. (---- unmodified theory [42], —— modified theory [48]).

REFERENCES

1. W. B. Russel J. Rheol. 24 287 (1980).
2. G. K. Batchelor Ann. Rev. Fluid Mech. 6 227 (1974).
3. J. Mewis and A. J. B. Spaull Adv. Coll. Inter. Sci. 6 173 (1976).
4. D. J. Jeffrey and A. Acrivos AIChE J. 22 417 (1976).
5. J. D. Goddard S.M. Archives 2 403 (1977).
6. J. Mewis J. Non-Newt. Fluid Mech. 6 1 (1979).
7. D. J. Evans and R. O. Watts Chem. Phys. 48 321 (1980).
8. E. J. Hinch and L. G. Leal J. Fluid Mech. 52 683 (1972).
9. D. H. Napper J. Coll. Inter. Sci. 58 390 (1977).
10. I. M. Krieger Adv. Coll. Inter. Sci. 3 111 (1972).
11. S. J. Willey and C. W. Macosko J. Rheol. 22 525 (1978).
12. T. A. Strivens J. Coll. Inter. Sci. 57 476 (1976).
13. I. Wagstaff and C. E. Chaffey J. Coll. Inter. Sci. 59 53, 63 (1977).
14. I. M. Krieger and M. Eguiluz Trans. Soc. Rheol. 20 29 (1976).
15. G. K. Batchelor J. Fluid Mech. 41 545 (1970).
16. G. K. Batchelor J. Fluid Mech. 83 97 (1977).
17. T. G. M. van de Ven and S. G. Mason J. Coll. Inter. Sci. 57 505, 517 (1976).
18. L. A. Spielman J. Coll. Inter. Sci. 33 562 (1970).
19. D. H. Berry and W. B. Russel J. Fluid Mech. (submitted 1982).
20. G. K. Batchelor and J. T. Green J. Fluid Mech. 56 375, 401 (1972).
21. L. N. McCartney and S. Levine J. Coll. Inter. Sci. 30 345 (1969).
22. W. B. Russel J. Fluid Mech. 85 209 (1978).
23. W. B. Russel Ann. Rev. Fluid Mech. 13 425 (1981).
24. H. Giesekus Rheol. Acta. 2 50 (1962).
25. F. L. Saunders J. Coll. Sci. 16 13 (1961).
26. J. D. Sherwood J. Fluid Mech. 101 609 (1980).
27. J. Stone-Masui and A. Watillon J. Coll. Inter. Sci. 28 187 (1968).
28. C. F. Tanford and J. G. Buzzell J. Phys. Chem. 60 225 (1956).
29. H. Brenner Inter. J. Multi. Flow 1 195 (1974).

30. P. J. Whitcomb and C. W. Macosko J. Rheol. 22 493 (1978).

31. G. Holzwarth Carbohydrate Research 66 173 (1978).

32. G. Chauvateau J. Rheol. 26 111 (1982).

33. H. A. Scheraga J. Chem. Phys. 23 1526 (1955).

34. H. de Hek and A. Vrij J. Coll. Inter. Sci. 84 409 (1981).

35. N. A. Frankel and A. Acrivos Ch. Eng. Sci. 22 847 (1967).

36. M. Doi and S. F. Edwards J. Chem. Soc. Far. Trans. II 74 560; 918 (1978).

37. S. Jain and C. Cohen Macromolecules 14 759 (1981).

38. S. Fesciyan and J. S. Dahler Macromolecules 15 517 (1982)

39. K. M. Zero and R. Pecora Macromolecules 15 87 (1982).

40. I. M. Krieger and P. A. Hiltner, in Polymer Colloids (ed. R. Fitch) Plenum, NY, 1971, p. 63.

41. S. Hachisu and Y. Kobayashi J. Coll. Inter. Sci. 46 470 (1974).

42. J. W. Goodwin and A. M. Khidher, in Colloid and Surface Science (ed. M. Kerker) Academic Press, NY, 1976, Vol. IV, p. 529.

43. I. Snook and W. van Megen J. Chem. Soc. London: Faraday Trans. II 72 216 (1976).

44. I. Snook and W. van Megen J. Coll. Inter. Sci. 57 40, 48 (1976).

45. P. A. Forsyth, S. Marcelja, D. J. Mitchell, and B. W. Ninham Adv. Coll. Inter. Sci. 9 37 (1978).

46. C. J. Barnes, D. Y. C. Chan, D. H. Everett, and D. E. Yates J. Chem. Soc. London: Far. Trans. II 74 136 (1978).

47. H. M. Lindsay and P. M. Chaikin J. Chem. Phys. 76 3774 (1982).

48. D. W. Benzing and W. B. Russel J. Coll. Inter. Sci. 83 163-178 (1981).

49. S. Mitaku, T. Ohtsuki, K. Okano, et al. Jap. J. Appl. Phys. 17 305 (1978); 17 627 (1978); 19 439 (1980); 20 509, (1981).

50. A. B. Glendinning and W. B. Russel J. Coll. Inter. Sci. (submitted 1982).

51. M. Doi "Rheology of Concentrated Macromolecular Solutions" in Advanced Seminar on the Theory of Dispersed Multiphase Flow, The Mathematics Research Center, University of Wisconsin, Madison, 1982, p. 35.

ACKNOWLEDGEMENTS

Work on which this review was based was supported by the National Science Foundation and the Xerox Education Fund.

NOMENCLATURE

a	radius of sphere or rod
A	Hamaker constant
$A(r)$	hydrodynamic function [20]
A_i	surface of ith particle
A_{ij}	surface of excluded volume for i-j interaction
$B(r)$	hydrodynamic function [20]
c_k, c_+, c_-	constants of integration in (31)
C_N	configuration of N particles
$\underline{C}_{ij}{}^{rt}$	tensors coupling stresslet to torque and force for i-j interaction
D_{ro}	rotary diffusion coefficient for isolated rod
$\underline{D}_r, \underline{D}_t$	rotational and translational diffusion tensors including hydrodynamic interactions
e	electronic charge
$\underline{e}'$	perturbation to local rate of strain due to presence of particle
$\underline{E}$	macroscopic rate of strain
$f(\underline{q})$	single particle orientation function including effect of interactions
$f_0(\underline{q})$	isolated particle orientation function
$\underline{f}_i$	linear force density along axis of ith rod
$\underline{F}_i$	total nonhydrodynamic force on ith particle
$g(r)$	radial dependence of perturbed pair density for spheres defined by (18)
$g(\underline{r}, \underline{q}_i, \underline{q}_2)$	effect of interactions on pair density for rods defined by (24)
$G(r)$	hydrodynamic function [16]
G'	shear modulus
$H(r)$	hydrodynamic function [16]
$\underline{I}$	isotropic unit tensor
kT	thermal energy
k, k_T	Huggins' coefficients for simple shear and elongation, respectively
ℓ	half length of rod

L	characteristic separation for charged spheres
n	average number density of particles
n_k, n_0	number densities of ions
$\underline{n}$	unit normal to surface
N	number of particles in representative volume
N_0	dimensionless number density of ions
N_1, N_2	first and second normal stress differences
P_1, P_N	probability densities for configurations of two and N particles, respectively
Pe	Peclét number
P_n	$O(Pe^n)$ portion of P
q	surface charge density on sphere
Q	dimensionless surface charge density
$\underline{q}_i, \underline{\dot{q}}_i$	orientation and rate of rotation for ith rod
$\underline{r}, r$	vector and scalar separation of the centers of mass of two particles
$\underline{\dot{r}}$	total relative velocity of centers of mass
s_1, s_2, s	positions along axis of rod relative to center of mass
$\underline{S}_i$	mechanical stress dipole for ith particle
$\underline{S}_{i0}, \underline{S}_{ij}'$	contributions to S_i for force- and torque-free ith particle, alone and interacting with jth particle
t	time
$\underline{T}_i$	nonhydrodynamic torque on ith particle
$\underline{U}$	relative velocity of centers of mass due to applied shear
V	representative volume or interaction potential
V_s	volume of solvent molecule
x, x_0	spatial variables
$\underline{x}, \underline{x}_i$	position vector and position of ith particle
z_k, z	ionic valences
α	ratio of electrostatic to thermal energy
β	function characterizing primary electroviscous effect [26] or dimensionless constant in expression for rotary diffusion coefficient in semi-dilute solutions
γ	shear rate
δ	minimum separation

$\delta(x)$	Dirac delta function
Δ	thickness of adsorbed polymer layer
ε	dielectric permittivity or slenderness parameter $(\ln 2\ell/a)^{-1}$
κ^{-1}	Debye length
ϕ	volume fraction
ψ_0	surface potential
μ, μ_0	suspension and solvent viscosities
$[\eta], [\eta]_T$	intrinsic viscosities in simple shear and elongation, respectively
$[\eta]_0$	zero-shear limit of intrinsic viscosity
ρ	minimum distance between axes of rods
ρ_0	charge density due to counterions
$\underline{\sigma}$	mechanical stress
$\underline{\tau}$	thermodynamic stress
$\underline{\Omega_i}$	rate of rotation of ith particle due to applied shear
χ	Flory-Huggins interaction parameter
$\langle\ \rangle$	indicates average

William B. Russel
Department of Chemical Engineering
Princeton University
Princeton, New Jersey 08544

Rheology of Concentrated Macromolecular Solutions

M. Doi

1. INTRODUCTION

Polymer solution is a kind of suspension. Since the size of polymers is much larger than the solvent molecules, the surrounding of the polymers can be regarded as a continuum Newtonian fluid. Furthermore in the usual theoretical treatment [1,2], the polymers are assumed to be made up of some frictional units which are large enough to be regarded as Brownian particles. Therefore the theory of polymer solutions is a theory of interacting Brownian particles.

In this paper we discuss some general ideas which were developed for polymer solutions but may be useful for the system of suspensions.

The paper is divided into two parts: in the first part, we discuss the classical theory of Kirkwood [3] for the dynamics of polymer solutions. Though old, this theory still has certain fresh aspects which are not well known and worth reviewing. In the second part, using this discussion, we study a particular system, i.e., the concentrated solution of rodlike polymers. Here an example of treating the concentrated system is shown.

ISBN 0-12-493120-0

2. GENERAL THEORY FOR POLYMER SOLUTIONS AND SUSPENSIONS

2.1 The Kirkwood theory

In a series of papers [3] Kirkwood gave a general theory for the dynamics of polymer solutions. The major purpose of the theory was to calculate the viscoelasticity of dilute polymer solutions. However his formulation is quite general and can be equally applied to suspensions of particles of any size and shape, and perhaps at any concentration. Various factors which are now thought to govern the dynamics of suspensions, such as the hydrodynamic interaction, the Brownian motion, and the potential force between particles are all included in his theory. Unfortunately, his original theory was written in the language of Riemannian geometry and had formidable appearance of mathematics. For this reason, his theory has been thought to be less useful for actual calculation, and the physical implication of the theory has not been well appreciated. However, as was shown by Fixman [4-6], the general structure of the Kirkwood theory is simple, and in a certain representation, the theory gives a convenient base for practical calculation. In the first part of this paper, I describe such modern version of the Kirkwood theory, and discuss some general conclusions derived from the theory.

2.2 Brownian particles interacting with soft potential

The Kirkwood theory considers a set of frictional units, called beads, immersed in a Newtonian fluid with viscosity η_s. The radii of the beads are assumed to be all equal to a and sufficiently small. This does not limit the applicability of his theory because, as shown in Fig.1, any solid objects can

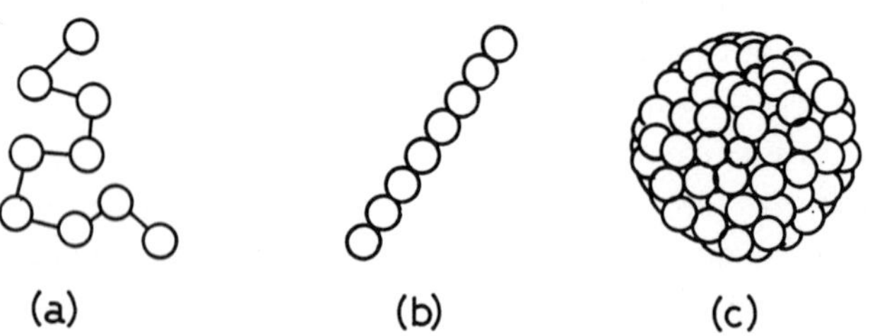

Fig.1 (a) flexible polymer, (b) rodlike polymer, (c) spherical polymer

be regarded as made up of such small beads.

The position of the m-th bead is denoted by $\mathbf{R}_m$ and the whole set of the position vectors $\mathbf{R}_1$, $\mathbf{R}_2$, ... are abbreviated as $\{\mathbf{R}\}$. Let $V(\{\mathbf{R}\})$ be the interaction potential between the beads. If the Brownian motion is neglected, the m-th bead feels the potential force

$$\mathbf{F}_m = -\partial V/\partial \mathbf{R}_m \tag{2.1}$$

and moves with certain velocity $\dot{\mathbf{R}}_m$. If the inertia of the beads is neglected, the velocity is given by the Stokes-Oseen law:

$$\zeta(\dot{\mathbf{R}}_n - \boldsymbol{\kappa}\cdot\mathbf{R}_n - \sum_{m\neq n} \mathbf{T}(\mathbf{R}_n - \mathbf{R}_m)\cdot\mathbf{F}_m) = \mathbf{F}_n \tag{2.2}$$

where $\boldsymbol{\kappa}^{\dagger}$ is the macroscopic velocity gradient, $\zeta \equiv 6\pi\eta_s a$, the Stokes friction constant, and $\mathbf{T}(\mathbf{r})$, the Oseen tensor:

$$\mathbf{T}(\mathbf{r}) = \frac{1}{8\pi\eta_s r}[\,\mathbf{I} + \frac{\mathbf{r}\mathbf{r}}{r^2}\,] \tag{2.3}$$

($\mathbf{I}$ denotes the unit tensor.) Defining the tensor $\mathbf{H}_{nm}$ as

$$\mathbf{H}_{nn} = \frac{\mathbf{I}}{\zeta}$$

$$\mathbf{H}_{nm} = \mathbf{T}(\mathbf{R}_n - \mathbf{R}_m) \qquad (n \neq m) \tag{2.4}$$

we can rewrite eq.(2.2) as

$$\dot{\mathbf{R}}_n - \boldsymbol{\kappa}\cdot\mathbf{R}_n = \sum_m \mathbf{H}_{nm}\cdot\mathbf{F}_m \tag{2.5}$$

Equations (2.1) and (2.5) are equivalent to the hydrodynamics in the Stokes approximation.

Now to take into account the Brownian motion of the beads, we introduce the probability density $\Psi(\{\mathbf{R}\};t)$ of finding the beads in the configuration $\{\mathbf{R}\}$ at time t. The conservation equation for the probability is written as

$$\frac{\partial\Psi}{\partial t} = -\sum_n \frac{\partial}{\partial \mathbf{R}_n}\cdot(\dot{\mathbf{R}}_n\Psi) \tag{2.6}$$

The essence of the Kirkwood theory is the proposition that, as a result of the Brownian motion, a new term $k_BT \ln\Psi$ must be added to the potential V: thus instead of eq.(2.1) $\mathbf{F}_m$ is now given as

$$\mathbf{F}_m = -\frac{\partial}{\partial \mathbf{R}_m}(k_B T \ln\Psi + V) \tag{2.7}$$

(The necessity of including the term $k_B T \ln\Psi$ was noted by several authors [7-9].)

Equations (2.5), (2.6) and, (2.7) lead to

$$\frac{\partial\Psi}{\partial t} = \sum_{n,m} \frac{\partial\Psi}{\partial R_n}\cdot\mathbf{H}_{nm}\cdot[k_B T\frac{\partial\Psi}{\partial \mathbf{R}_m} + \Psi\frac{\partial V}{\partial \mathbf{R}_m}] - \sum_n \frac{\partial}{\partial R_n}\cdot\boldsymbol{\kappa}\cdot\mathbf{R}_n \tag{2.8}$$

This equation determines $\Psi(\{\mathbf{R}\};t)$ for given velocity gradient $\boldsymbol{\kappa}(t)$.

To discuss the rheological properties, we need toknow the microscopic expression for the stress tensor, for which Kirkwood gave the following formula:

$$\boldsymbol{\sigma}_{tot} = \boldsymbol{\sigma} + \eta_s(\boldsymbol{\kappa} + \boldsymbol{\kappa}^+) + P\mathbf{I} \tag{2.9}$$

$$\boldsymbol{\sigma} = -\sum_m \langle \mathbf{R}_m \mathbf{F}_m \rangle / \Omega \tag{2.10}$$

where P is the pressure, Ω the volume of the system, and <...> is the average with respect to $\Psi(\{\mathbf{R}\};t)$:

$$\langle \ldots \rangle \equiv \int d\{\mathbf{R}\}\Psi(\{\mathbf{R}\};t)\ldots \tag{2.11}$$

Equation (2.10) agrees with the formula given recently by Batchelor [10].

2.3 The effect of constraints

In the above discussion, the position vectors $\mathbf{R}_n$ are assumed to be independent. However, except for certain models of polymers, $\mathbf{R}_n$ cannot be regarded as independent variables. For example, the position vectors of the beads constitutinga solid particles as shown in Fig.1 are subject to the condition that the distance between any pairs of beads remain constant. Such conditions are generally written as a set of functional relations:

$$C_a(\{\mathbf{R}\}) = 0 \qquad a = 1,2,\ldots \tag{2.12}$$

To take into account this constraint Kirkwood introduced the generalized coordinates $\{Q\} \equiv Q_1, Q_2, \ldots$ which specify the configuration of the beads under the constraint (2.12). The velocity $\dot{\mathbf{R}}_m$ is expressed by the velocity of the generalized coordinate $\dot{Q}_a$ as

$$\dot{\mathbf{R}}_m = \sum_a \frac{\partial \mathbf{R}_m}{\partial Q_a} \dot{Q}_a \tag{2.13}$$

The force $\mathbf{F}_m$ now represents both the potential force and the constraining force. By considering the work necessary to displace Q_a, we get

$$\sum_m \mathbf{F}_m \cdot \frac{\partial \mathbf{R}_m}{\partial Q_a} = -\frac{\partial}{\partial Q_a}(k_B T \ln\Psi + V) \tag{2.14}$$

Equations (2.5), (2.13) , and (2.14) determine $\dot{Q}_a$ and $\mathbf{F}_m$ as a function of $\{\mathbf{R}\}$ and Ψ. In the generalized coordinate space, the conservation equation is written as

$$\frac{\partial \Psi}{\partial t} = -\frac{1}{\sqrt{g}} \frac{\partial}{\partial Q_a} [\sqrt{g}\, \dot{Q}_a \Psi] \tag{2.15}$$

where g is the determinant of the matrix g_{ab} defined by

$$g_{ab} \equiv \sum_m \frac{\partial \mathbf{R}_m}{\partial Q_a} \cdot \frac{\partial \mathbf{R}_m}{\partial Q_b} \tag{2.16}$$

The explicit form of eq.(2.15) is quite complicated and is not written here because it is not needed in the following discussion.

The constraint can be treated by an alternative method. We regard $\{\mathbf{R}\}$ as independent variables, and add a constraining force to the right hand side of eq.(2.7):

$$\mathbf{F}_m = -\frac{\partial}{\partial \mathbf{R}_m} (k_B T \ln\Psi + V) + \sum_a \lambda_a \frac{\partial C_a}{\partial \mathbf{R}_m} \tag{2.17}$$

The unknown parameters λ_a are chosen such that $\dot{\mathbf{R}}_m$ satisfy the following conditions.

$$\sum_m \frac{\partial C_a}{\partial \mathbf{R}_m} \cdot \dot{\mathbf{R}}_m = 0 \tag{2.18}$$

Equations (2.5), (2.17) and (2.18) determine $\dot{R}_m$, F_m and λ_a. This method was first introduced by Fixman [4,5], who succeeded in calculating the viscoelasticity of polymers with internal constraints.

In either formulation, the stress tensor is given by eq.(2.10). Since $\mathbf{F}_m$ is obtained as a linear function of $\boldsymbol{\kappa}$, the microscopic expression for the stress is generally written as

$$-\sum_m R_{m\alpha}F_{m\beta} = \left[\tilde{\sigma}^{(E)}_{\alpha\beta}(\{Q\},\ln\Psi) + \tilde{\eta}_{\alpha\beta\mu\nu}(\{Q\})\kappa_{\mu\nu}\right]\Omega \qquad (2.19)$$

Hence

$$\sigma_{\alpha\beta} = \sigma^{(E)}_{\alpha\beta} + \sigma^{(V)}_{\alpha\beta} \qquad (2.20)$$

where

$$\sigma^{(E)}_{\alpha\beta} = \langle\tilde{\sigma}^{(E)}_{\alpha\beta}\rangle \quad ; \quad \sigma^{(V)}_{\alpha\beta} = \langle\tilde{\eta}_{\alpha\beta\mu\nu}\rangle\kappa_{\mu\nu} \qquad (2.21)$$

We shall call $\boldsymbol{\sigma}^{(E)}$ the elastic stress and $\boldsymbol{\sigma}^{(V)}$ the viscous stress. The distinction between these two stresses is important, and will be discussed in the next section.

2.4 The elastic stress and the viscous stress

The viscous stress is proportional to the current velocity gradient $\boldsymbol{\kappa}(t)$, while the elastic stress does not include $\boldsymbol{\kappa}(t)$. Although the actual stress is the sum of the two, each component can be determined by experiment. For example, the contribution of the two stresses in steady shear flow can be determined by measuring the stress immediately after stopping the shear flow (see Fig.2). The viscous stress is given by the instantaneous drop in the stress when the flow is stopped, and

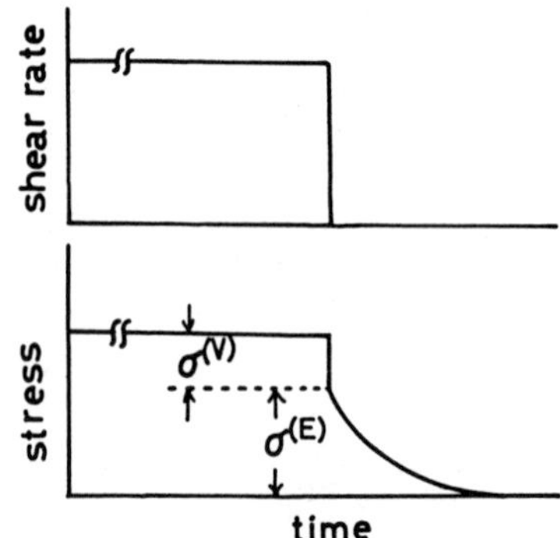

Fig.2 Experimental determination of the elastic stress and the viscous stress

the elastic stress is obtained by subtracting the viscous stress from the total stress in the steady state.

In concentrated polymer solutions, the instantaneous drop in the stress is quite small [11], which means that the elastic stress is essential in such system. On the other hand in suspensions of large particle size, the stress vanishes immediately, which indicates the dominant contribution of the viscous stress [12,13].

A more fundamental difference between the viscous stress and the elastic stress becomes clear from the energetic consideration.

Consider a hypothetical deformation $\delta\varepsilon_{\alpha\beta}$ which displaces the material point at r_α to $r_\alpha + \delta\varepsilon_{\alpha\beta}\, r_\beta$. (In this paper summation is implied over the repeated Greek indices.) If the deformation is done in a short time interval δt, the velocity gradient is given as

$$\kappa_{\alpha\beta} = \delta\varepsilon_{\alpha\beta}/\delta t \tag{2.22}$$

We consider the limit of instantaneous deformation by putting $\delta t \to 0$ while keeping $\delta\varepsilon_{\alpha\beta}$ small but finite. This deformation displaces the beads at $\mathbf{R}_m$ to $\mathbf{R}_m + \dot{\mathbf{R}}_m \delta t$. The velocity $\dot{\mathbf{R}}_m$ can be calculated by solving eqs.(2.5), (2.17) and (2.18), but as $\boldsymbol{\kappa}$ is very large, we may neglect the first term in the right hand side of eq.(2.17). Alternatively, the velocity can be determined by the following variational principle.

We regard $\dot{\mathbf{R}}_m$ as independent variables and consider the following function of $\{\dot{\mathbf{R}}\}$:

$$W(\{\dot{\mathbf{R}}\}) = \sum_{m,n} (\dot{\mathbf{R}}_m - \boldsymbol{\kappa}\cdot\mathbf{R}_m)\cdot(\mathbf{H}^{-1})_{mn}\cdot(\dot{\mathbf{R}}_n - \boldsymbol{\kappa}\cdot\mathbf{R}_n) \tag{2.23}$$

where $(\mathbf{H}^{-1})_{mn}$ is the inverse of $\mathbf{H}_{mn}$:

$$\sum_m \mathbf{H}_{lm}\cdot(\mathbf{H}^{-1})_{mn} = \mathbf{I}\delta_{ln} \tag{2.24}$$

The real velocity is given by the $\{\dot{\mathbf{R}}\}$ which minimize W subject to the condition (2.18).

The function W represents the energy dissipated to the

solvent fluid due to the hypothetical motion $\{\dot{R}\}$. The above formula is a special representation of the principle of the minimum energy dissipation known in low Reynolds number hydrodynamics.

It is easy to prove the following relations.

(i) The minimum value of W gives the viscous stress as

$$\text{Mini } W = \tilde{\eta}_{\alpha\beta\mu\nu}\,\kappa_{\beta\alpha}\,\kappa_{\nu\mu}\,\Omega \tag{2.25}$$

or from eq.(2.21),

$$\sigma^{(V)}_{\alpha\beta}\,\kappa_{\beta\alpha} = \langle \text{Mini } W \rangle/\Omega \tag{2.26}$$

(ii) The elastic stress is related to the variation of the scalar $\mathcal{A}$ which is defined by

$$\mathcal{A} = \int d\{Q\}\sqrt{g}\,(k_B T \Psi \ln\Psi + \Psi V) \tag{2.27}$$

as

$$\delta\mathcal{A} = \sigma^{(E)}_{\alpha\beta}\,\delta\varepsilon_{\beta\alpha}\,\Omega \tag{2.28}$$

In the ideal elastic materials, the stress is given by the variation in the free energy under a hypothetical strain. Equation (2.28) is a generalization of such relation for the viscoelastic materials. The scalar $\mathcal{A}$ plays the role of the free energy. Since the system we are considering is not in equilibrium, we call $\mathcal{A}$ the dynamic free energy.

As the viscous stress is negligibly small in concentrated polymer solutions, the stress in such systems can be calculated from the variation in the dynamic free energy. The existence of the dynamic free energy has been presumed by various authors. One of the conventional theory of the polymeric liquids assumed such energy [2, 14, 15]. Coleman also proposed a phenomenological theory which asserts such free energy [16]. Recently Hassager [17] has shown how to use such free energy for the variational calculation in the fluid mechanics of polymeric liquids.

2.5 Variational formulation of the Kirkwood theory

The Kirkwood theory can be formulated in the form of the variational principle. This formulation states the mathematical structure of the Kirkwood theory in a simple form. Also it gives a convenient base for various approximate calculations.

In this formulation, we regard $\dot{\mathbf{R}}_m$ as a function of $\{\mathbf{R}\}$, and consider the functional defined by

$$\mathcal{R} = \frac{1}{2}\mathcal{W} + \dot{\mathcal{A}} \tag{2.29}$$

$$\mathcal{W} \equiv \int d\{\mathbf{R}\}\Psi \sum_{m,n} (\dot{\mathbf{R}}_m - \boldsymbol{\kappa}\cdot\mathbf{R}_m)\cdot(\mathbf{H}^{-1})_{mn}\cdot(\dot{\mathbf{R}}_n - \boldsymbol{\kappa}\cdot\mathbf{R}_n) \tag{2.30}$$

and

$$\dot{\mathcal{A}} \equiv \int d\{\mathbf{R}\}\,[k_BT\dot{\Psi}\ln\Psi + k_BT\dot{\Psi} + \dot{\Psi}V] \tag{2.31}$$

where $\dot{\Psi}$ is defined by

$$\dot{\Psi} \equiv -\sum_m \frac{\partial}{\partial \mathbf{R}_m}\cdot(\dot{\mathbf{R}}_m\Psi) \tag{2.32}$$

By simple calculation we can show:

(i) The minimization of $\mathcal{R}$ for all variation of $\dot{\mathbf{R}}_m$ subject to the constraint (2.18) determines the time evolution of Ψ as

$$\frac{\partial}{\partial t}\Psi(\{\mathbf{R}\};t) = \dot{\Psi}(\{\mathbf{R}\}) \tag{2.33}$$

(ii) The minimum value of $\mathcal{R}$ is related to the stress as

$$\text{Mini }\mathcal{R} = \sigma^{(E)}_{\alpha\beta}\kappa_{\beta\alpha} + \frac{1}{2}\eta_{\alpha\beta\mu\nu}\kappa_{\beta\alpha}\kappa_{\nu\mu} + [\text{terms independent of }\kappa] \tag{2.34}$$

Several applications of this formulation will be published elsewhere.

In closing this section it is worthwhile to stress again that in the above formulation, no condition has been imposed on the concentration of the particles. Therefore the theory will apply for concentrated suspensions as well.

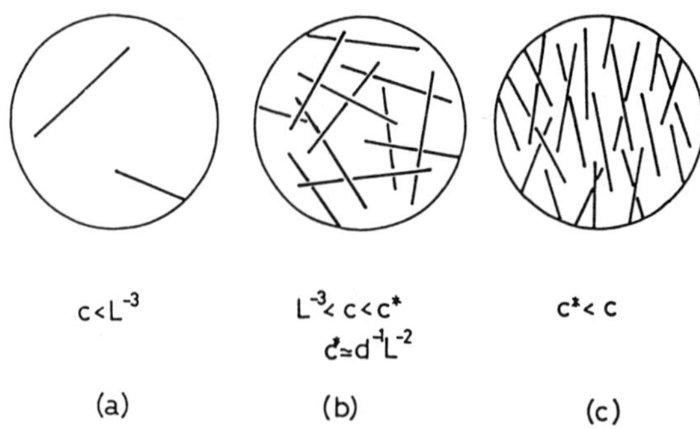

Fig.3 Solutions of rodlike polymers at various concentrations

3. RODLIKE POLYMERS IN DILUTE SOLUTIONS

As an example of the above formulation, we now discuss the solution of rodlike polymers. We consider rodlike polymers of length L and diameter d immersed in a Newtonian fluid. Let c be the number of polymers in unit volume. Fig.3 illustrates the microscopic view of such systems at various concentrations: If $cL^3<1$ (Fig.3(a)), the interaction between polymers is weak and their motion is independent of each other. If $cL^3>1$ (Fig.3(b)), the motion of individual polymers is hampered by the surrounding ones, and one expects strong effect of molecular interaction. If $c>c^*\simeq 1/dL^2$, the system forms a liquid crystalline phase in which polymers orient toward certain direction (Fig.3(c)) [18, 19]. In this section we discuss the dilute solution ($cL^3<1$). The concentrated solution is discussed in the next section.

In the dilute solution, it is sufficient to consider an isolated polymer in an infinite fluid. Following the Kirkwood theory we assume that the polymer is made up of $N \equiv L/d$ beads, connected along a straight line with separation d (see Fig.1(b)). The beads are numbered from -N/2 to N/2. Let $\mathbf{u}$ be the unit vector along the polymer and $\mathbf{R}_G$ be the position of the center of mass, then

$$\mathbf{R}_m = md\mathbf{u} + \mathbf{R}_G \qquad \dot{\mathbf{R}}_m = md\dot{\mathbf{u}} + \dot{\mathbf{R}}_G \tag{3.1}$$

From eqs.(2.3), (2.4) and (3.1) we have

$$\mathbf{H}_{mn} = (1 - \delta_{mn})h_{mn}(\mathbf{uu} + \mathbf{I}) \tag{3.2}$$

where

$$h_{mn} = \frac{1}{8\pi\eta_s d|m - n|} \tag{3.3}$$

In eq.(3.2), the term $\mathbf{I}\delta_{mn}/\zeta$ is neglected since such term turns out negligible for large N. The inverse of $\mathbf{H}_{mn}$ is given as

$$(\mathbf{H}^{-1})_{mn} = g_{mn}(\mathbf{I} - \frac{\mathbf{uu}}{2}) \tag{3.4}$$

where

$$\sum_{m=-N/2}^{N/2} h_{km}g_{mn} = \delta_{kn} \tag{3.5}$$

Substituting eqs.(3.1) and (3.4) into eq.(2.30), we have

$$\mathcal{W} = \int d^2\mathbf{u}\, d^3\mathbf{R}_G \Psi\Big[\zeta_r^{(0)}[(\dot{\mathbf{u}} - \boldsymbol{\kappa}\cdot\mathbf{u})^2 - \tfrac{1}{2}(\mathbf{u}\cdot\boldsymbol{\kappa}\cdot\mathbf{u})^2] + 2\,\zeta_{t/\!/}^{(0)}[(\dot{\mathbf{R}}_G - \boldsymbol{\kappa}\cdot\mathbf{R}_G)^2 - \tfrac{1}{2}(\mathbf{u}\cdot(\dot{\mathbf{R}}_G - \boldsymbol{\kappa}\cdot\mathbf{R}_G))^2]\Big] \tag{3.6}$$

where

$$\zeta_r^{(0)} = d^2 \sum_{m,n} g_{mn}mn \quad ; \quad \zeta_{t/\!/}^{(0)} = \frac{1}{2}\sum_{m,n} g_{mn} \tag{3.7}$$

Hence

$$\mathcal{R} = \int d^2\mathbf{u}\, d^2\mathbf{R}_G \Psi\Big[\frac{\zeta_r^{(0)}}{2}(\dot{\mathbf{u}} - \boldsymbol{\kappa}\cdot\mathbf{u})^2 - \frac{\zeta_r^{(0)}}{4}(\mathbf{u}\cdot\boldsymbol{\kappa}\cdot\mathbf{u})^2 + \dot{\mathbf{u}}\cdot\frac{\partial}{\partial\mathbf{u}}(k_BT \ln\Psi + V_e) + \zeta_{t/\!/}^{(0)}(\dot{\mathbf{R}}_G - \boldsymbol{\kappa}\cdot\mathbf{R}_G)^2 - \frac{\zeta_{t/\!/}^{(0)}}{2}(\mathbf{u}\cdot(\dot{\mathbf{R}}_G - \boldsymbol{\kappa}\cdot\mathbf{R}_G))^2 + \dot{\mathbf{R}}_G\cdot\frac{\partial}{\partial\mathbf{R}_G}[k_BT \ln\Psi + V_e]\Big] \tag{3.8}$$

where $V_e(\mathbf{u},\mathbf{R}_G)$ is the potential due to an external field. Since $\mathbf{u}$ is a unit vector, $\dot{\mathbf{u}}$ must satisfy the condition

$$\mathbf{u}\cdot\dot{\mathbf{u}} = 0 \tag{3.9}$$

Minimizing $\mathcal{R}$ under this condition, we get

$$\dot{\mathbf{u}} = \frac{1}{\zeta_r^{(0)}}\nabla[k_BT\ \ln\Psi + V_e] + \boldsymbol{\kappa}\cdot\mathbf{u} - (\mathbf{u}\cdot\boldsymbol{\kappa}\cdot\mathbf{u})\mathbf{u} \tag{3.10}$$

$$\dot{\mathbf{R}}_G = \frac{1}{2\zeta_{t/\!/}^{(0)}}(\mathbf{I} + \mathbf{uu})\cdot\frac{\partial}{\partial \mathbf{R}_G}[k_BT\ \ln\Psi + V_e] + \boldsymbol{\kappa}\cdot\mathbf{R}_G \tag{3.11}$$

with

$$\nabla \equiv (\mathbf{I} - \mathbf{uu})\cdot\frac{\partial}{\partial \mathbf{u}} \tag{3.12}$$

The diffusion equation for Ψ is thus obtained as

$$\frac{\partial\Psi}{\partial t} = D_r^{(0)}\nabla\cdot[\nabla\Psi + \frac{\Psi}{k_BT}\nabla V_e] - \nabla\cdot[(\boldsymbol{\kappa}\cdot\mathbf{u} - (\mathbf{u}\cdot\boldsymbol{\kappa}\cdot\mathbf{u})\mathbf{u}\Psi]$$
$$+ \frac{1}{2}D_{t/\!/}^{(0)}\frac{\partial}{\partial \mathbf{R}_G}\cdot(\mathbf{I} + \mathbf{uu})\cdot[\frac{\partial\Psi}{\partial \mathbf{R}_G} + \frac{\Psi}{k_BT}\frac{\partial V_e}{\partial \mathbf{R}_G}] - \frac{\partial}{\partial \mathbf{R}_G}\cdot\boldsymbol{\kappa}\cdot\mathbf{R}_G\Psi \tag{3.13}$$

where

$$D_r^{(0)} = k_BT/\zeta_r^{(0)} \quad ; \quad D_{t/\!/}^{(0)} = k_BT/\zeta_{t/\!/}^{(0)} \tag{3.14}$$

is the rotational, and the translational diffusion constant respectively.

Since the minimum of $\mathcal{R}$ is obtained as

$$\text{Min}\ \mathcal{R} = \int[\frac{\zeta_r^{(0)}}{4}(\mathbf{u}\cdot\boldsymbol{\kappa}\cdot\mathbf{u})^2 + \boldsymbol{\kappa}:\mathbf{u}\ \nabla(k_BT\ \ln\Psi + V_e)]\Psi d^2u\, d^3\mathbf{R}_G$$
$$+ [\text{terms independent of } \boldsymbol{\kappa}] \tag{3.15}$$

the elastic stress and the viscous stress are obtained as

$$\boldsymbol{\sigma}^{(E)} = \frac{1}{\Omega}\int d^2\mathbf{u}\, d^3R_G\Psi[\mathbf{u}\nabla(k_BT\ \ln\Psi + V_e)]$$
$$= 3ck_BT\langle\mathbf{uu} - \frac{\mathbf{I}}{3}\rangle + c\langle\nabla V_e\mathbf{u}\rangle \tag{3.16}$$

$$\sigma^{(V)} = c\frac{\zeta_r^{(0)}}{2} \langle \mathbf{uuuu} \rangle : \boldsymbol{\kappa} \tag{3.17}$$

To calculate g_{mn} we approximate h_{mn} as

$$h_{mn} \simeq \bar{h}\delta_{mn} \tag{3.18}$$

with

$$\bar{h} = 2\int_1^{N/2} dm\, h_{m0} = \frac{1}{4\pi\eta_s d} \ln(N/2) \tag{3.19}$$

Hence

$$g_{mn} \simeq \delta_{mn}/\bar{h} \tag{3.20}$$

and

$$\zeta_r^{(0)} = 2d^2\int_0^{N/2} dm \frac{m^2}{\bar{h}}$$

$$\simeq \frac{\pi\eta(Nd)^3}{3}/\ln(N/2)$$

$$\simeq \frac{\pi\eta_s L^3}{3}/\ln(L/d) \tag{3.21}$$

$$\zeta_{t\parallel}^{(0)} = N/2\bar{h} \simeq 2\pi\eta_s L/\ln(L/d) \tag{3.22}$$

These results agree with those of Kirkwood [20, 21]. Detailed calculation of the rheological properties has been done by various authors [2, 9, 22, 23, see also the footnote of p.522 in ref.2].

4. RODLIKE POLYMERS IN CONCENTRATED SOLUTION

4.1 Interaction between the rodlike polymers

We now consider the concentrated solutions of rodlike polymers shown in Fig.3(b) and (c). It is of course not possible to treat such systems rigorously following the general formulation of Kirkwood. Here we take an approach of the mean field theory. Thus instead of the probability density Ψ for the whole configuration of the polymers, we consider the probability density

$f(\mathbf{u};t)$ that an arbitrary chosen polymer is in the direction $\mathbf{u}$. We then derive a time evolution equation for $f(\mathbf{u};t)$ by considering the motion of a certain test polymer under the influence of the other polymers. It may not be appropriate to give this theory as an example of the application of the Kirkwood theory because this theory is not derived from the Kirkwood theory, rather it was constructed on the basis of some physical argument. However the general relations given in section 2 are essential in this theory. (Constructing the mean field theory directly on the basis of the Kirkwood theory is an interesting problem, but has not been successful.)

Now in the concentrated solution of rodlike polymers, we have to take into account two types of interaction: one is the hydrodynamic interaction and the other is the interaction due to molecular potential. As the molecular potential, we assume a hard repulsive potential of rigid rod. The effect of this potential on the dynamics of polymers has two features quite different from each other.

First the repulsive potential affects the thermodynamic functions in equilibrium. For example the osmotic pressure of the polymers are written as

$$\Pi = ck_BT(1 + \frac{\pi}{4}cdL^2 + \dots) \tag{4.1}$$

The correction term cdL^2 comes from the decrease in the volume available for each polymer due to the mutual volume exclusion of the polymers. We call this effect the excluded volume effect.

Second the repulsive potential imposes an obvious constraint that two polymers cannot pass through crossing each other. This constraint should be distinguished from the excluded volume effect because the constraint is independent of the volume of the polymer. (For example, in the limiting case of $d \to 0$, the excluded volume effect vanishes, while the constraint still remains.) We call the effect due to this constraint the entanglement effect.

We shall now discuss these effects in detail.

4.2 The hydrodynamic interaction

First we consider the hydrodynamic interaction. Again we assume that the polymers are made up of N beads and denote the position of the m-th bead ($-N/2 < m < N/2$) of the a-th polymer as $\mathbf{R}_{am}$. The energy dissipation function W per unit volume is written as

$$W = \sum(\dot{\mathbf{R}}_{am} - \boldsymbol{\kappa}\cdot\mathbf{R}_{am})\cdot(\mathbf{H}^{-1})_{am,bn}\cdot(\dot{\mathbf{R}}_{bn} - \boldsymbol{\kappa}\cdot\mathbf{R}_{bn}) \tag{4.2}$$

To reduce the many body problem into a single body problem, we neglect the correlation in the motion of different polymers, and approximate W as

$$W = \sum(\dot{\mathbf{R}}_{am} - \boldsymbol{\kappa}\cdot\mathbf{R}_{am})\cdot(\mathbf{H}^{-1})_{am,an}\cdot(\dot{\mathbf{R}}_{an} - \boldsymbol{\kappa}\cdot\mathbf{R}_{an}) \tag{4.3}$$

Furthermore the matrix element $(\mathbf{H}^{-1})_{am,an}$, which actually depends on the configuration of the whole polymers, is averaged over the configuration of the other polymers.

$$(\mathbf{H}^{-1})_{am,an} \simeq \langle(\mathbf{H}^{-1})_{am,an}\rangle' \equiv (\mathbf{H}'^{-1})_{mn} \tag{4.4}$$

where $\langle\ldots\rangle'$ means the average over the configuration of the other polymers. The tensor $\mathbf{H}'^{-1}$ can be calculated by the diagramatic method used by Edwards and Freed [27, 28]. Alternatively, the same result can be derived from the following qualitative argument (see [29]).

In concentrated solution, the hydrodynamic perturbation caused by a motion of a certain polymer does not reach far away as in the dilute solution, but decays quickly in a certain characteristic length ξ_H because the perturbation is screened by the other polymers. The screening length ξ_H is given as [27 - 29]

$$\xi_H \simeq (cd^2L)^{-1/2} \tag{4.5}$$

If the approximations of eqs.(4.3) and (4.4) are used, the energy dissipation function is written in the same form as in the dilute solution theory:

$$W = c \sum_{m,n} (\dot{\mathbf{R}}_m - \boldsymbol{\kappa}\cdot\mathbf{R}_m)\cdot(\mathbf{H}'^{-1})_{mn}\cdot(\dot{\mathbf{R}}_n - \boldsymbol{\kappa}\cdot\mathbf{R}_n) \tag{4.6}$$

where

$$\mathbf{H}'_{mn} = (1 - \delta_{mn})h'_{mn}(\mathbf{I} + \mathbf{uu}) \tag{4.7}$$

Due to the hydrodynamic screening h'_{mn} becomes negligible, if the distance between the beads m and n is larger than ξ_H. Hence

$$h'_{mn} \simeq \begin{cases} 1/8\pi\eta_s d\,|n-m| & \text{for } |n-m|\,d < \xi_H \\ 0 & \text{for } |n-m|\,d > \xi_H \end{cases} \tag{4.8}$$

or in the approximation (3.18), we have

$$h'_{mn} = \bar{h}'\delta_{mn} \quad ; \quad \bar{h}' = \frac{1}{4\pi\eta_s d}\ln(\xi_H/d) \tag{4.9}$$

Repeating the same procedure as described in section 3, we arrive at:

$$W = c\zeta'_r[(\dot{\mathbf{u}} - \boldsymbol{\kappa}\cdot\mathbf{u})^2 - \tfrac{1}{2}(\mathbf{u}\cdot\boldsymbol{\kappa}\cdot\mathbf{u})^2] + [\text{terms including } \dot{\mathbf{R}}_G] \tag{4.10}$$

with

$$\zeta'_r \simeq \frac{\pi\eta_s L^3}{3} / \ln(\xi_H/d) \tag{4.11}$$

Comparing eq.(4.11) with eq.(3.21), we note that the energy dissipation function is little affected by the hydrodynamic interaction. Especially, the expression for the viscous stress is nearly the same as in dilute solution:

$$\boldsymbol{\sigma}^{(V)} = \frac{c}{2}\zeta'_r\langle\mathbf{uuuu}\rangle:\boldsymbol{\kappa} \tag{4.12}$$

Equation (4.12) is consistent with Batchelor's result [30] for the elongational viscosity in the limit of high elongational rate.

4.3 The entanglement effect

Due to the entanglement effect the rotational Brownian motion of each polymer becomes very slow in concentrated solution. This effect can be quantitatively discussed by the tube

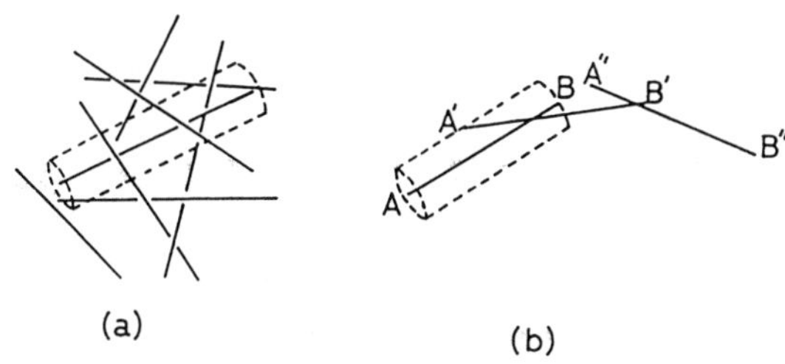

Fig.4 The tube model for concentrated solutions of rodlike polymers.

model [25, 26].

The tube model assumes that in the situation shown in Fig.3(b) and (c), the motion of each polymer is restricted in a tube like region made of the surrounding polymers. Suppose that at time $t = 0$ the test polymer is in a certain tube of direction $\mathbf{n}$. As long as the test polymer stays in this tube, its direction $\mathbf{u}(t)$ is restricted within a small angle $\varepsilon \simeq a_t/L$ around $\mathbf{n}$, where a_t is the radius of the tube. If the test polymer moves out of the tube by diffusion, it can change the direction by a small angle ε(see Fig.4). Although ε is small, repetition of this process can cause a large change in $\mathbf{u}(t)$. If the average time necessary for the test polymer to get out of a certain tube is τ_t, then the mean square displacement of $\mathbf{u}(t)$ in a time interval t is estimated as

$$\langle(\mathbf{u}(t) - \mathbf{u}(0))^2\rangle \simeq \frac{t}{\tau_t}\varepsilon^2 \tag{4.13}$$

Thus the Brownian motion of $\mathbf{u}(t)$ is described by an effective rotational diffusion constant:

$$D_r \simeq \varepsilon^2/\tau_t \tag{4.14}$$

The time τ_t is estimated by the time necessary for the polymer to move the distance L by diffusion:

$$\tau_t \simeq L^2/D_{t/\!/} \tag{4.15}$$

where $D_{t\parallel}$ is the translational diffusion constant of the test polymer along the tube. Since the translation along the tube is almost free, $D_{t\parallel}$ is nearly equal to $D_{t\parallel}^{(0)}$, the diffusion constant in dilute solution:

$$D_{t\parallel} \simeq D_{t\parallel}^{(0)} = \frac{k_B T \ln(L/d)}{2\pi\eta_s L} = \frac{1}{6} L^2 D_r^{(0)} \tag{4.16}$$

From eqs.(4.14) - (4.16), we have

$$D_r \simeq \varepsilon^2 D_r^{(0)} \simeq (a_t/L)^2 D_r^{(0)} \tag{4.17}$$

The tube radius a_t is estimated by the distance between the test polymer and its nearest neighbour. For the test polymer in the direction $\mathbf{u}$, this distance is estimated as [26]

$$a_t[\mathbf{u};f(\mathbf{u};t)] = [cL^2 \int d^2\mathbf{u}_1 f(\mathbf{u}_1;t)\, |\mathbf{u}\times\mathbf{u}_1|\,]^{-1} \tag{4.18}$$

From eqs.(4.17) and (4.18), we finally get

$$D_r = \nu_1 D_r^{(0)} (cL^3)^{-2} [\, \frac{4}{\pi} \int d^2\mathbf{u}_1 f(\mathbf{u}_1;t)\, |\mathbf{u}\times\mathbf{u}_1|\,]^{-2} \tag{4.19}$$

where ν_1 is a certain numerical constant whose precise value cannot be determined by the present theory.

Equation (4.19) gives the effective rotational diffusion constant of the polymers in the direction $\mathbf{u}$. Note that D_r is much smaller than $D_r^{(0)}$ in the concentrated solution for which $cL^3 >> 1$.

4.4 The excluded volume effect

Due to the excluded volume effect, the rodlike polymers form the liquid crystalline phase above a certain critical concentration c*. The statistical mechanical theory for the formation of the liquid crystalline phase was given by Onsager and Flory [18, 19].

According to Onsager [18] the free energy $\mathcal{A}$ per unit volume is given as a functional of f(u) as

$$\mathcal{A} = ck_B T[\int d^2\mathbf{u}\, f \ln f + cdL^2 \int d^2\mathbf{u}_1\, d^2\mathbf{u}_2\, |\mathbf{u}_1\times\mathbf{u}_2|\, f(\mathbf{u}_1)f(\mathbf{u}_2)] \tag{4.20}$$

The equilibrium distribution function is given by the condition that $\mathcal{A}$ to be mininized for all variation of f, i.e., $\delta\mathcal{A}/\delta f = 0$, which gives

$$f_{eq} = \text{constant} \times \exp[-V_{SCF}/k_B T] \qquad (4.21)$$

where

$$V_{SCF} = 2cdL^2 k_B T \int d^2\mathbf{u}_1 \, |\mathbf{u}\times\mathbf{u}_1| \, f_{eq}(\mathbf{u}_1) \qquad (4.22)$$

is the mean field potential acting on the test polymer. Onsager showed that above a certain critical concentration $c^* \propto 1/dL^2$, eq.(4.21) has an nonisotropic solution corresponding to the liquid crystalline phase.

Now in the nonequilibrium case, it is natural to assume the same mean field potential for the motion of the test polymer:

$$V_{SCF} = 2cdL^2 k_B T \int d^2\mathbf{u}_1 \, |\mathbf{u}\times\mathbf{u}_1| \, f(\mathbf{u}_1;t) \qquad (4.23)$$

This potential acts to aline the test polymer in the same direction as the surrounding ones.

4.5 The rheological constitutive equation

From the above discussion, the motion of the test polymer can be regarded as a Brownian motion in the potential field V_{SCF} with the effective rotational diffusion constant D_r. Hence the time evolution equation for f(u;t) is obtained as a natural generalization of eq.(3.13):

$$\frac{\partial f}{\partial t} = \nabla D_r \cdot [\nabla f + \frac{f}{k_B T} \nabla V_{SCF}] - \nabla\cdot[\boldsymbol{\kappa}\cdot u - (\mathbf{u}\cdot\boldsymbol{\kappa}\cdot\mathbf{u})u]f \qquad (4.24)$$

The expression for the viscous stress is already obtained in eq.(4.12). The expression for the elastic stress is obtained from the variation of $\mathcal{A}$ for a hypothetical deformation. The result is written in the similar form as eq.(3.16):

$$\sigma^{(E)}_{\alpha\beta} = 3ck_B T\langle u_\alpha u_\beta - \tfrac{1}{3}\delta_{\alpha\beta}\rangle + \langle[\tfrac{\partial}{\partial u} V_{SCF}]_\alpha u_\beta\rangle \qquad (4.25)$$

Equations (4.24), (4.25) and (4.12) give the rheological constitutive equation: for given $\boldsymbol{\kappa}(t)$, $f(\mathbf{u};t)$ is solved from eq.(4.24), and then the stress is calculated from eqs.(4.25) and (4.12). Some results of the constitutive equation have already been published [24, 26, 31, 32]. In Fig.5, the concentration dependence of the zero shear rate viscosity is shown. This behaviour is in qualitative agreement with experiments [33-36].

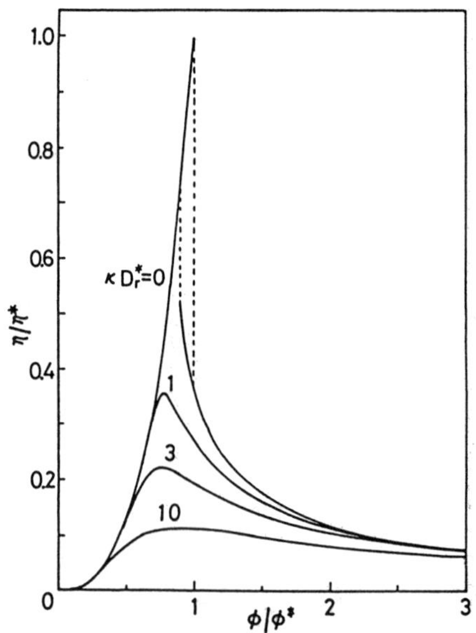

Fig.5 The steady state viscosity $\eta(\kappa)$ of rodlike polymers

REFERENCES

1. Yamakawa, H., Modern theory of Polymer Solutions, Harper and Row, New York, 1971.
2. Bird, R. B., O. Hassager, R. C. Armstrong, C. F. Curtiss, Dynamics of Polymeric Liquids, vol.2, Kinetic Theory, John Wiley & Sons, 1977.
3. Kirkwood, J. G., Rec. trac. chim., 68 (1949), 649. See also Kirkwood, J. G., Macromolecules, in Documents of Modern Physics, Gordon Breach, New York, 1967.
4. Fixman, M. and J. Kovac, J. Chem. Phys., 61 (1974), 4939, 4950; ibid., 63 (1975), 935.
5. Fixman, M. and G. T. Evans, J. Chem. Phys., 64 (1976), 3474; ibid., 68 (1978), 195.
6. Fixman, M., J. Chem. Phys., 69 (1978), 1527, 1538.

7. Kuhn, W. and H. Kuhn, Helv. Chim. Acta, 37 (1944), 97; ibid., 38 (1945), 1533; ibid., 39 (1946), 71.
8. Fraenkel, G. K., J. Chem. Phys., 20 (1952), 642.
9. Saito, N., J. Phys. Soc. Japan, 6 (1951), 297, 302.
10. Batchelor, G. K., J. Fluid Mech., 41 (1970), 545.
11. Bird, R. B., R. C. Armstrong and O. Hassager, Dynamics of Polymeric Liquids, vol.1, John Wiley & Sons, 1977.
12. Cheng, D. C. H. and F. Evans, Brit. J. Appl. Phys., 16 (1955), 1599.
13. Gadara-Maria, F. and A. Acrivos, J. Rheology, 24 (1980), 799.
14. Green, M. S. and A. V. Tobolsky, J. Chem. Phys., 14 (1946), 80.
15. Lodge, A. S., Elastic Liquids, Academic Press, New York, 1964.
16. Astarita, G. and G. Marrucci, Principles in Non-Newtonian Fluid Mechanics, MacGraw Hill, London, 1974.
17. Hassager, O., J. Non-Newtonian Fluid Mech., 9 (1981), 321.
18. Onsager, L., Ann. N.Y. Acad.Sci., 51 (1949), 627.
19. Flory, P. J., Proc. Roy. Soc. London, Ser. A, 234 (1956), 73.
20. Riseman, J. and J. G. Kirkwood, J. Chem. Phys., 17 (1949), 442.
21. Kirkwood, J. G. and P. L. Auer, J. Chem. Phys., 19 (1951), 281.
22. Hinch, J. and L. G. Leal, J. Fluid Mech., 52 (1972), 683; ibid., 76 (1976), 187.
23. Stewart, W. E. and J. P. Sorensen, Trans. Soc. Rheol., 16 (1972), 1.
24. Doi, M., J. Polymer Sci., 19 (1981), 229.
25. Doi, M., J. Phys. Paris, 36 (1975), 607.
26. Doi, M. and S. F. Edwards, J. Chem. Soc. Faraday Trans. II, 74 (1978), 568, 918.
27. Edwards, S. F. and K. F. Freed, J. Chem. Phys., 61 (1974), 1189.

28. Freed, K. F. and S. F. Edwards, J. Chem. Phys., 61 (1974), 3626.
29. Wang, F and B. H. Zimm, J. Polymer Sci., 12 (1974), 1619.
30. Batchelor, G. K., J. Fliud Mech., 46 (1971), 813.
31. Kuzuu, N. and M. Doi, Polymer J., 12 (1980), 883.
32. Jain, S. and C. Cohen, Macromolecules, 14 (1981), 759.
33. Papkov, S. P., U. G. Kulichikhin, V. D. Kalmykova and A. Y. Malkin, J. Polymer Sci. Phys. Ed., 12 (1974), 1753.
34. Kiss, G. and R. S. Porter, J. Polymer Sci. Phys. Ed., 18 (1980), 361.
35. Asada, T., T. Tanaka and S. Onogi, Polymer Preprints Japan, (1981), 1604.
36. Chu, S. G., S. Venkatraman, G. C. Berry and Y. Einaga, Macromolecules, 14 (1981), 939.

This research was supported in part by a Grant-in-Aid from the Ministry of Education, Science, and Culture of Japan.

Masao Doi
Department of Physics
Faculty of Science
Tokyo Metropolitan University
Setagaya, Tokyo, Japan

A Survey of Some Results in the Mathematical Theory of Fluidization

G. M. Homsy

INTRODUCTION

Fluidization refers to the phenomena which occur when a collection of particles is immersed in a fluid stream moving antiparallel to gravity at velocities sufficiently large that the drag exerted on the particles exceeds their net buoyant weight and they are thus free to move. The fluid may be either gas or liquid and for simplicity the particles will be taken to have the same density, shape, and size. The fluidized state has certain advantages when used as a reaction medium in which the particles themselves may be reactive or catalytic, or as a fluid-solids contacting device in which the solids are chemically inert but other transport processes are taking place, e.g. drying or heating. Thus fluidized beds have been under study and processes utilizing them have been under development for over fifty years. In spite of this development, there are many mechanical aspects of the motions of the individual phases which are still not well understood, and the theory may be considered to be in a primative state of development relative to more established areas of fluid mechanics. Nonetheless the mathematical theory of fluidization has progressed over the last decade to the point where there can be considerable confidence placed in continuum description of the motions of the phases. The framework for such a description is the modelling of the two-phase mixture as interpenetrating continua. We will briefly discuss some of this progress.

ISBN 0-12-493120-0

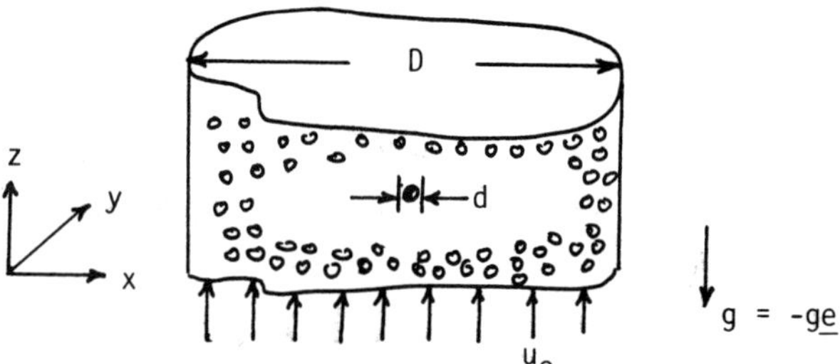

Figure 1. Definition Sketch

Figure 1 serves as a definition sketch and sets the coordinate system. An assemblage of particles of size d and density ρ_s is contained within a vessel of macroscopic dimension D. Fluid of viscosity μ and density ρ_f is introduced at the base with average velocity u_0 and flows upward in the vertical (z) direction. Gravity acts on the mixture and the coordinate system is as shown. The volume fraction of space occupied by fluid is the void fraction ε. The volume fraction of particles is thus $\phi = 1 - \varepsilon$. We are generally interested in the case $D \gg d$ and motions of length scale much larger than d, although not necessarily on the order of D, so that a continuum approach is appropriate.

BASIC EQUATIONS

Because the basic phenomenon involves a particulate phase which is not neutrally buoyant, and because this phase is typically characterized by relatively large density, and thus non-negligible inertia, the mathematical description is necessarily more complex than for other fluid - particle systems such as neutrally buoyant and slowly sedimenting suspensions. In turn, it is also of a relatively phenomenological nature, as we shall see.

Because of the large discrepancy in phase densities and velocities, it is necessary to write balances laws for each of the phases. Denoting $\underline{u}$, $\underline{v}$, ε as the continuum (volume averaged) fluid velocity, solids velocity and void fraction respectively, the basic equations thought to describe the fluidized state are

$$\frac{\partial \varepsilon}{\partial t} + \nabla \cdot (\varepsilon \underline{u}) = 0 \quad (1)$$

$$\frac{\partial (1-\varepsilon)}{\partial t} + \nabla \cdot ((1-\varepsilon)\underline{v}) = 0 \quad (2)$$

$$\varepsilon \rho_f \frac{Du}{Dt} = -\nabla(\varepsilon p^f) + \nabla \cdot \underline{T}^f + \underline{I} + \varepsilon \rho_f g \quad (3)$$

$$(1-\varepsilon)\, \rho_s \frac{Dv}{Dt} = -\nabla((1-\varepsilon)p^s) + \nabla \cdot \underline{T}^s - \underline{I} + (1-\varepsilon)\, \rho_s g \quad (4)$$

Here p is the pressure, $\underline{T}$ the extra stress, $\underline{I}$ the interphase force, and the subscripts f,s denote fluid and solid respectively. These equations have been discussed by Anderson & Jackson (1967), Drew & Segel (1971), Ishii (1975), and Homsy et al. (1980), among others.

A constitutive equation for $\underline{I}$, has been proposed of the form

$$\underline{I} = p^f \nabla \varepsilon - \frac{\varepsilon f(\varepsilon)\mu}{d^2}(\underline{u}-\underline{v}) - \varepsilon c(\varepsilon)\rho_f \frac{D(\underline{u}-\underline{v})}{Dt} \quad (5)$$

The material functions $f(\varepsilon)$, $c(\varepsilon)$ are the drag and virtual mass coefficients, respectively. More general forms of the drag term are possible if the particle Reynolds number is not small. See Drew et al (1979) for a discussion of the relative phase acceleration.

It is possible to manipulate equations (3-5) in a number of ways. Two forms will be of use in what follows. In the first, we display the buoyant force on the particles in a clear manner: equation (3) is unchanged, and Equations (3) and (4) may be combined to read:

$$(1-\varepsilon)\left[\rho_s \frac{Dv}{Dt} - \rho_f \frac{Du}{Dt}\right] = -\nabla((1-\varepsilon)(p^s-p^f)) + \frac{f(\varepsilon)\mu}{d^2}(\underline{u}-\underline{v})$$
$$+c(\varepsilon)\frac{D(\underline{u}-\underline{v})}{Dt} + (1-\varepsilon)(\rho_s-\rho_f)g + \underline{\nabla} \cdot \underline{T}^{s'} . \quad (6)$$

Here $\underline{T}^{s'}$ is a modified extra stress tensor for which constitutive equations must be given, (p^s-p^f) is a collisional pressure due to stresses transmitted by direct particle contacts and hence is referred to as a contact stress.

A second useful form of the equations may be obtained by adding

(1)-(2) and (3)-(4) to obtain balance laws for the mixture, taken as an effective fluid. These read

$$\nabla \cdot \bar{\underline{u}} = 0, \tag{7}$$

$$\bar{\rho} \frac{D\bar{u}}{Dt} = -\nabla \bar{p} + \nabla \cdot \bar{\underline{T}} + \bar{\rho} g, \tag{8}$$

where $\bar{p}$, $\bar{u}$ are the volume-averaged density and velocity, respectively, and $\bar{T}$ the effective extra stress. Eqs. (7)-(8) are sufficient to describe neutrally buoyant suspensions, and when appended to equation (2) and a simplified version of equation (4), have been used to study sedimentation in the quasi-static limit: see e.g. Acrivos & Herbolzheimer (1979).

Constitutive equations are required for $T^{s'}$, T^f in order to close these equations; these are usually taken to be of the Newtonian form:

$$\underline{T}^{s'} = \beta(\varepsilon)(\nabla \cdot \underline{v})\underline{\underline{I}} + 2\,\eta(\varepsilon)(\underline{d} - \tfrac{1}{3}(\text{trace } \underline{d})\underline{\underline{I}}) \tag{9}$$

$$\underline{d} = \frac{1}{2}\,(\nabla \underline{v} + \nabla \underline{v}^T). \tag{10}$$

Similar expressions for $\underline{T}^f$ apply.

The mathematical theory of fluidization thus consists of seeking solutions to these balance laws.

STEADY SOLUTIONS

It has been noted by many authors that these equations admit a particularly simple steady solution, referred to as homogeneous fluidization. This state is described by

$$\begin{aligned} \varepsilon &= \varepsilon_0 \\ \underline{u} &= u_0 \underline{e} \\ \underline{v} &= \underline{0}\,. \end{aligned} \tag{11}$$

The momentum equations reduce to

$$\frac{dp^f}{dz} = \left(\frac{f(\varepsilon_0)\mu u_0}{d^2} - \rho_f g\right), \tag{12}$$

$$(1-\varepsilon_0)\,\frac{d(p^s - p^f)}{dz} = \frac{f(\varepsilon_0)\,\mu\, u_0}{d^2} - (1-\varepsilon_0)(\rho_s - \rho_f)g \quad . \tag{13}$$

The point at which the contact stress has no gradient is that for which the drag balances the net weight of the particles. This is known as the "minimum fluidization velocity" and will be referred to as the critical velocity in what follows. It is given by

$$u_c = \frac{(1-\epsilon_0)(\rho_s - \rho_f) g d^2}{f(\epsilon_0)\,\mu} \quad . \tag{14}$$

Further features of this steady solution are discussed by Anderson and Jackson (1968) and Homsy *et al.* (1980).

SCALING

The equations may be scaled using the following characteristic quantities

velocity $\sim u_0$ (or u_c)

length $\sim d$

time $\sim d/u_0$

pressure $\sim \rho_s u_0^2$

It is then found that the equations contain three parameters,

$Re = \frac{u_0 \rho_f d}{\mu}$ particle Reynolds number

$R = \rho_f/\rho_s$ density ratio

$Fr = gd/u_0^2$ particle Froude number

and the dimensionless material functions

$f(\epsilon)$ drag coefficient

$c(\epsilon)$ virtual mass coefficient

$\hat{\eta} = \eta/\mu$, $\hat{\beta} = \beta/\mu$ rheological parameters

We expect that $f(\epsilon) \sim O(1)$; the magnitude of the other material functions is unknown.

LINEAR INSTABILITY THEORY

It is natural to determine the stability of the steady solution. The linearized version of the dimensionless equations is:

$$\frac{\partial\varepsilon}{\partial t} + \varepsilon_0 \; \nabla\cdot\underline{u} + \frac{\partial\varepsilon}{\partial z} = 0 \qquad (15)$$

$$-\frac{\partial\varepsilon}{\partial t} + (1-\varepsilon_0)(\nabla\cdot\underline{v}) = 0 \qquad (16)$$

$$(1-\varepsilon_0)\frac{D\underline{v}}{Dt} = \frac{R}{Re}\left[(\hat{\beta}+4\hat{\eta})\nabla.((\nabla\cdot\underline{v})\underline{\underline{I}}) + \eta\nabla\cdot(\underline{d}) + f(\varepsilon_0)(\underline{u}-\underline{v}) + f'(\varepsilon_0)\varepsilon\underline{e}\right]$$

$$+ R(1-\varepsilon_0)\left[\frac{\partial\underline{u}}{\partial t} + \frac{\partial\underline{u}}{\partial z}\right] + M\,\nabla\varepsilon + Rc(\varepsilon_0)\frac{D(\underline{u}-\underline{v})}{Dt} + (1-R)Fr\varepsilon\underline{e} \qquad (17)$$

The perturbed version of equation (3) serves to determine the pressure; it is not of immediate interest in the present discussion. Here we have set $M = -\varepsilon_0 \left.\frac{d(p^s-p^f)}{d\varepsilon}\right|_{\varepsilon_0}$. By taking the divergence of eq. (17) and using (15)-(16), the following evolution equation for $\varepsilon(\underline{x},t)$ may be derived:

$$A\varepsilon_{tt} + 2B\varepsilon_t + C\varepsilon_z - 2D\nabla^2\varepsilon_t + E\varepsilon_{zt} + F\varepsilon_{zz} - M\nabla^2\varepsilon = 0 \qquad (18)$$

A,B, etc. are positive constants, being combinations of the coefficients in eq. (17). Specific formulae are given in Didwania & Homsy (1982). The constant A can be set to one without loss of generality. ε_{tt} is associated with the inertia of the particles. The other associations are as follows: B is the gravity coefficient, C that associated with drag, D with viscous dissipation, E,F with the inertia of the fluid, and M the coefficient of the contact stress. The term $M\nabla^2\varepsilon$ also appears in treatments which model an elastic response of the solid phase, and in particular occurs in the treatment of magnetically stabilized systems; Rosenswaig (1979); thus M may be considered as a generalized elasticity.

Eq. (18) admits solutions of the form

$$\varepsilon = \hat{\varepsilon}\exp\left(\sigma t + i\underline{k}\cdot\underline{x}\right) \qquad (19)$$

and we shall be particularly interested in the dispersion relation $\sigma(k)$. This is given as the roots of the quadratic associated with eq. (18), i.e.

$$A\sigma^2 + 2B\sigma + ik_zC + 2D|k|^2\sigma + Eik_z\sigma - Fk_z^2 + M|k|^2 = 0 \qquad (20)$$

The following subcases are of interest.

(i) Inertialess quasitatic systems (A = E = F = M = 0)

$$\sigma = \frac{ik_z C}{2(B+D|k|^2)} \tag{21}$$

Thus in systems in which buoyancy, drag and damping are the only forces, only neutral waves are possible.

(ii) Inertialess fluid, no damping, no contact stresses.

$D = E = F = M = 0;\ A = 1$

$$\sigma = -B \pm \sqrt{B^2 - ik_z C} \tag{22}$$

This result indicates that instabilities are possible if they are periodic in the vertical direction, $k_z \neq 0$. A sketch of the growth constant as a function of wave number is given in Figure 2, and indicates that

$$\sigma_r \sim k_z^2, \qquad k_z \to 0$$

$$\sigma_r \sqrt{k_z}, \qquad k_z \to \infty \tag{23}$$

The behavior at high wave number is unacceptable, as it indicates a breakdown of the continuum theory, and is in disagreement with experiment: Jackson (1963), Anderson and Jackson (1968), El-Kaissy & Homsy (1976). The inclusion of viscous damping $D \neq 0$ modifies this result with

$$\sigma = -(B+D|k|^2) \quad \pm \sqrt{(B+D|k|^2)^2 - Cik_z} \tag{24}$$

The properties of (24) are

$$\sigma_r \sim 0(|k|^2),\ |\underline{k}| \to 0,$$

$$\sigma_r \sim 0(1),\ |\underline{k}| \to \infty$$

For D sufficiently large, the growth constant is a maximum at a finite wave number, as may be established by direct evaluation of (24); this behavior also sketched in Figure 2. The addition of damping furthermore indicates that the growth rate is a maximum for $\underline{k} = (0,0,k_z)$, i.e. for wave-fronts perpendicular to the direction of the mean flow. In fact, it is easy to establish that

$$\sigma(|k|) = \sigma(k_z) - a\,k_x^2 - b\,k_y^2 + 0\,(k_x^4 + k_y^4). \tag{25}$$

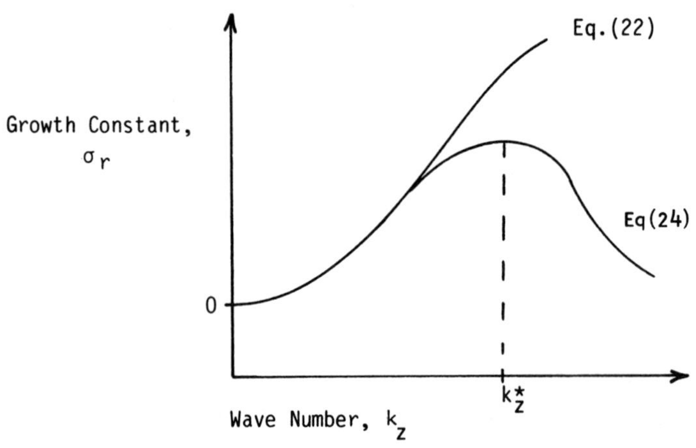

Figure 2. Stability Characteristics of the Uniform State.

(iii) Effect of contact stresses (elasticity)

Consider the situation without viscous damping and an inertialess fluid, D = E = F = 0, A = 1.

$$\sigma = -B \pm \sqrt{B^2 - ik_z C + M|\underline{k}|^2}. \tag{26}$$

It may be shown that elasticity can stabilize the otherwise unstable mode if

$$M > C^2/4 \tag{27}$$

a result due to Rosenswaig (1978), but implicit in the work of Anderson & Jackson (1968). When M is due simply to contact stresses, it cannot stabilize the steady solution for all values of u_0, since there is a void fraction ε_0 above which this term will vanish. If however the term M is a truly independent parameter, as in the case of magnetic beds at or near magnetic saturation, then the steady solution may be stabilized when the condition (27) may be met for sufficiently strong fields or low flows. For further details, see Rosenswaig (1979).

Experiments have been conducted in which growth constant, wave number, and phase velocity have been measured, and when used to fit the material functions $c(\varepsilon)$, $M(\varepsilon)$, $\hat{\eta}(\varepsilon)$, $\hat{\beta}(\varepsilon)$ appearing in the full solution of eq. (20), yield results which are of reasonable magnitude and internally consistent over a wide range of particle size, fluid velocity, and void fraction: see Homsy et al. (1980).

WEAK NON-LINEAR WAVES

Experiments by Didwania & Homsy (1981) have shown that the one-dimensional wave train characterized by wave-vector $\underline{k}_c = (0,0,k_z^*)$ becomes unstable to oblique waves having a two-dimensional structure. Thus the primary wave becomes unstable in the non-linear regime by resonance with side bands, and the following resonance conditions between a triad of waves may be easily established: see e.g. Philips (1974).

$$\underline{k}_1 = \underline{k}_2 + \underline{k}_3 \tag{28}$$

$$\sigma_{i1} = \sigma_{i2} + \sigma_{i3} \tag{29}$$

By assuming that the resonance is between the first harmonic and the side-bands $\underline{k}_{2,3} = (\pm\gamma,\ 0,\ k_z^*)$, (an assumption well-supported by experimental evidence), eq. (28) is satisfied, and eq. (29) reads

$$\sigma_i(0,2k_z^*) = \sigma_i(-\gamma,k_z^*) + \sigma_i(+\gamma,k_z^*) \tag{30}$$

Since the full dispersion relation is known as the solution of eq. (20), the resonance condition may be solved for γ as a function of the parameters A,B,C, etc. The resulting values of γ are in good agreement with those obtained experimentally: for details see Didwania & Homsy (1982).

VORTEX INSTABILITIES

It has been known for some time that if fluidization occurs under conditions for which the distributor has only a small pressure drop relative to that of the bed, motions are observed which are periodic in the horizontal and which decay in the vertical. In "two-dimensional" beds these take the form of two-dimensional cellular flows while in three dimensions they are associated with a toroidal vortex near the distributor. These have been analyzed in two dimensions by Medlin, Wong and Jackson (1974). Now two fundamental solutions to eq. (18) exist of the form

$$\varepsilon = \exp(\sigma t + i\,(k_x x + k_y y) + rz) \tag{31}$$

where r and σ are related through eq. (18). Unlike the case of traveling waves, when the eigenmodes of eq. (18) were of interest, these solutions are determined by the relevant boundary conditions at the base and top of the fluidization mixture. It is easy to show that in the absence of any coupling of perturbations in void fraction with the other dynamic variables, (fluid & solid velocities and pressure), all solutions of the form of eq. (31) are stable. If the distributor has a finite resistance to flow comparable to the pressure drop across the bed, however, then pressure perturbations in the clear fluid below the distributor plate couple to those in the bed and result in boundary conditions which admit solutions of the form of eq. (31). The analysis is algebraically complicated however, because of the need to determine the fundamental solutions of eqs. (15-17) and the linearized dimensionless version of eq. (2), since all flow quantities are coupled at the boundaries. For details, see Medlin _et al_. (1974).

A key parameter of the theory which emerges is the aforementioned ratio of pressure drops,

$$\delta \equiv \frac{Khu_0}{\rho_s gH}\left[\frac{1}{1 + (1-\varepsilon_0)(1-R)}\right] \tag{32}$$

where K and h are the permeability and thickness of the distributor, and H the height of the fluidized mixture. The results of the theory are sketched schematically in Figure 3. Figure 3 indicates how the growth constant σ depends upon horizontal wave number and the parameter δ. Two features are of interest. First, for fixed values of the other parameters, there is always a cut-off wave number which shifts to longer wave lengths as δ is increased. Second, there is a "critical" value of δ above which there is complete stability. These features depend upon the material functions in a manner which is more complicated than in the case of the traveling wave instabilities, but which may be established by numerical analysis. It is found that experiments in which the cut-off wave numbers are measured are in good agreement with the theory in which values of these parameters are chosen consistent with those values measured by wave propagation: see Agarwal, Hudson & Jackson (1980).

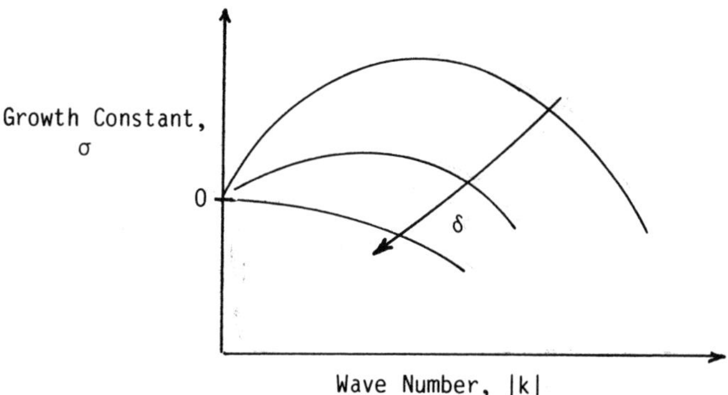

Figure 3. Schematic of the Results for the Vortex Instability.

A physical mechanism for the instability may be seen as follows. For small δ, pressure fluctuations in the clear fluid below the distributor are communicated to the fluidized mixture through the distributor. In particular, an increase in pressure in manifested by an increase in fluid velocity, which in turn produces a local increase in void fraction through an accompanying solids movement due to drag. Now an increase in void fraction results in a lowered resistance to flow in the fluidized mixture (decreased friction coefficient, $f(\varepsilon)$ or equivalently increased permeability) in the vertical plane above the pressure disturbance. Under conditions of fixed total flow, this results in a preferential flow of fluid above this point and a partial defluidization of the areas out of phase with the positive pressure fluctuation, with the clear possibility of the motion becoming self-sustaining. As the plate pressure drop increases, this coupling decreases, until a critical condition is reached where the coupling between pressure fluctuations and the velocity and voidage fluctuations they produce is too weak to drive an instability. In the limit in which there is no coupling, we recover the result that all eigensolutions of the form of eq. (31) are stable.

We see that this instability is related to the Saffman-Taylor-Chouke instability in displacements in porous media. In the former case the flow resistance is decreased by an increase in voidage; in the latter it is produced by a decrease in flow resistance due to preferential flow of a second, less viscous fluid.

STEADY TRAVELLING WAVE SOLUTIONS

There is general interest in the propagation and stability of fronts or shocks across which there is a rapid variation in void fraction. This situation has been analyzed rigorously only in the case of little or no inertia associated with either phase. It is convenient to use the equivalent set of governing equations

$$\nabla \cdot u = 0 \tag{7}$$

$$\rho \frac{D\overline{u}}{Dt} = -\nabla\overline{p} + \nabla \cdot \overline{\underline{T}} + \rho g \tag{8}$$

and for the particle phase (with $\phi = 1 - \varepsilon$),

$$\frac{\partial \phi}{\partial t} + \nabla \cdot (\phi \underline{v}) = 0 \tag{33}$$

$$\underline{v} - \underline{u} = v_s(\phi)\underline{e} \tag{34}$$

Thus the slip velocity is taken as simply the Stokes hindered settling velocity in the direction of gravity. It is well-known that shock-like solutions of eqs. (7-8, 32-33) exist in which

$$\overline{u} = \text{constant},$$
$$\phi = \text{constant}$$
$$\nabla\overline{p} = \text{constant}$$

on either side of the shock, and the shock speed c may be easily determined as

$$c = \frac{(v_s^- - v_s^+)(\phi^+)}{(\phi^+ - \phi^-)} \tag{35}$$

a result first due to Slis *et al*. (1959).

Such shocks may be produced by a sudden increase or decrease in the velocity of the fluid, and when care is taken to ensure uniform distribution of the fluid, and when the relative accelerations are low such that eq. (34) is justified, the existing experiments are in good agreement with eq. (35); see Didwania & Homsy (1981).

RAYLEIGH-TAYLOR INSTABILITIES

Considerable interest in gravitational instabilities of the Rayleigh-Taylor type has accrued in connection with the question of bubble stability, and essentially identical treatments have been given by Rice & Wilhelm (1958), Upson & Pyle (1973) and Clift *et al*. (1974). In the simplest of treatments, the mixture is taken as an effective fluid and eqs (7-8) are solved subject to conditions at the "interface" between the effective fluids:

$$[[\underline{u}]] = 0 \tag{36}$$

$$\left[\left[p + \frac{1}{Re} w_z\right]\right] = B\eta \tag{37}$$

$$\frac{\partial \eta}{\partial t} = w \tag{38}$$

$$[[\underline{e} \cdot \underline{T} \cdot \underline{t}]] = 0 \tag{39}$$

In these equations, w is the vertical velocity, η the interfacial location and $\underline{e}$, $\underline{t}$ the vectors normal and tangent to the interface. The bracket denotes a jump. Equation (36) expresses continuity of velocity, eq. (38) with $w = \underline{u} \cdot \underline{e}$ and B the Bond number gives the normal stress balance, eq. (38) is the kinematic condition and eq. (39) the tangential stress balance. It is well known for the semi-infinite fluids with no surface tension, normal modes of the form

$$\exp\ (\sigma t + i(k_x\ x + k_y y) \pm |k|z) \tag{40}$$

exist and that the dispersion relation depends strongly upon the viscosities of the phases: see e.g. Chandrashekhar (1961). For effective viscosities consistent with those measured in the wave propagation and vortex instability cases the theory predicts a high degree of instability rarely seen in experiment. The most complete analysis of the full two-fluid equations together with a discussion of the relevant experiments has been given recently by Didwania & Homsy (1981). The details are complicated, but the conclusion is that closures of the form of eqs. (9-10) cannot be correct for systems of relatively low voidage; and that finite yield-stresses exist, especially for weakly fluidized systems.

CONCLUSIONS

From the above survey, we may draw the following conclusions:

1. The mathematical properties of the simplest of solutions have been established and verified by experiment, namely:

(i) There exists an exact steady solution of the full equations (the homogeneous state) and it is unstable to waves with wave vectors in the direction of the mean flow.

(ii) The steady solution may be stabilized by an "elasticity" of an independent origin.

(iii) Travelling shock-like solutions exist in the quasi-static limit of no appreciable particle inertia.

2. By close quantitative interplay between theory and experiment, the magnitude of the material functions appearing in the two-fluid equations has been established. The agreement is most firmly established in the following cases:

(i) The propagation and secondary instabilities of instability waves.

(ii) The magnetic stabilization of fluidized beds.

(iii) The existence and suppression of vortex instabilities.

(iv) Rayleigh-Taylor instabilities of unstable shocks.

We do comment, however, that the majority of these problems are linear, and no (non-trivial) nonlinear problems have been solved to date.

REFERENCES

1. Acrivos, A. and E. Herbolzheimer, 1979, J. Fluid Mech. 92, 435.
2. Agarwal, G.P. Hudson, J.L. and R. Jackson, 1980, I.&E.C. Fund. 19, 59.
3. Anderson, T.B. and R. Jackson, 1967, I&EC Fund. 6, 527.
4. Anderson, T.B. and R. Jackson, 1968, I&EC Fund. 7, 12.
5. Chandrasekhar, S. Hydrodynamic & Hydromagnetic Stability, 1961, Oxford Univ. Press.
6. Clift, R., Grace, J.R. & M.E. Weber, 1974, I&EC Fund. 13, 45.
7. Didwania, A. & G.M. Homsy, 1981, I&EC Fund 20, 318.
8. Didwania, A. & G.M. Homsy, 1981, Int. J. Multiphase Flow 7, 563.
9. Didwania, A. & G.M. Homsy, 1982, Resonant Side-Band Instabilities in Wave Propagation in Fluidized Beds, to appear in J. Fluid Mech.

10. Drew, D. & L. Segel, 1971, Stud. App. Math. 50, 205.
11. Drew, D., L. Cheng & R. Lahey, 1979, Int. J. Multiphase Flow, 5, 233.
12. El-Kaissy, M. & G.M. Homsy, 1976, Int. J. Multiphase Flow, 2, 379.
13. Homsy, G.M., El-Kaissy, M. & A. Didwania, 1980, Int. J. Multiphase Flow, 6, 305.
14. Ishii, 1975, Thermo-Fluid Dynamic Theory of Two Phase Flow, Eyrolles, Paris.
15. El-Kaissy, M. & G.M. Homsy, 1976, Int. J. Multiphase Flow, 2, 379.
16. Medlin, J. Wong, H.W., & R. Jackson, 1974, I.&E.C. Fund. 13, 247.
17. Philips, O., 1974, Ann. Rev. Fluid Mech. 6, 93.
18. Rice, W.J. & R.H. Wilhelm, 1958, AIChEJ, 4, 423.
19. Rosenswaig, R., 1979, I.&E.C. Fund. 18, 260.
20. Slis, P.L., Willemse, Th.W. & H. Kramers, 1959, App. Sci. Res. A8, 209.
21. Upson, P.C. & L.D. Pyle, 1973, Proc. Int. Symp. Fluid. Appls. Toulouse, France.

I wish to acknowledge the support of the U.S. Department of Energy and the Particulates & Multiphase Processing Section of the National Science Foundation for the support of the research reported upon in references (7,8,9,12,13) and of the preparation of this paper.

G. M. Homsy
Department of Chemical Engineering
Stanford University
Stanford, California 94305

Instability in Settling of Suspensions Due to Brownian Effects

R. Caflisch and G. Papanicolaou

It has been known for a long time that initially uniform suspensions of hydrophobic colloidal particles sediment in a manner that is not compatible with the elementary theory of Brownian motion. That is, the particles do not sediment to reach an equilibrium concentration that is linearly increasing with depth into the container but instead the concentration forms nearly constant layers and sharp transitions to higher densities as the depth increases. Recent experimental work by Siano [1] confirms and delineates quantitatively this phenomenon. References to earlier work can also be found in [1].

From Siano's work one may conclude that the principal mechanism for this effect must still be Brownian motion and not electrostatic forces, boundary influences etc. However, since the elementary theory of noninteracting Brownian particles is not suitable one must account for interactions.

We shall describe briefly our work [2] which we believe gives a theoretical explanation of Siano's experiments. The main point of [2] is that starting from a suitable description of the problem by a Fokker-Planck equation (equation (2) below) and using a scaling that fits reasonably well with Siano's experiment we derive an effective equation for the evolution of the concentration $\phi(t,x)$ which has the form

$$\phi_t = \alpha\nabla^2(\sigma\phi) - \varepsilon^2\alpha^3\nabla^2[\sigma_1\nabla^2\phi + \sigma_2(\nabla\phi)^2] \qquad (1)$$

ISBN 0-12-493120-0

where α, σ, σ_1 and σ_2 are functions of ϕ that are defined below and ε is a small dimensionless parameter. This equation, which arises in the Cahn-Hilliard theory of spinodal decomposition, was proposed (in a somewhat different form) by Siano [1] as a possible governing equation. We have succeeded in deriving this equation from a more basic description of the problem and in identifying the parametric functions in (1) in terms of quantities that are natural in that description.

The fact that (1) can account qualitatively for the observed form of the concentration is a consequence of the properties of σ, σ_1 and σ_2. Namely, $1+\phi\sigma'/\sigma$ becomes negative for a certain range of values of ϕ but σ_1 and σ_2 are such that the overall equation (1) is restabilized. The uniform state ϕ = constant is not however stable and new, nonuniform, steady states arise [3].

Let us now give a description of the formulation presented in [2] along with the relevant scaling that leads to (1).

We assume that the suspension is confined to a container and so its average velocity $u(t,x)$ is zero. Let $f(t,x,\xi)$ denote the phase space volume density of particles, i.e. the average volume of particles at x (per unit volume) going with velocity ξ (per unit volume in velocity space) at time t. We shall assume that f satisfies the Fokker-Planck equation

$$f_t + \xi\cdot\nabla f + \nabla_\xi\cdot\left[\left(g\gamma\left(1-\frac{\rho_f}{\rho_p}\right) - M^{-1}6\pi\mu\theta a\xi\right)f\right] = D\delta\nabla_\xi^2 f \tag{2}$$

in which ∇ is the x-gradient and ∇_ξ is the ξ-gradient. The average volume fraction or concentration of the particles is

$$\phi = \int f d\xi \tag{3}$$

while the average number density n and average particle velocity v are given by

$$n = \phi \,/\, \tfrac{4}{3}\pi a^3 , \qquad \phi v = \int \xi f d\xi \ . \tag{4}$$

In (2) - (4) appear several parameters defined as follows:

$$
\begin{aligned}
a &= \text{radius of particles} \\
\mu &= \text{viscosity of the fluid} \\
\rho_p &= \text{mass density of a particle} \\
\rho_f &= \text{mass density of the fluid} \\
M &= \tfrac{4}{3}\pi a^3 \rho_p = \text{particle mass} \\
g &= \text{gravitational acceleration vector pointing downward} \\
D &= \frac{kT}{M}\,\frac{6\pi\mu a}{M} = \text{velocity diffusion coefficient (derived by Einstein) in which} \\
k &= \text{Boltzmann's constant and} \\
T &= \text{temperature of the suspension assumed constant.}
\end{aligned}
\tag{5}
$$

There are also several dimensionless parametric functions appearing in (2) which depend on the concentration ϕ

$$\gamma(\phi) = \left(1 - \frac{\rho_p\phi + (1-\phi)\rho_f}{\rho_p}\right) \Big/ (1 - \rho_f/\rho_p) = 1 - \phi \tag{6}$$

modifies the gravitational acceleration to account for the additional buoyancy of the fluid due to the suspended particles.

$\theta(\phi)$ modifies the Stokes drag on a moving particle (hindrance effects due to concentration)

$\delta(\phi)$ modifies the velocity diffusion coefficient.

We also define σ by

$$\sigma(\phi) = \delta/\theta \quad , \tag{7}$$

and this is the same function σ appearing in (1).

Equation (2) is a reasonable model for describing the suspension provided one has a good idea what the function $\theta(\phi)$ and $\delta(\phi)$ look like in their dependence on concentration. It is through these functions that interaction effects come in and make (2) a nonlinear equation. As far as θ is concerned there are experiments as well as theory available. The theory [4] is confined to low concentrations. The experiments give for θ expressions of the form

$$\theta(\phi) = (1 + c_1\phi^{1/3})(1 + c_2\phi) \tag{8}$$

where c_1 and c_2 are positive constants depending on the size of the particles [5,6] (small particles $c_1 = 0$ [4]).

The behavior of δ as a function of the concentration does not seem to be known experimentally or theoretically. In [2] we argue that $\delta(\phi)$ should behave like the selfdiffusion coefficient for interacting Brownian particles. Note that (2) is a Fokker-Planck equation in phase space describing interacting Ornstein-Uhlenbeck particles (particles whose velocity is an Ornstein-Uhlenbeck process). We therefore assume that $\delta(\phi)$ decreases with concentration. This is crucial in order that (1) have the properties we want.

We are at present looking into various simplified problems that may lead to a theoretical determination of properties of $\delta(\phi)$.

Typical parameter values from Siano's experiments are

$$\begin{aligned} a &= .5 \times 10^{-4} \text{ cm} \\ \mu &= .01 \text{ dynes sec cm}^{-2} \\ \rho_p &= 1.05 \text{ gr cm}^{-3} \\ \rho_f &= 1.00 \text{ gr cm}^{-3} \\ T &= 293 \text{ deg K} \\ \phi &= .001 \ (.1\,\% \text{ concentration}) . \end{aligned} \tag{9}$$

With these values and the constants $k = 1.38 \times 10^{-6}$ ergs/°K and $g = 980$ cm/sec we find that the Stokes settling velocity for a single particle is

$$v_{ST} = .3 \times 10^{-6} \text{ cm/sec} \tag{10}$$

while the RMS velocity of the particles is

$$v_{RMS} = .3 \text{ cm/sec} . \tag{11}$$

We nondimensionalize (2) using 1 cm and 10^5 sec (days) as the laboratory reference space and time scales. Three dimensionless parameters appear which lead to the scaled equation

$$\varepsilon f_t + \xi \cdot \nabla f + \nabla_\xi \cdot \left[\left(b\gamma \hat{z} - \frac{c}{\varepsilon} \theta \xi \right) f \right] = \frac{c}{\varepsilon} \delta \nabla_\xi^2 f \tag{12}$$

where $\varepsilon \sim 10^{-5}$ and b and c are dimensionless numbers of order one.

We have analyzed (12) in two ways. First we carried out a linearized stability analysis about the equilibrium solution

$$f_o(\xi) = \phi_o (2\pi\sigma)^{-3/2} \exp\{-(\xi - v_o)^2 / 2\sigma\} \tag{13}$$

where

$$v_o = \hat{z} b\gamma / c\theta \qquad \text{and} \qquad \sigma = \delta/\theta \quad . \tag{14}$$

In this calculation we do not use the fact that ε is small. We calculate explicitly the dispersion relation. We find [2] that if

$$1 + \frac{\sigma'}{\sigma} \phi_o = 1 + \left(\frac{\delta'}{\delta} - \frac{\theta'}{\theta} \right) \phi_o < 0 \tag{15}$$

then there are wavenumbers k which grow exponentially (i.e. are linearly unstable). For k small the dispersion relation (in suitable variables) has the form

$$\lambda = (d_1 - 1)k^2 - d_1 \left(\frac{3}{2} d_1 - 1 \right) k^4 + O(k^6) \tag{16}$$

where

$$d_1 = -\frac{\sigma'}{\sigma} \phi_o = \left(\frac{\theta'}{\theta} - \frac{\delta'}{\delta} \right) \phi_o \quad .$$

The second way we analyze (12) is as a full nonlinear problem with exact equilibrium solution (13) and the small parameter ε. The method we use is a variant of the usual Chapman-Enskog procedure in kinetic theory. It is after this is done that equation (1) emerges (with some other simplifications) and

$$\begin{aligned} \alpha &= \frac{1}{c\theta} \\ \sigma_1 &= \sigma'\sigma\phi + \frac{3}{2}\sigma'^2\phi^2 \\ \sigma_2 &= \frac{1}{2}\frac{\alpha}{\alpha}(\sigma+\sigma'\phi)^2 + \sigma'^2\phi - 2\sigma'\sigma + \sigma''\left(\frac{3}{2}\sigma'\phi^2 - \frac{1}{2}\sigma\phi\right) \quad . \end{aligned} \tag{18}$$

In conclusion, we have shown how starting from a Fokker-Planck equation (2) one can derive (1) and interpret Siano's experiments via the instability and restabilization properties that (1) displays. The clustering effects seen in the suspensions by Siano do seem to be due to Brownian interaction effects. However they could not be explained from a model that assumes spatial diffusion for the particles (interacting Brownian particles) because for such a model one could not come up with an equation like (1). It is necessary to start with a model of diffusion in phase space such as (2).

At least two important tasks remain to be carried out. One, to obtain a reliable theory for the function $\delta(\phi)$ appearing in (2). Two, to obtain quantitative agreement between the equilibrium pattern observed in [1] and those deduced from (1). We do not have such agreement at present and this may be due to a number of reasons such as neglect of gravity in (1), neglect of bottom effects, etc.

REFERENCES

1. Siano, D.B., Layered sedimentation in suspensions of monodisperse spherical colloidal particles, J. Colloid. Interface Sci. 68 (1979) 111-127.
2. Caflisch, R. and G. Papanicolaou, Dynamic theory of suspensions with Brownian effects, SIAM J. Appl. Math., to appear.
3. Novic-Cohen, A. and L.A. Segel, Nonlinear aspects of the Cahn-Hilliard equation, submitted to Physica D.
4. Batchelor, G.K., Sedimentation in a dilute suspension of spheres, J. Fl. Mech. 52 (1972) 245-268.
5. Barnea, E. and J. Mizrahi, A generalized approach to fluid dynamics of particulate systems I, Chem. Eng. J. 5 (1973) 171-189.
6. Oliver, D.R., Chem. Eng. Sci. 15 (1961) 230.

R. Caflisch
Department of Mathematics
Stanford University
Stanford, CA 94305

G. C. Papanicolaou
Courant Institute
New York University
New York, NY 10012

Enhanced Sedimentation in Vessels Having Inclined Walls. The Boycott Effect

A. Acrivos, R. H. Davis, and E. Herbolzheimer

1. INTRODUCTION.

The separation of particles from a solid-liquid suspension by gravity settling is an important step in many physical operations. The process is often very slow especially when the particles are small, and hence there exists a need for constructing simple devices which could accomplish this separation rapidly. One such class of devices, known commercially as "supersettlers," consists of settling vessels with inclined walls where the retention times can, in principle, be as low as a few minutes rather than hours.

This phenomenon of enhanced sedimentation was first described by Boycott [3] who observed that "when...blood is put to stand in narrow tubes, the corpuscles sediment a good deal faster if the tube is inclined than when it is vertical." This increase in the settling rate results primarily from the fact that--c.f. Figure 1--when the walls are vertical, a particle within the suspension has to travel an O(H) distance before settling whereas, when the walls are inclined, the settling distance becomes O(b). Thus, according to this simple argument the relative augmentation in the rate of settling should be O(H/b) which, in principle, could be made arbitrarily large by decreasing the spacing between the two walls of the vessel.

THEORY OF DISPERSED MULTIPHASE FLOW

ISBN 0-12-493120-0

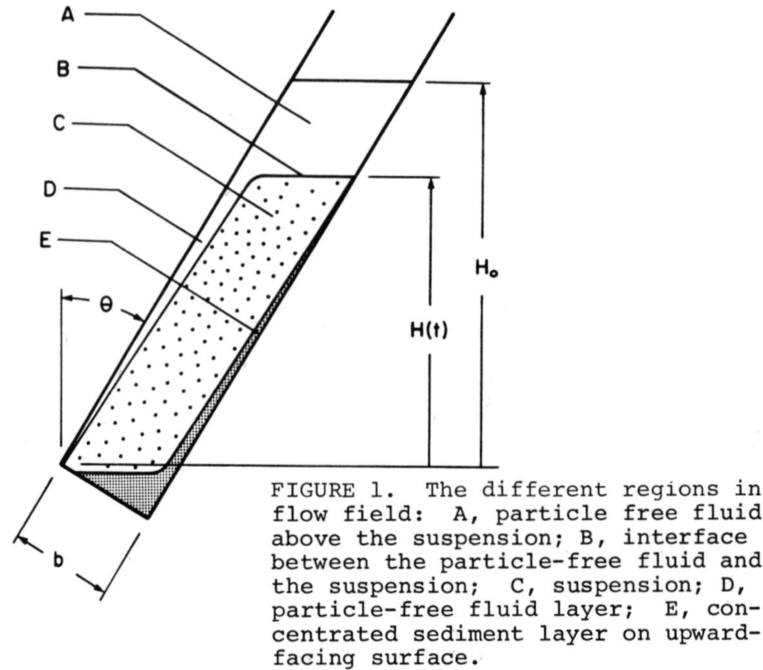

FIGURE 1. The different regions in flow field: A, particle free fluid above the suspension; B, interface between the particle-free fluid and the suspension; C, suspension; D, particle-free fluid layer; E, concentrated sediment layer on upward-facing surface.

The enhancement in the sedimentation rate can also be viewed as resulting from the fact that whereas, in a channel with vertical walls particles can only settle onto the bottom, in a tilted channel such particles can also settle onto the upward-facing wall as shown in Figure 1. These particles then form a thin sediment layer which rapidly slides down towards the bottom of the vessel under the action of gravity. Thus, one can interpret the increase in the settling rate as due to an increase in the surface area available for settling. Of course, since the suspension is incompressible, coincident with the removal of particles from the suspension, there must also occur an accompanying production of clarified fluid. This takes place at the top of the suspension and along the downward facing wall where a thin clear-fluid layer is formed --c.f. Figure 1. This particle-free layer is buoyant compared to the bulk suspension, and the fluid within it flows rapidly to the top of the vessel. Also, this layer reaches a steady-state thickness in a relatively short amount of time because the fluid that is being entrained into the layer exerts a drag force on the particles at the interface that counter-balances the normal component of the gravitational force on these particles.

The augmentation in the settling rate predicted by the above arguments holds true only so long as the flow in the channel remains laminar. Under some conditions, however, waves are seen to form at the interface between the suspension and the layer of particle-free fluid underneath the downward facing wall. These waves grow as they travel up the vessel and often break before reaching the top of the suspension. This phenomenon of wave-break-up is detrimental to the efficient operation of such systems because fluid that has already been clarified is remixed with the suspension. Such waves have been observed both in this study and in earlier investigations (Pearce [12], Probstein, et al. [14], and Mund, et al. [10]). As a first step to investigate in detail the conditions for wave-formation and break-up, a linear-stability analysis has been performed to predict the maximum amplification rates of such waves. A prerequisite to this stability analysis is the development of the base state velocity profiles for the laminar flow regime which have already been presented recently by Acrivos and Herbolzheimer [1] and by Herbolzheimer and Acrivos [8]. The analysis of [1] is restricted to situations where the thickness of the clear-fluid slit beneath the downward facing surface of the vessel is much less than the spacing between the plates, whereas more narrow channels in which the clear-fluid layer occupies an appreciable fraction of the channel-width are considered in [8]. Reference [1] also includes a detailed description of the "Boycott effect" discussed above, as well as a summary of many of the earlier papers on the subject.

In what follows we shall first summarize the laminar flow theory presented in [1] and [8] and then briefly consider the linear stability analysis. Finally, experimental results for both stable and unstable flows will be compared to the theoretical predictions.

2. THEORY AND EXPERIMENTS FOR LAMINAR FLOW.

The analysis of [1] and [8] considered the settling of a suspension composed of a Newtonian fluid and identical solid spheres under conditions of vanishingly small particle Reynolds numbers and developed solutions of the equations of continuum mechanics by analytical techniques. A key feature of this analysis was to treat the suspension as an effective fluid and express the ensemble-averaged fluid velocity, $\vec{u}_f$, and particle velocity, $\vec{u}_p$, in terms of the bulk average velocity, $\vec{u}$, and the average slip velocity, $\vec{u}_s$, defined as

$$\vec{u} = (1-c)\vec{u}_f + c\vec{u}_p, \tag{2.1a}$$

$$\vec{u}_s = \vec{u}_p - \vec{u}, \tag{2.1b}$$

where c is the local volume-fraction of solids. The slip velocity was then assumed to be in the direction of gravity and equal in magnitude to the vertical settling velocity of the particles. This assumption is valid if we restrict our attention to flows in which the particle Reynolds number is negligibly small--a condition that is satisfied in most systems of practical interest owing to the small size and settling velocity of the suspended particles.

Furthermore, the analysis was restricted to laminar flows and concluded that, aside from the vessel geometry, a complete description of the settling process is governed by two dimensionless parameters: R, a sedimentation Reynolds number, and Λ, the ratio of a sedimentation Grashof number to R, where

$$R \equiv \frac{\rho H v_o}{\mu} \quad \text{and} \quad \Lambda \equiv \frac{H^2 g(\rho_s-\rho)c_o}{v_o \mu} \tag{2.2}$$

with ρ being the density of the fluid, ρ_s that of the solids, μ the viscosity of the fluid, H the vertical height of the suspension, g the gravitational constant, c_o the initial volume fraction of solids, and v_o the particle settling velocity. Since in most cases of interest, Λ is $O(10^6)-O(10^8)$, while R is only $O(1)-O(10^2)$, an asymptotic analysis for determining the details of the motion was developed in [1] and [8] by taking $\Lambda \gg 1$ with $R\Lambda^{-1/3} \leq O(1)$. In particular, this asymptotic solution gave explicit expressions for the thickness of the clear-fluid slit that forms beneath the downward-facing wall of the vessel--cf. Figure 1--and for the velocity profile both within this slit and within the adjacent portion of the particle-filled region.

A simple, but important, result shown in [1] is that, when $\Lambda >> 1$ and when the solids concentration is initially uniform throughout the domain (batch settling), or when the feed concentration is steady (continuous settling), $c = c_o$ everywhere within the bulk suspension throughout the duration of the settling process, except in the particle-free fluid above the suspension and underneath the downward-facing wall, and in the concentrated sediment on the upward-facing wall and on the bottom of the vessel.

Another important result of this asymptotic analysis of [1] is that it confirms the validity of the expressions for the clarification rate and for the rate of descent of the top of the suspension given by an elementary kinematic model by Ponder [13] and independently by Nakamura and Kuroda [11]--the so-called PNK theory--provided that the flow is laminar and that $\Lambda >> 1$. These expressions are, for the parallel plate geometry of Figure 1,

$$S(t) = \frac{v_o b}{\cos\theta}\left(1 + \frac{H}{b}\sin\theta\right) \quad , \qquad (2.3)$$

$$\frac{dH}{dt} = -v_o\left(1 + \frac{H}{b}\sin\theta\right) \quad , \qquad (2.4)$$

where θ is the angle of inclination of the plates from the vertical and b is the channel width.

Batch experiments described in Section 4 of [1] under the following set of conditions: $2 \le H_o/b \le 8$, $0^o \le \theta \le 50^o$, $0.76 \le R_o \le 3.04$, $4.8 \le 10^5 \le \Lambda_o \le 1.5\times10^7$, $0.01 \le c_o \le 0.10$ (where H_o refers to the initial height of the suspension, and similarly for R_o and Λ_o), gave results in very good agreement with these theoretical predictions. One particularly sensitive test of the theory involves measuring the thickness of the clear fluid slit beneath the downward facing wall of the vessel. For the parallel-plate geometry it was predicted in [1] that, after a short initial time period, $\delta(x)$, the thickness of this layer as a function of x, the distance along the downward-facing surface, should be given by

$$\delta(x) = \Lambda^{-1/3}(3x\tan\theta)^{1/3} , \qquad (2.5)$$

where both δ and x have been made dimensionless with H. Figure 5 of [1] compared experimental measurements of this thickness with the above prediction, and the agreement is seen to be very good.

More recently, a continuous settling system has been fabricated, and experiments have been performed under the following set of conditions: $3 \leq H/b \leq 20$, $0^o \leq \theta \leq 40^o$, $0.1 \leq R \leq 9.8$, $6.0\times10^5 \leq \Lambda \leq 6.0\times10^7$, $0.005 \leq c_o \leq 0.05$. As long as the flow remained laminar, these experiments also gave results in good agreement with the analysis of [1]. However, rather than forming a sharp, horizontal interface separating the top of the suspension from the clear-fluid above (as was the case in batch experiments), the suspension was observed to spread near the top of the vessel. This was accompanied by a continuous decrease in the solids concentration with height in that region. It is believed that this spreading resulted from the polydispersity of the sieved particles, with the smaller particles accumulating at the top of the suspension. A detailed analysis of this effect is presented in the recent paper by Davis, et al. [4].

The final important result shown in [1] is that, in the clear-fluid slit beneath the upper wall, the velocity in the up-channel direction is given by

$$u = \Lambda^{1/3}(\tilde{y}\tilde{\delta} - \frac{1}{2}\tilde{y}^2)\cos\theta , \qquad (2.6)$$

where the velocity, u, has been rendered dimensionless by v_o, $\tilde{\delta}\equiv\Lambda^{1/3}\delta$, $\tilde{y} \equiv \Lambda^{1/3} y$, and y is the dimensionless distance from the wall. In the adjacent suspension region there exists a boundary layer flow arising from the large interface velocity given by (2.6) with $\tilde{y} = \tilde{\delta}$ and the relatively large distance of the lower wall from this interface. It is these velocity profiles along with (2.5) that form the basis for the linear stability analysis presented in section 3.

The analysis of [1] is restricted to situations where the thickness of the clear fluid slit beneath the downward facing surface of the vessel is much less than the spacing between the plates, i.e. for $b/H \gg \Lambda^{-1/3}$. All of the experiments described above satisfy these conditions. However, it is clear from the PNK theory that the enhancement of the settling rate is highest for very large values of the aspect ratio, H/b. In order to test this prediction, the analysis of [1] was extended to the case $H/b \geq O(\Lambda^{1/3})$, and a series of batch laboratory experiments were performed using a very high aspect ratio vessel, i.e. $H/b = O(10^2)$. The results of this study are given in [8]. For batch settling, one finding of major interest is that, surprisingly, the clear fluid layer which forms beneath the downward facing plate attains a steady shape only along the lower portion of the channel while, in contrast, its thickness is independent of distance along the plate and increases with time for locations along the channel above some critical point. In other words, when viewed as a function of distance along the channel, the thickness of the clear-fluid layer becomes discontinuous. This interesting behavior is sketched in Figure 3 of [8]. Because of this transient behavior, the PNK formula overestimates the rate of descent of the top of the suspension; however, the corresponding results for the volumetric settling rate still hold under the conditions considered in [8].

The batch experiments described in [8] were performed for the following range of parameters: $b = 1$ cm, $H_o \approx 90$ cm, $0^o \leq \theta \leq 45^o$, $0.01 \leq c_o \leq 0.025$, $1.7\times10^7 \leq \Lambda_o \leq 3.5\times10^7$, and $1.8 \leq R_o \leq 2.1$. In all cases, a somewhat smoothed discontinuity in the thickness of the clear fluid layer was indeed observed, and the location of the discontinuity was in excellent agreement with the theory, as was the prediction of the large enhancement in the settling rate.

Of greater practical interest than batch settling is the continuous operation of inclined settlers. Even though, as mentioned earlier, the thickness of the clear-fluid layer remains transient in the upper portion of the vessel for batch systems, the analysis of [8] predicts that high aspect ratio settlers can be operated continuously provided the feed and withdrawal locations are chosen properly (a thorough discussion of the feed and withdrawal conditions that should lead to a steady operation of such continuous settlers is found in section 4 of [8]). One such possibility is for the feed to enter at the bottom of the vessel and for the clarified fluid to be removed from the top of the vessel so that a net flow of fluid up the channel is maintained. Such a setup has been recently built, and steady operation has indeed been achieved for ranges of the parameters similar to those of the batch experiments described above.

Another case of practical interest is where both the feed and withdrawal are located at the top of the vessel; then, as shown in [8] for high aspect ratio vessels and by Herbolzheimer [6] for low aspect ratio vessels, two steady solutions for $\delta(x)$, the thickness of the clear fluid layer, are possible. In the first solution, the thickness of the clear-fluid layer vanishes at the bottom of the vessel and then grows with x--as is always observed in batch settling experiments--whereas in the second solution, the thickness of the suspension layer vanishes at the bottom of the vessel and then grows with x. This second mode of operation was first observed by Probstein and Hicks [14], and later by Leung and Probstein [9], who argued that the second flow regime may be advantageous in suppressing the effects of any remixing of the particle-free fluid and the suspension, which might occur if the interface between these regions became unstable. These authors also included the presence of the thin sediment layer on the lower wall in their analysis.

Since the expressions for the velocity field and for the clear-fluid layer thickness in narrow tilted channels are somewhat cumbersome, these will not be written down here and the reader is referred to [8] for details.

3. THEORY AND EXPERIMENTS FOR UNSTABLE FLOW

It has been noted earlier that, in certain cases, the interface between the clear-fluid slit and the suspension becomes wavy. Evidently, the occurrence of such an instability limits the efficiency of the inclined settling process especially when the waves break. This phenomenon of wave formation and growth is similar to the well known problem of a falling liquid film on an inclined surface which has also been observed to become unstable. In order to investigate its origins, a linear stability analysis was performed for the laminar flow profiles derived in [1] and [8]. The details of this analysis can be found in the thesis of Herbolzheimer [6] and in the recent papers by Herbolzheimer [7] and by Davis, et al. [5].

The theoretical development proceeds along the lines of the classical works by Yih [16] and by Benjamin [2] who studied the instability of the falling liquid film. The disturbances to the base-state flow are assumed to be two-dimensional and infinitesimally small and are expressed as superpositions of normal modes of the general form

$$\Psi(x,y,t) = \psi(y) \exp i (\bar{\alpha}x - \bar{\omega}t) , \tag{3.1a}$$

$$\Phi(x,y,t) = \phi(y) \exp i (\bar{\alpha}x - \bar{\omega}t) , \tag{3.1b}$$

where, since the waves are spatially (rather than temporally) growing, the wave number of the disturbance, $\bar{\alpha}$, is treated as complex, while the frequency, $\bar{\omega}$, is a real constant; also Ψ and Φ are the normal-mode disturbance stream functions in the clear-fluid layer and in the adjacent suspension layer, respectively. A similar expression applies for the displaced position of the interface separating these two regions. The analysis is a local (quasi-parallel and quasi-steady) one because the base-state profiles vary parametrically with x and t but only slowly compared to the variations in the disturbances.

When (3.1a,b) are substituted into the linearized momentum equations for the disturbances, the well known Orr-Sommerfeld equation results together with the associated boundary conditions of no-slip at the walls and matching of velocity and stress at the interface. This system represents an eigenvalue problem with $\bar{\alpha}$ as the complex eigenvalue. If the solution of this problem yields a positive value for $\bar{\alpha}_I$, the imaginary part of the eigenvalue, then the disturbances are damped as they travel up the vessel, whereas, if $\bar{\alpha}_I$ is negative then these same disturbances grow and the system is unstable.

The Orr-Sommerfeld equations were solved first asymptotically for small frequencies (this corresponds to long wavelengths since the wave speed remained O(1)), and then these results were supplemented with a numerical analysis because the most highly amplified waves were found to occur for moderate values of the frequency and the Reynolds number. For the steady flow profile derived in [1] the results of the linear theory [6,7] showed that small disturbances always grow, i.e. that the presumably steady flow in the narrow slit is always unstable. However, the rate of growth of these disturbances depends on R, Λ , and the geometrical and physical properties of the system, so that under many circumstances these do not grow to a visible size over the length of the vessel. Thus, in many of the experiments described earlier, the flow appeared to be stable. The local exponential amplification factor for the wave growth is $-\bar{\alpha}_I = -\alpha_I/\delta$, where α_I is the imaginary part of the wave number scaled with the clear-fluid layer thickness. This factor is proportional to $-\alpha_I(\tan\theta)^{-1/3}$-- c.f.(2.5)--and the numerical results for the latter are plotted in Figure 2 for the most unstable mode as a function of θ and for various values of the local Reynolds number, Rx. Clearly, according to the predictions of the theory the waves have a maximum aplification that generally occurs in the θ range 5°-15°.

Using the linear stability analysis as a guide, experiments, for which a visible observation of the instability was expected, were performed in our laboratory, both on a batch and a continuous basis, under the following set of conditions: $8 \leq H_o/b \leq 15$, $5^o \leq \theta \leq 50^o$, $4 \leq R_o \leq 25$, $7\times10^6 \leq \Lambda \leq 2\times10^8$, $0.005 \leq c_o \leq 0.15$. The experimental data were found to follow the general trends predicted by the linear theory. In particular, the distance along the channel at which waves first became visible, x_i, was measured, because the inverse of this distance provides an experimental indication of the average growth rate of the waves. As an example of these measurements, $1/x_i$ is plotted in Figure 3 as a function of θ for $c_o = 0.05$ and various values of Rx (determined midway up the channel) from which it is evident that the shapes of the curves are very similar to the corresponding theoretical predictions for the maximum amplification rates shown in Figure 2.

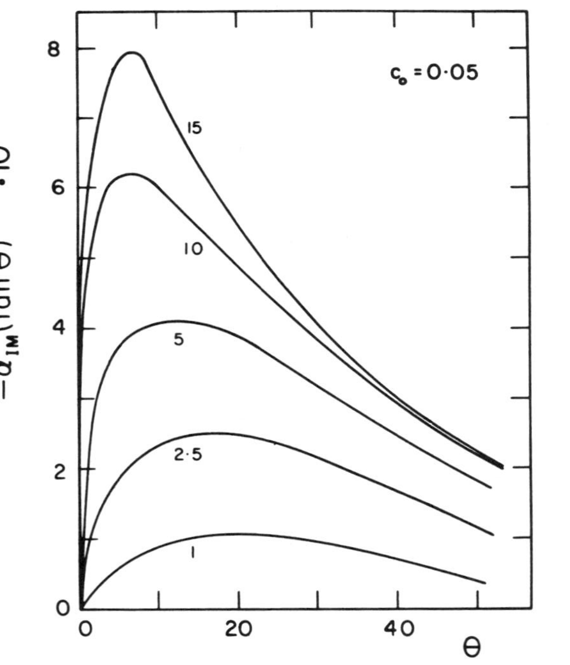

FIGURE 2. The parameter $-\alpha_I(\tan\theta)^{-1/3}$ for the most unstable mode vs. θ for $c_o = 0.05$ and Rx = 1, 2.5, 5, 10 15.

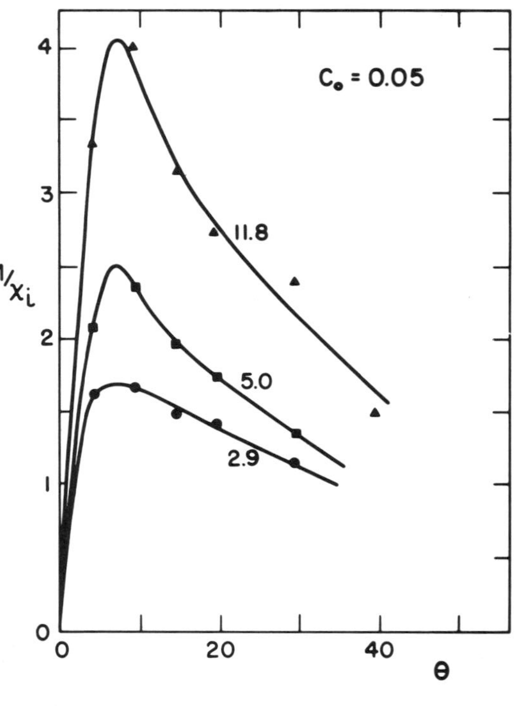

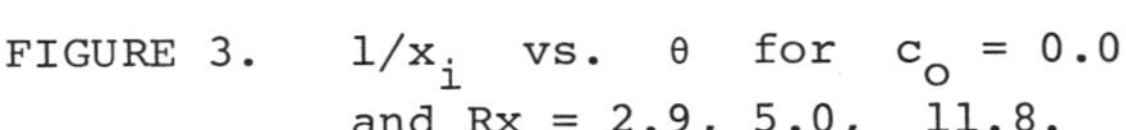

FIGURE 3. $1/x_i$ vs. θ for $c_o = 0.05$ and Rx = 2.9, 5.0, 11.8.

The linear stability analysis and experiments described above have been extended [5,6] to determine the amplification rates of waves in tall, narrow channels with $H/b \geq O(\Lambda^{1/3})$. This case is of particular interest since, in light of (2.3), such vessels give very high enhancement in the clarification rate, and it is important to know, therefore, under what conditions, if any, a stable operation of these vessels can be achieved. It was found that, unlike the case of low aspect ratio vessels, which were shown to be unstable for all values of the Reynolds number, the analysis for these high aspect ratio vessels led to a critical Reynolds number, R_c, below which the system can be expected to be stable to infinitesimal disturbances. In general, it was shown [5] that R_c increased with increasing H/b and increasing θ. Thus, the same geometric factors that lead to the highest enhancement in the settling rate also give the most stable operating conditions--a surprising result that can be very useful in the future design of "supersettlers."

This exciting prediction was tested in the laboratory using the bottom-feeding continuous high aspect ratio vessel described in section 2, and the results were in very good agreement with the linear theory. This can be seen in Figures 4 and 5 where the amplification rates of the waves are plotted for theory and experiment, respectively, as functions of θ under the conditions $\Lambda\cos^2\theta = 1.3\times10^7$, $\tilde{b}\cos^{1/3}\theta = 2.4$ (where $\tilde{b} \equiv \Lambda^{1/3}b/H$), $c_o = 0.01$, and for various values of the local Reynolds number with both the theoretical amplification rates and the local Reynolds numbers being computed at the location midway up the channel. Compared to the low aspect ratio case (Figures 2 and 3) there is dramatic stabilization for $\theta > 20^\circ$.

4. SUMMARY

The preceding represents a synopsis of our investigation of the observation that sedimentation of a suspension in a settling vessel having inclined walls can occur, in principle, much faster than in a corresponding vertical system, especially when the aspect ratio (height over width) of these vessels is large. A recently developed theoretical model--

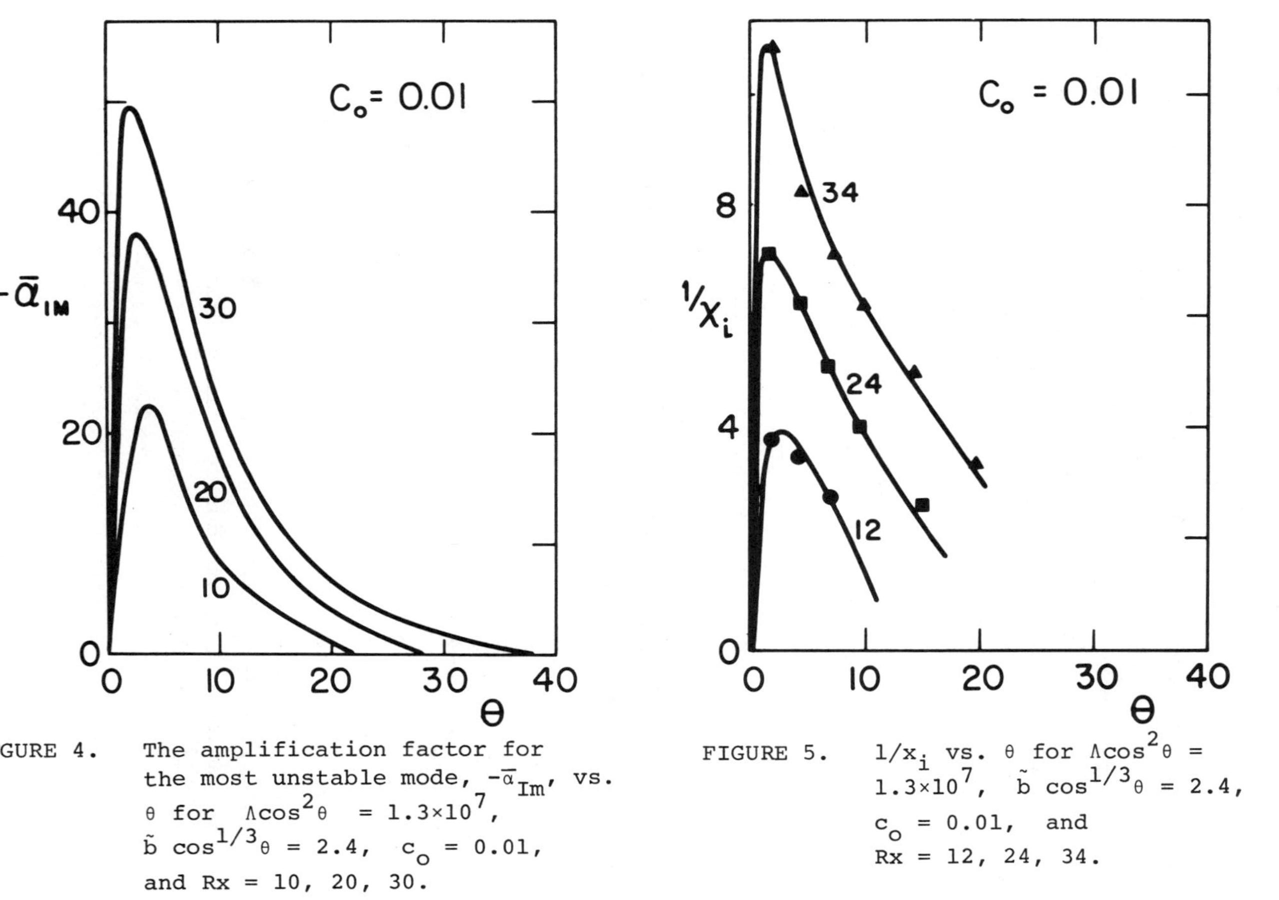

FIGURE 4. The amplification factor for the most unstable mode, $-\bar{\alpha}_{Im}$, vs. θ for $\Lambda \cos^2\theta = 1.3\times10^7$, $\tilde{b} \cos^{1/3}\theta = 2.4$, $c_o = 0.01$, and Rx = 10, 20, 30.

FIGURE 5. $1/x_i$ vs. θ for $\Lambda \cos^2\theta = 1.3\times10^7$, $\tilde{b} \cos^{1/3}\theta = 2.4$, $c_o = 0.01$, and Rx = 12, 24, 34.

based on the principles of continuum mechanics--which offers a basic understanding of the mechanisms responsible for this phenomenon as well as a detailed description of the resulting flow patterns is discussed. This model was tested in batch and continuous experiments with very good success so long as the flow remained laminar. Under some conditions, however, the interface between the suspension and the layer of particle free fluid underneath the downward-facing wall became wavy and broke with a resulting decrease in the operating efficiency of the vessel. As a first stop to investigate in detail the conditions for wave-formation and growth, a linear-stability analysis was performed to predict the maximum amplification rates of such waves. Experiments in our laboratory proved consistent with the general trends set forth by the linear theory. In particular, it was found that tall, narrow vessels not only gave the largest enhancement in the settling rate but also significantly increased the stability of such systems.

REFERENCES

1. Acrivos, A. and E. Herbolzheimer, 1979. Sedimentation in settling tanks with inclined walls. J. Fluid Mech. 92, 435-457.
2. Benjamin, T.B., 1957. Wave formation in laminar flow down an inclined plane. J. Fluid Mech. 2, 554-574.
3. Boycott, A.E., 1920. Sedimentation of blood corpuscles. Nature 104, 532.
4. Davis, R.H., E. Herbolzheimer, and A. Acrivos, 1982. The sedimentation of polydisperse suspensions in vessels having inclined walls. Int. J. Multiphase Flow (to appear).
5. Davis, R.H., E. Herbolzheimer, and A. Acrivos. Wave formation and growth during sedimentation in narrow tilted channels (in preparation).
6. Herbolzheimer, E., 1980. Enhanced sedimentation in settling vessels having inclined walls. Ph.D. dissertation, Stanford University.
7. Herbolzheimer, E. The stability of the flow during sedimentation beneath inclined surfaces (in preparation).

8. Herbolzheimer, E., and A. Acrivos, 1981. Enhanced sedimentation in narrow tilted channels. J. Fluid Mech. 108, 485-499.
9. Leung, W.F., and R.F. Probstein, 1982. Lamella and tube settlers. I. Model and operation. Ind. Eng. Chem. Process Des. Dev. (to appear).
10. Mund, H., E. Rubin and A. Nir, 1979. Sedimentation of concentrated suspensions under inclined surfaces (personal communication).
11. Nakamura, N. and K. Kuroda, 1937. La cause de l'acceleration de la vitesse de sedimentation des suspensions dans les recipients inclines. Keijo J. Med. 8, 256-296.
12. Pearce, K.W., 1962. Settling in the presence of downward facing surfaces. Proceedings of the Third Congress of European Federation of Chemical Engineering, London, 30-39.
13. Ponder, E., 1925. On sedimentation and rouleaux formation. Q.J. Expt'l. Physiol. 15, 235-252.
14. Probstein, R.F. and R. Hicks, 1978. Lamella settlers: a new operating mode for high performance. Ind. Water Engng. 15, 6-8.
15. Probstein, R.F., D. Yung, and R. Hicks, 1977. A model for lamella settlers. Presented at "Theory, Practice, and Process Principles for Physical Separations," Engng. Foundation Conference, Asilomar, CA.
16. Yih, C.-S., 1963. Stability of liquid flow down an inclined plane. Phys. Fluids 6, 321-334.

This research was supported in part by contract EPRI-RP-314-1 with the Electric Power Research Institute in Palo Alto, California, and by the National Science Foundation under Grants ENG 78-16928 and CPE 81-08200.

Andreas Acrivos
Robert H. Davis
Department of Chemical Engineering
Stanford University
Stanford, CA 94305

Eric Herbolzheimer
Department of Chemical Engineering
California Institute of Technology
Pasadena, CA 91125

Simple Kinetic Theory of Brownian Diffusion in Vapors and Aerosols

B. Dahneke

1. INTRODUCTION.

The various processes important in the behavior and effects of aerosol systems can be listed in several general categories, as illustrated in Table 1. Throughout the various categories of this type of list, several processes recur because they are fundamental in controlling or influencing many of the properties of aerosol systems. This paper will describe and extend the theory of some of these basic processes, namely, those which depend on the transport and deposition of vapors and particles by Brownian diffusion. This topic is important because of the broad influence of diffusional transport on the behavior and effects of aerosol systems. Moreover, this topic is also useful as an illustration of a secondary theme of this paper, namely, that the study of aerosol systems requires a broad range of disciplines to understand and characterize the behavior and effects of aerosol systems. This secondary theme is also illustrated by the contents of Table 1 and by its title.

The central focus of this paper is a simple kinetic theory of aerosol and vapor diffusion. I will describe this theory and apply it to several example problems that illustrate its utility. This theory can be generally described as a mean-free-path type theory that contains neither the rigor nor complexity of a Chapman-Enskog type theory [2]. Its simplicity thus provides the advantages of ease of formulation

ISBN 0-12-493120-0

Table 1. Aerosols: a kingdom of many subjects.

	Basic processes important to the behavior and effects of aerosols:	Scientific disciplines required to understand and describe these processes:
1.	Aerosol formation (nucleation, condensation, evaporation, coagulation, chemical reactions, entrainment, combustion, emissions, sea spray, volcanoes.)	Transport phenomena, thermodynamics, surface chemistry & physics, kinetic theory & stat. mech., chemistry & chemical kinetics, mathematics.
2.	Aerosol stability/dynamics (sedimentation, coagulation, evaporation, condensation, chemical reactions; filtration, precipitation, sampling, deposition, etc.)	Thermodynamics, transport phenomena, kinetic theory & stat. mech., mechanics, dynamics, electricity & magnetism, mathematics.
3.	Atmospheric aerosol effects (visibility, meteorology & climate alteration, pollution, toxicity & public health.)	Light scattering & absorption, chemistry, chemical kinetics, combustion, transport phenomena, surface chemistry & physics, photochemistry, analytical methods, mathematics, epidemiology, biology & biochemistry.
4.	Sampling & removal of aerosols (detection & measurement of particle size, shape, composition, charge, activity, mass, surface, reactivity; deposition, filtration, precipitation, classification.)	Light scattering, optics, electronics, analytical methods, statistics, stat. mech., transport phenomena, mechanics, dynamics, electricity & magnetism, electro-optics, mathematics.

and interpretation. The importance of these advantages is illustrated by a lesson from the history of the kinetic theory of gases. Upon development of the rigorous kinetic theory of gases by Chapman and Enskog [2], the previously unknown second order effect called thermal diffusion was first predicted. Within approximately one year this predicted phenomena was experimentally demonstrated by Chapman and Dootson [1]. Nevertheless, twenty-five years later Furth [9] published a simple mean-free-path theory of thermal diffusion that provides, through its simplicity, a clear physical picture of the thermal diffusion process not available from the complex theory.

In view of this historical lesson, the starting point in the present development of a kinetic theory of Brownian diffusion in aerosol systems will be a simple mean-free-path theory that provides a clear physical picture of the processes it describes. Once the important underlying physical processes have been identified and quantified as much as possible by means of the simple theory, more exact theory can be sought using the results of the simple theory as a guide.

The theory presented herein utilizes earlier results and follows to large extent the pattern of similar theories that have preceded it, most notably those due to Maxwell [10], Schmoluchowski [12] and Fuchs [6,7,8].

2. DROPLET/PARTICLE GROWTH BY CONDENSATION/EVAPORATION.

In an article on Diffusion written in 1877 for the Encyclopedia Britanica, Maxwell [10] described the basis of the theory of droplet (and particle) growth by condensation or evaporation. Since the fundamental laws of diffusion also describe the transport of other particles besides vapor molecules, Maxwell's theory also applies to the process of aerosol coagulation, diffusional charging of particles and diffusion controlled chemical reactions on a particle surface.

For stationary diffusion of vapor to or from a droplet of radius r, the droplet growth rate in molecules/sec is given by Maxwell's equation

$$I_M = 4\pi r\, D(n_\infty - n_s) \qquad (2.1)$$

where D is the diffusion coefficient of the vapor molecules and n_∞ and n_s the number concentrations of the vapor molecules at infinity and at the droplet surface. The latter concentration is equated to that of the saturated vapor at the droplet temperature and radius for the case of droplet growth by condensation or evaporation and to zero for the case of aerosol coagulation or diffusion controlled reaction.

Maxwell also considered the temperature of a droplet undergoing stationary growth by condensation or evaporation. The heat conduction rate to the droplet is

$$Q_M = 4\pi r \kappa (T_\infty - T_o) \tag{2.2}$$

where κ is the thermal conductivity of the gaseous medium, T_∞ the gas temperature at infinity and T_o the temperature of the droplet. The heat transfer and droplet growth rates are not independent in the stationary case but are related according to

$$I_M = Q_M/L \tag{2.3}$$

where L is the molecular latent heat of vaporization. The values of n_∞ and T_∞ and the above three expressions specify the stationary growth rate of a droplet. Thus, Maxwell's expressions provide a complete description of the stationary growth of large droplets and his results apply in general to the diffusional transport to or from a spherical body.

However, since continuum expressions are used to obtain Maxwell's results, they fail when the mean-free-path of the diffusing particles is not negligible compared to the sphere radius. Fuchs [6,7] derived a useful correction in 1934 that accounts for the kinetic behavior of the diffusing particles. In this theory the diffusing particles within the distance Δ of the sphere surface are regarded as being in a vacuum. This model is reasonable since Δ is of the order of the particle mean-free-path so that within Δ of the sphere surface the motion of the diffusing particles follows straight line trajectories as in a vacuum. Fuchs assumed that beyond Δ the diffusing particles obeyed continuum behavior. By equating the continuum transport at the limiting sphere surface of radius $r+\Delta$ to the net molecular transport at the sphere surface, Fuchs obtained the correction factor

$$\beta = I/I_M = (1+\Delta/r)/[1+4D(1+\Delta/r)/(\delta\bar{c}r)] \tag{2.4}$$

where I is the actual droplet growth rate, I_M the continuum result due to Maxwell, δ the evaporation-condensation coefficient, $\bar{c} = (8\ kT/\pi m)^{\frac{1}{2}}$ the mean thermal velocity of the diffusing particles, k is Boltzmann's constant, T the absolute system temperature and m the mass of the diffusing particles. This expression and the application of Fuchs' limiting sphere model to other problems have been useful. The model suffers, however, because the distance Δ is not specified in the theory but must be assumed, adjusted empirically or estimated by independent theory.

In the following sections a simple kinetic theory of the diffusion process is presented which gives a correction factor identical to that of Fuchs with the exception that all parameters are specified by the theory. No parameter similar to the Δ of Fuchs' theory needs to be adjusted to empirical data or determined by additional theory. This simple kinetic theory has also been used to determine particle coagulation rates and particle deposition on a tube wall by convective diffusion and heat conduction to a sphere. The result for heat conduction to a sphere, stated below, is also essential to an accurate description of the growth of small droplets by condensation or evaporation. The derivation of such an expression has apparently not been previously reported.

3. SIMPLE KINETIC THEORY OF VAPOR DIFFUSION.

A net diffusion of vapor molecules generally accompanies a concentration gradient because the thermal motions of molecules in the direction from higher to lower concentration are not balanced by the thermal molecule motions in the opposite direction. For uniform concentration and temperature, the equilibrium velocity distribution of Maxwell requires the one-way diffusional flux of vapor molecules, i.e., the one-way flux of molecules due to their thermal motions, $j = \frac{1}{4}\bar{c}n$. For the case of non-uniform vapor concentration, I assume the local one-way flux in the y direction, where y is the direction of the concentration gradient,

$$j_+ = \tfrac{1}{4}\,\bar{c}\,n(y - \ell_D) \tag{3.1}$$

while the flux in the opposite direction is

$$j_- = \tfrac{1}{4}\,\bar{c}\,n(y + \ell_D) \tag{3.2}$$

where $n(y)$ is the local number concentration of the diffusing molecules and ℓ_D the average distance from the control surface at which the molecules crossing the surface are "concentration equilibrated". The quantity ℓ_D will be regarded as the diffusional mean-free-path of the vapor molecules.

I further assume that near the control surface the Taylor series expression

$$n(y \pm \ell_D) = n(y) \pm \ell_D \left.\frac{\partial n}{\partial y}\right)_y + \tfrac{1}{2}\,\ell_D^{\,2} \left.\frac{\partial^2 n}{\partial y^2}\right)_y \tag{3.3}$$

is adequate. Then the net flux through the control surface is given by

$$j = j_+ - j_- = -\tfrac{1}{2}\,\bar{c}\,\ell_D \left.\frac{\partial n}{\partial y}\right)_y . \tag{3.4}$$

Comparison with the continuum expression

$$j = -D \left.\frac{\partial n}{\partial y}\right)_y \tag{3.5}$$

gives

$$\ell_D = 2D/\bar{c} . \tag{3.6}$$

Thus, the diffusional mean-free-path ℓ_D is specified so that the net diffusional flux given by the kinetic expressions (3.1) and (3.2) is forced to agree with that of the continuum law (3.5). This feature of the present theory is essential. The inaccuracies caused by the usual assumptions of simple mean-free-path theories are largely removed by the use of the correct value of D in (3.6), which forces the value of ℓ_D to correspond to the actual diffusion rate for the local conditions. When a simple mean-free-path theory is used to *predict* the value of the diffusion coefficient, thermal conductivity or viscosity based on some molecular interaction law, such agreement is not forced and the resulting expressions may contain substantial error.

4. THE KINETIC BOUNDARY CONDITION FOR VAPOR CONDENSATION.

As assumed in Fuchs' limiting sphere model wherein vapor molecules within Δ of the droplet surface behave as in a vacuum, I assume here that the continuum diffusion law is not valid within a distance of the order of the mean-free-path of a surface. In terms of the simple kinetic theory of the previous section, within the region near a surface the outward flux of vapor molecules (particles) is diminished in the non-equilibrium state because the presence of an adsorbing or partially adsorbing surface excludes particles which would otherwise particpate in the outward diffusion process. (The continuum expression is valid near a completely reflecting wall.) Since the combination of outward and inward fluxes (3.1) and (3.2) result in the continuum expression (3.5), the partial disruption of the outward flux by the presence of the adsorbing wall causes the diffusion process near the wall to obey a law other than this continuum law. Beyond the distance of order ℓ_D the continuum expressions (3.4) and (3.5) are adequate.

Although the continuum expression is not valid near the wall, the kinetic expression (3.2) does describe the flux of particles towards the wall. The net flux of particles captured by the wall is obtained by multiplying the quantity (3.2) by the particle sticking probability δ, the condensation-evaporation coefficient for the case of vapor molecules, and subtracting the flux due to particle evaporation from the wall. Thus, the net flux to the wall at $y = 0$ is

$$j = \tfrac{1}{4}\,\delta\bar{c}[n(y=+\ell_D) - n_s(T_o)] \qquad (4.1)$$

where $n_s(T_o)$ is the saturation number concentration at the wall temperature T_o. Generally, for the deposition of an aerosol and for diffusion controlled chemical reactions, $n_s=0$.

Even though the continuum expressions (3.5) and the diffusion equation which derives from it are not valid within ℓ_D of a wall, they can still be used in this region to retain the convenience of continuum theory wherein a single law describes the concentration distribution over the entire field of interest. If the correct boundary condition is imposed so as to require the correct flux at the wall, the concentration field obtained from solution of the continuum expressions will be distorted near the wall to just compensate for inaccuracy in the continuum expressions. The continuum expression (3.5) and the distorted concentration field then give the correct flux at the wall.

I therefore impose the kinetic boundary condition at the wall $y = 0$

$$\tfrac{1}{4}\ \delta\bar{c}[n(y=+\ell_D) - n_s(T_o)] = D\left.\frac{\partial n}{\partial y}\right)_{y=0} . \qquad (4.2)$$

By use of (3.6), the kinetic boundary condition can be stated in terms of the dimensionless coordinate $x = y/L'$

$$n(x=Kn_D) - n_s(T_o) = 2Kn_D/\delta\left.\frac{\partial n}{\partial x}\right)_{x=0} \qquad (4.3)$$

where L' is a characteristic system length and $Kn_D = \ell_D/L' = 2D/(\bar{c}L')$.

5. APPLICATION: DIFFUSIONAL TRANSPORT TO A SPHERE.

When the stationary diffusion equation

$$D\nabla^2 n = 0 \qquad (5.1)$$

is solved for a spherical system with boundary conditions

$$n = n_\infty \quad \text{as} \quad \rho \to \infty \qquad (5.2)$$

$$n(\rho = r[1 + Kn_D]) - n_s(T_o, r) = 2r\ Kn_D/\delta\left.\frac{\partial n}{\partial \rho}\right)_{\rho=r} \qquad (5.3)$$

the resulting concentration field gives the diffusional transport of particles to the sphere (particles/sec)

$$I = 4\pi r\ D\ \beta\ (n_\infty - n_s) \qquad (5.4)$$

where

$$\beta = (Kn_D + 1)/(2Kn_D(Kn_D+1)/\delta + 1) \qquad (5.5)$$

and $Kn_D = 2D/\bar{c}r$. In the limit $Kn_D \to 0$, $\beta = 1$ and (5.4) becomes the continuum result of Maxwell. In the opposite limit $Kn_D >> 1$, $\beta = \delta\bar{c}r/4D$ so that (5.4) gives the correct result for free-molecular diffusion.

At small values of Kn_D the β of (5.5) agrees exactly with the β obtained by Monchick and Reiss [11] by use of a first-order perturbation theory which is valid for small Kn_D.

Wright [15] pointed out that if the Δ of Fuchs' theory is chosen to cause Fuchs' β to agree with the β of Monchick and Reiss, the value

$$\Delta = 2D/\bar{c} = \ell_D$$

is obtained and the β of Fuchs' theory agrees exactly with that of the present theory. Thus, the present theory gives, in a single derivation, the earlier results of Fuchs, of Monchick and Reiss and of Wright.

To complete the solution of the problem of stationary growth of small droplets by condensation or evaporation, I simply state the result for the correction factor β_q for the stationary heat conduction to a sphere of radius r. That is,

$$\beta_q = Q/Q_M \tag{5.6}$$

where Q is the actual heat transfer rate and Q_M the continuum result (2.2).

$$\beta_q = (Kn_K + 1)/(2\ Kn_K(Kn_K + 1)/\alpha + 1) \tag{5.7}$$

where $Kn_K = 2\kappa/(\bar{c}'n'c_v r)$, n' is the number concentration of the suspending gas molecules, $\bar{c}'$ their mean thermal velocity, c_v their molecular specific heat at constant volume and α the thermal accommodation coefficient. Although no derivation is given here, the derivation is exactly parallel to the one outlined above for mass transport.

An interesting consequence of the present theory is the relationship between δ and α required by the conservation equation (2.3). That is, for stationary condensation or evaporation

$$\beta I_M = \beta_q\ Q_M/L \tag{5.8}$$

so that δ and α are not independent.

The correction of Maxwell's results for the influence of non-negligible mean-free-path of the suspending gas molecules is thus complete, in so far as the simple kinetic theory described here is accurate. Determination of the accuracy of the theory must await adequate data, such as the measured data of Davis and Ray [4]. The data presently available strongly support the theory.

Finally, I point out that in addition to the obvious cases of small sphere radius or reduced system pressure, the correction factor β or β_q can also become important when the coefficient δ or α is sufficiently small. For example, if $Kn_D = .01$ and the sticking probability δ takes the values 1.0, .01 and .001, then β obtains the values 0.99, 0.33 and .048. Since molecular sticking probabilities may be even smaller, the kinetic correction factor β may sometimes be extremely important in systems where Kn_D is quite small. Such corrections would therefore be essential in the understanding of, say, diffusion controlled reactions on the surface of airborne particles.

6. BROWNIAN DIFFUSION OF FLUID-BORNE PARTICLES.

Particles suspended in a fluid medium undergo thermal or Brownian motions due to fluctuations in their bombardment by surrounding fluid molecules. Because the motion of the fluid molecules is random, the frequency, magnitudes and directions of the Brownian impulses experienced by a particle are also random. As with the thermal motion of vapor molecules treated in the preceding sections, the distribution of velocities of the individual particles of uniform mass M comprising a fluid-borne suspension is given by Maxwell's famous law:

$$N(u,v,w) = (M/2\pi kT)^{3/2}\, e^{-M(u^2+v^2+w^2)/2kT}. \qquad (6.1)$$

This equilibrium velocity distribution, which may be superimposed on a drift velocity due to, say, gravity or hydrodynamic flow, causes an average thermal speed $\bar{c}$ of the particles

$$\bar{c} = \sqrt{\frac{8kT}{\pi M}} \qquad (6.2)$$

and also causes the one-way effusion of particles (particles/sec) through a control surface of unit area

$$j = \tfrac{1}{4}\, n\, \bar{c} \qquad (6.3)$$

where n is the particle number density. Expressions (6.1), (6.2) and (6.3) are exact for equilibrium suspensions of particles in liquids or gases.

In the simple kinetic theory of dilute gases, the gas molecules are pictured as following straight line trajectories at constant velocities between collisions with other molecules (see, for example, Chapman and Cowling [2]). The concept of the mean-free-path λ of the gas is therefore an easy one which characterizes well the actual properties of dilute gases such as viscosity, thermal conductivity and diffusion. Moreover, the dimensionless Knudsen number $Kn = \lambda/L$, where L is the characteristic length of the system, reveals when the gas can be treated as a continuum fluid ($Kn \ll 1$), when the gas can be treated as discrete molecules acting essentially independently of one another ($Kn \gg 1$) or when theories incorporating both qualities are required to characterize the gas behavior ($Kn \sim 1$).

The motion of aerosol or colloid particles is not as simple to characterize. Because of their relatively large size and mass, these particles experience a large number of collisions per unit time with surrounding fluid molecules while they are not generally accelerated to a significant degree by a single collision. Thus, the equilibrium motion of fluid-borne particles over time scales of practical interest can be regarded as continuously changing in both particle speed and direction of motion, such that the probability distribution (6.1) is maintained as a time or ensemble distribution.

This picture of the Brownian motion of fluid-borne particles provides no obvious path length which can be identified with the gas molecule mean-free-path. Nevertheless, it is indeed possible to define a particle mean-free-path ℓ which is similar in its function and usefulness to the gas molecule mean-free-path λ . After Fuchs [7,8], I define the particle mean-free-path ℓ as being proportional to the average path length travelled by a particle undergoing Brownian motion before its motion is completely diverted into a direction perpendicular to the initial particle motion. While this definition can be used to provide a unique value of the particle mean-free-path, its evaluation is not as straightforward as the mean-free-path of a gas molecule. However, the evaluation of ℓ is, in fact, similar to the evaluation of λ .

The gas mean-free-path λ is usually evaluated by combining a measured quantity of the gas (generally the gas viscosity η) with a simple kinetic theory expression for this quantity, and solving for λ. A similar procedure is used to obtain the particle mean-free-path from the particle diffusion coefficient and a simple kinetic theory expression like the one derived in Section 3. As before, ℓ_D denotes the particle mean-free-path determined by use of the particle diffusion coefficient.

In the equilibrium case for which no concentration gradients exist among the particles suspended in a fluid, expressions (6.1), (6.2), and (6.3) are valid. Of course, no net diffusion of particles can occur in this case, the Brownian diffusion of particles in any direction, given by (6.3), being exactly balanced by an equal flux of particles in the opposite direction.

When a particle concentration gradient exists, the system is in a nonequilibrium state and (6.1), (6.2) and (6.3) are no longer exact. However, the fluid-borne particles remain in thermal equilibrium with their suspending medium so that (6.1) still applies as the probable velocity distribution of each individual particle and (6.2) still gives the mean thermal velocity of each particle. It also follows that (6.3) applies in a "local equilibrium" sense that will be described below.

Consider first the case in which a y-direction gradient in the particle concentration occurs. The particle diffusion flux resulting from this gradient is described by Fick's law

$$j = -D \frac{\partial n}{\partial y} \qquad (6.4)$$

where the particle diffusion coefficient D is given by Einstein's [5] well known result generalized here to particles of any shape

$$D = kT/f \qquad (6.5)$$

where k is Boltzmann's constant, T the absolute fluid temperature and f the coefficient of friction of the particle in the fluid. For example, a spherical particle of radius a in an infinite fluid of viscosity η has the friction coefficient

$$f = 6\pi\eta a/C_s \ ,$$

where C_s is the slip correction factor, and the diffusion coefficient

$$D = kTC_s/(6\pi\eta a) \ .$$

In analogy to the derivation of Section 3 for vapor molecules, the particle diffusion flux can be described in terms of the particle mean-free-path ℓ_D. Consider a control surface located in the plane $y = 0$. From (6.3) we assume the one-way flux of particles through the surface in the $-y$ direction is given by

$$j_-(y=0) = \tfrac{1}{4}\,\bar{c}\; n(y=\ell_D) \ . \tag{6.6}$$

This expression simply describes the one-way flux of particles across the control surface in the -y direction as depending not on the particle concentration at the control surface y=0, but at the location $y=\ell_D$ because this is the average plane of "origin" of particles crossing the surface. That is, particles crossing the control surface in the -y direction are, on the average, equilibrated to the local concentration at $y=\ell_D$ rather than at $y=0$. A Taylor's series expansion for $n(\ell_D)$ gives, as in Section 3,

$$j_-(0) = \tfrac{1}{4}\,\bar{c}\left\{ n(0) + \ell_D \left.\frac{\partial n}{\partial y}\right)_{y=0} + \tfrac{1}{2}\,\ell_D{}^2 \left.\frac{\partial^2 n}{\partial y^2}\right)_{y=0} \right\} . \tag{6.7}$$

Simultaneously, particles cross the control surface in the opposite direction according to

$$j_+(0) = \tfrac{1}{4}\bar{c}\left\{ n(0) - \ell_D \left.\frac{\partial n}{\partial y}\right)_{y=0} + \tfrac{1}{2}\,\ell_D{}^2 \left.\frac{\partial^2 n}{\partial y^2}\right)_{y=0} \right\} . \tag{6.8}$$

Combination of (6.7) and (6.8) gives the net flux of particles through the control surface in the $+y$ direction.

$$j(0) = j_+(0) - j_-(0) = -\tfrac{1}{2}\,\bar{c}\,\ell_D \left.\frac{\partial n}{\partial y}\right)_{y=0} \tag{6.9}$$

This result is Fick's law (6.4). We therefore obtain, by comparison with (6.4),

$$D = \tfrac{1}{2}\,\bar{c}\,\ell_D \tag{6.10}$$

and combination of (6.5) and (6.10) gives

$$\ell_D = 2kT/(\bar{c}f) . \tag{6.11}$$

Example values of ℓ_D are shown in Fig. 1 for unit density spheres of radius a in water and air media at NTP. Because of the high value of f in liquid media, ℓ_D is extremely small for liquid-borne particles.

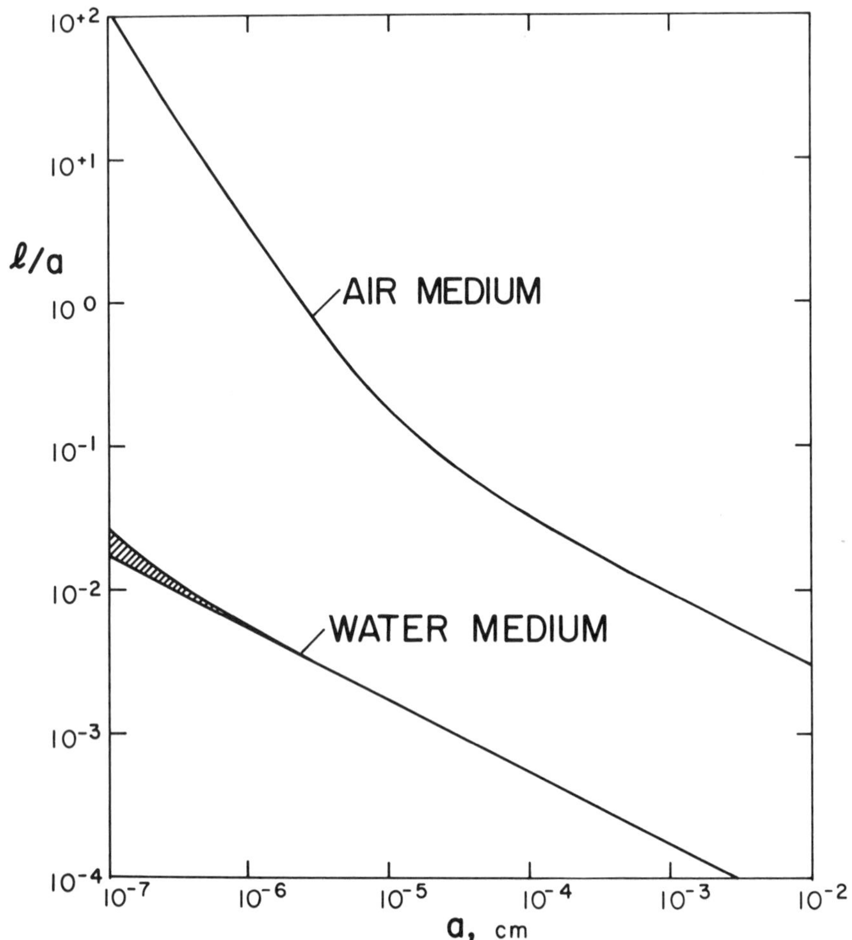

Fig. 1. Calculated ratios $\ell/a = 2kT/\bar{c}fa$ versus particle radius a for unit density spheres in air and water media at NTP.

Although the foregoing argument has only demonstrated the validity of the combination (6.6) and (6.11) for the one-way diffusional flux of particles when Fick's law applies and the Taylor's series expansion of n(y) is dominated by its leading terms, I shall assume throughout that this combination always provides an adequate approximation of the one-way diffusional flux of particles.

The particle diffusion Knudsen number is defined by

$$Kn_D = \ell_D/L = 2kT/(\bar{c}fL) \tag{6.12}$$

where L is the characteristic system length. In liquids Kn_D is always small because of the extremely small values of ℓ_D. In aerosols Kn_D is usually negligible, with a few important exceptions where the characteristic system length is sufficiently small or the aerosol gas sufficiently rarefied to make Kn_D significant or even large. Alternatively, when the particle sticking probability is small, partice deposition rates can be substantially attenuated at even small values of Kn_D. In these cases the normal boundary condition used in calculating diffusional aerosol deposition must be replaced by a kinetic expression. The following sections define this expression and use it in two example calculations: aerosol coagulation and aerosol deposition in a fine capillary. Additional systems for which the kinetic boundary condition is required because Kn_D is not negligible include diffusional deposition of aerosols on ultra-fine fibers and various rarefied aerosol systems such as diffusional sampling systems for the upper atmospheric aerosol. Since, for spherical particles of radius a,

$$Kn_D = kTC_s/(3\pi\eta a\bar{c}L)$$

the particle Knudsen number can become large in rarefied systems where C_s becomes large.

7. THE KINETIC BOUNDARY CONDITION FOR PARTICLE DEPOSITION.

The rate at which fluid-borne particles deposit by diffusion onto a collector surface depends not on the particle concentration at the surface $y = 0$, where the concentration of fluid-borne particles is generally very small, but rather on the distribution in particle concentration over small separations from the surface. The simplest treatment of particle deposition is by use of the simple kinetic theory of Section 6. In their fluctuating motion towards the surface, the depositing particles appear, on the average, to come from among those in local equilibrium at $y = \ell_D$. If we assume the fraction δ of all particles striking the surface is captured, the flux of particles depositing on the surface is given by the kinetic boundary condition

$$-j(y=0) = \tfrac{1}{4}\,\delta\,\bar{c}\;n(y=\ell_D) \; . \tag{7.1}$$

The continuum law of diffusion (6.4) is valid at sufficiently large separations from the surface. Near the surface this law no longer describes the diffusion process correctly. In terms of the simple kinetic theory of Section 6, the net diffusional flux at separation y from the surface

$$j(y) = \tfrac{1}{4}\,\bar{c}\;[n(y-\ell_D) - n(y+\ell_D)] \; ,$$

in agreement with Fick's law at large separations, becomes distorted for separations $y \lesssim \ell_D$ because of the absence of fluid-borne particles in the range $y \leq 0$ available to the diffusion process. Thus, the one-way flux of particles in the direction away from the wall, normally given at large separations by $\bar{c}\;n(y-\ell_D)/4$, is diminished while the particle flux towards the wall remains unchanged. The particle concentration field near the wall therefore differs from that predicted by Fick's law. Fuchs [8] provides a more rigorous discussion of this subject.

Rewriting the above expression in terms of the dimensionless distance $x = y/L$, we obtain

$$j(x) = \tfrac{1}{4}\,\bar{c}\;[n(x-Kn_D) - n(x+Kn_D)] \; .$$

Thus, in the special cases when Kn_D is significant, we can expect Fick's law to be inadequate within dimensionless separations of the order Kn_D from the adsorbing surface. When Kn_D is negligible, Fick's law is adequate at all separations.

In spite of its deficiency near a collector surface when Kn_D is significant, we can still assume Fick's law is valid at all separations from the surface in order to describe the diffusion process at all separations with a simple law of diffusion. We must, however, impose the proper condition at the surface to insure the solution n(y) gives the correct deposition flux. We therefore require

$$D \left.\frac{\partial n}{\partial y}\right)_{y=0} = \tfrac{1}{4}\,\delta\,\bar{c}\;n(y=\ell_D)\ . \tag{7.2}$$

Since Fick's law is valid for $y > \ell_D$, this combination extrapolates Fick's law into the surface region so that the correct particle flux to the surface is insured even though Fick's law may not be valid near the surface. Inaccuracy in Fick's law near the surface is compensated by a distortion of the concentration field n(y) so that the correct flux of particles to the surface is always maintained.

When $Kn_D \to 0$ the kinetic boundary condition (7.2) becomes the usual boundary condition for the deposition of fluid-borne particles onto a surface. This result is easily demonstrated by writing (7.2) in terms of the dimensionless coordinate x

$$(2\ Kn_D/\delta) \left.\frac{\partial n}{\partial x}\right)_{x=0} = n(x=Kn_D) \tag{7.3}$$

and taking the limit as $Kn_D \to 0$. This gives the usual boundary condition

$$n(x=0) = 0 \tag{7.4}$$

which is generally satisfied with (6.4) in aerosols and liquid suspensions of particles (except when δ is very small). There exist, however, some important aerosol systems for which (7.4) is not valid and the more general boundary condition (7.3) must be used with (6.4). These include the diffusional coagulation of aerosols and diffusional deposition of aerosols in a fine capillary, problems which will be treated in Sections 8 and 9. Because Kn_D is extremely small in liquid suspensions (7.4) is usually adequate in such systems. Use of (7.3) would only be necessary when $\delta \ll 1$ which is not uncommon when colloid suspensions are strongly stabilized.

The mathematical form of (7.3) makes this boundary condition more difficult to use than (7.4). Certain similarity transformations are no longer possible using (7.3). Moreover, this boundary condition does not satisfy the form required by Sturm-Liouville theory so that (7.3) must be approximated to apply this method of solution for $n(x)$. An example approximation is obtained by expanding $n(x)$ in a Taylor's series about $x = 0$.

$$2\,Kn_D/\delta \left.\frac{\partial n}{\partial x}\right)_{x=0} = n(x=0) + Kn_D \left.\frac{\partial n}{\partial x}\right)_{x=0} + \tfrac{1}{2} Kn_D^2 \left.\frac{\partial^2 n}{\partial x^2}\right)_{x=0} + \ldots$$

Truncation of the series after the first order term in Kn_D gives the boundary condition

$$Kn_D(2/\delta - 1) \left.\frac{\partial n}{\partial x}\right)_{x=0} \cong n(x=0) \tag{7.5}$$

which satisfies the required form for Sturm-Liouville theory. This approximation of (7.3) is a "slip flow" approximation since it is only valid for sufficiently small Kn_D. As with (7.3), the mathematical form of (7.5) serves to demonstrate how non-continuum effects can be significant whenever Kn_D is sufficiently large or δ is sufficiently small.

8. COAGULATION.

The fundamental expression of coagulation theory describes the rate of deposition of species 2 particles (of uniform radius r_2 and mass m_2) onto a species 1 particle (of radius r_1 and mass m_1). To obtain this expression for the rate of deposition we need to solve the diffusion equation in spherical coordinates which derives from Fick's law

$$\frac{\partial(n_2 r)}{\partial t} = D\,\frac{\partial^2(n_2 r)}{\partial r^2} \tag{8.1}$$

with the initial and boundary conditions

$$\begin{aligned} n_2 &= N_2 \quad \text{when } t = 0,\ r > R \\ n_2 &= N_2 \quad \text{when } t > 0,\ r \to \infty \\ n_2(r=R[1+Kn_D],t) &= (2RKn_D/\delta)\left.\frac{\partial n_2}{\partial r}\right)_{r=R} \quad \text{when } t > 0 \end{aligned} \tag{8.2}$$

where $R = r_1 + r_2$ is the separation of centers at which the spheres touch (adsorb or coalesce). The other quantities pertinent to this system are $D = D_1 + D_2$, as shown by

Smoluchowski [12], so that $D = kT/f_1 + kT/f_2 = kT/f$ where $f = f_1 f_2/(f_1 + f_2)$, $Kn_D = 2kT/(\bar{c} f R)$ and $\bar{c} = \sqrt{\bar{c}_1^2 + \bar{c}_2^2}$. The latter boundary condition is taken from (7.3).

To solve this initial boundary value problem we use the Laplace transform of $n_2(r,t)$, which we denote as $\tilde{n}_2(r,s)$. The transform of (8.1), incorporating the initial condition of (8.2) is

$$s\,\tilde{n}_2\,r - N_2\,r = D\,\frac{\partial^2(\tilde{n}_2 r)}{\partial r^2}$$

with the boundary conditions of (8.2) transformed to

$$\tilde{n}_2 = N_2/s \quad \text{when } r \to \infty$$

and

$$\tilde{n}_2(r{=}R[1{+}Kn_D],s) = (2R\,Kn_D/\delta)\left.\frac{\partial \tilde{n}_2}{\partial r}\right|_{r=R}.$$

The solution is

$$\tilde{n}_2(r,s) = N_2/s\left\{1 - C(s)\,R/r\;e^{-\sqrt{s/D}\,(r-R)}\right\} \tag{8.3}$$

where

$$C(s) = \left\{(2\,Kn_D/\delta)(1+\sqrt{s/D}\,R) + (1+Kn_D)^{-1}e^{-\sqrt{s/D}\,Kn_D r}\right\}^{-1} \tag{8.4}$$

The concentration field of the species 2 particles around the species 1 particle is therefore given by

$$n_2(r,t) = \frac{1}{2\pi i}\int_{a'-i\infty}^{a'+i\infty} e^{st}\,\tilde{n}_2(r,s)\,ds \tag{8.5}$$

where i is the imaginary unit $\sqrt{-1}$ and a' is any real constant chosen to the right in the complex plane of all singularities of $\tilde{n}_2(r,s)$.

The complex integral (8.5) can be easily evaluated in two limiting cases. In the continuum limit $Kn_D \to 0$, $C(s) = 1$ and (8.3) is easily inverted using standard reference tables of Laplace transforms.

$$n_2(r,t) = N_2[1 - R/r\,\text{erfc}(\psi)] \tag{8.6}$$

where

$$\mathrm{erfc}(\psi) = 1 - \mathrm{erf}(\psi)$$

$$\mathrm{erf}(\psi) = \frac{2}{\sqrt{\pi}} \int_o^{\psi} e^{-x^2} dx$$

and

$$\psi = (r-R)/\sqrt{4Dt} \; .$$

The deposition rate (particles/sec) of species 2 particles onto the species 1 particle is obtained from (8.6).

$$-J = 4\pi R^2 D \left.\frac{\partial n_2}{\partial r}\right)_{r=R} = 4\pi RDN_2(1 + R/\sqrt{\pi Dt}) \qquad (8.7)$$

This is the famous result of Smoluchowski [12], which is similar to Maxwell's result (2.1). The coefficient

$$K_o = 4\pi RD \qquad (8.8)$$

is the continuum coagulation constant. At time intervals of usual interest, $R << \sqrt{\pi Dt}$ so that the quantity $R/\sqrt{\pi Dt}$ can usually be neglected. The total coagulation rate between all the type 1 and type 2 particles (number/cm^3/sec) is therefore given by $K_o N_1 N_2$, where N_1 is the concentration of type 1 particles.

The complex integral (8.5) is also easily evaluated in the opposite limit, $Kn_D \to \infty$. In this "free particle" limit $C(s) = 0$ and we obtain

$$n_2(r,t) = N_2 \; . \qquad (8.9)$$

The rate of deposition of species 2 particles onto a species 1 particle is obtained from (7.1)

$$-J = \pi\delta\bar{c}\, R^2\, N_2 \; . \qquad (8.10)$$

Note from (8.9) and (8.10) that the concentration field and rate of deposition are stationary for $t > 0$ compared to the continuum case where the stationary condition was obtained only for $t >> R^2/\pi D$. Thus, the "induction time" required to obtain the stationary condition is maximum in the continuum limit $Kn_D \to 0$.

At intermediate values of Kn_D the evaluation of (8.5) is not trivial. Fortunately, evaluation of this complex integral is not required at intermediate Kn_D because the stationary solution can be obtained by use of a simple theorem (see, for example, Wylie [16]).

$$\lim_{s \to 0} s\tilde{n}_2(r,s) = \lim_{t \to \infty} n_2(r,t).$$

Thus, the stationary solution is

$$n_2(r) = N_2 \left\{ 1 - \frac{Kn_D + 1}{1 + 2Kn_D(Kn_D+1)/\delta} R/r \right\}. \quad (8.11)$$

This result gives the stationary deposition rate

$$-J = 4\pi R^2 D \left.\frac{\partial n_2}{\partial r}\right)_{r=R} = \frac{4\pi RD(Kn_D+1)}{1 + 2\ Kn_D(Kn_D+1)/\delta} N_2 . \quad (8.12)$$

Evaluation of these expressions at the extreme limits $Kn_D \to 0$ and $Kn_D \to \infty$ gives the previous stationary results for the continuum and free particle cases. The results (8.11) and (8.12) are also valid for the stationary state at intermediate values of Kn_D. Since the induction period has the maximum duration $t >> R^2/\pi D$ in the case of the continuum limit $Kn_D \to 0$, the results (8.11) and (8.12) are generally valid over the entire time range of practical interest in aerosol systems.

The concentration distribution (8.11) gives the stationary concentration jump at the surface $r = R$.

$$n_2(R) = N_2 \left\{ 1 - \frac{Kn_D + 1}{1 + 2Kn_D(Kn_D+1)/\delta} \right\}$$

Even for a perfectly adsorbing surface $\delta = 1$ this expression predicts a non-zero concentration jump at non-zero values of Kn_D.

The total rate of coagulation of species 2 particles onto all the species 1 particles is $-J = K N_1 N_2$ where

$$K = 4\pi RD \frac{Kn_D+1}{1 + 2\ Kn_D(Kn_D+1)/\delta} \quad (8.13)$$

is the coagulation rate constant valid at all values of Kn_D.

Let the quantity β be defined as the ratio

$$\beta = K/K_o .$$

By (8.8),

$$K_o = 2(r_1+r_2)^2 kT/(3\eta r_1 r_2)$$

since $C_{s1} = C_{s2} = 1$ must apply in the limit $Kn_D \to 0$. Thus,

$$\beta = \frac{(Kn_D+1)\ C_{s1}C_{s2}\ (r_1/C_{s1} + r_2/C_{s2})}{[1 + 2\ Kn_D(Kn_D+1)/\delta](r_1 + r_2)} . \tag{8.14}$$

When $r_1 = r_2$,

$$\beta = \frac{(Kn_D + 1)\ C_s}{1 + 2Kn_D(Kn_D+1)/\delta} \tag{8.15}$$

where $C_s = C_{s1} = C_{s2}$.

The correction factor can be separated into two independent factors $\beta = \beta_1\beta_2$ that correct K_o for non-negligible Kn and for non-negligible Kn_D, respectively. Individually, the factor

$$\beta_1 = C_{s1}C_{s2}(r_1/C_{s1} + r_2/C_{s2})/(r_1+r_2)$$

in (8.14) corrects K_o for non-negligible Kn while the remaining factor

$$\beta_2 = (Kn_D+1)/[1+2Kn_D(Kn_D+1)/\delta]$$

corrects K_o for non-negligible Kn_D.

The factor β_1 is easily derived since this factor simply corrects K_o for the influence of C_s on D when Kn is significant. This correction has been derived by several authors. Fuchs [7,8] also derived an expression for β_2 for the case of coagulation of equal size particlcs. Fuchs' result is obtained using the stationary form of (8.1) with the boundary conditions

$$n_2(r) = N_2 \quad \text{when} \quad r \to \infty$$

$$4\pi\ (R+\delta')^2\ D\ \left.\frac{\partial n_2}{\partial r}\right)_{r=R+\delta'} = \pi R^2\ \bar{c}\ n_2(r=R+\delta').$$

Since the stationary current of particles is the same through every surface $r = \text{constant}$, the latter condition is equivalent to the corresponding condition of the present analysis (the last of conditions (8.2) with $\delta = 1$), with the exception that Fuchs used a value of δ' different from ℓ_D. Fuchs thus obtained the correction

$$\beta_2 = (1+\delta'/R) / \{1 + \frac{\sqrt{\pi}}{2}(\ell_B/R)(1+\delta'/R)\}$$

which is similar in form to the β_2 of the present theory.

Fuchs defined $\delta'/\sqrt{2}$ as the mean distance from the surface of a central sphere $r = R = 2a$ reached by a large number of particles of radius a leaving the central sphere surface and obtaining the average path length $\ell_B = 8D/(\pi\bar{c})$. This quantity is multiplied by $\sqrt{2}$ to account for motion of the central sphere. Fuchs derived the result

$$\delta'/R = \frac{\sqrt{2}}{3\ell_B/R} \{(1 + \ell_B/R)^3 - (1 + \ell_B^2/R^2)^{3/2}\} - 1 .$$

The product $\beta = \beta_1\beta_2$ evaluated by the present theory (8.14) with $\delta = 1$ and using Fuchs' expression for β_2 are compared in Table 2. The results show a maximum difference of less than 4 percent. Fuchs claims that his expression provides only a rough estimate of the proper correction. The good agreement of experimental data with the present result and Fuchs' expression suggests, however, that his correction is better than a rough approximation.

The data of Table 2 were calculated with the assumptions that $r_1 = r_2 = 0.22$ μm, $T = 25°C$, $\delta = 1$, $\eta = 1.84 \times 10^{-4}$ gm/cm/sec and $\rho_1 = \rho_2 = 0.917$ gm/cm^3. The variation in Kn_D and in $Kn = \lambda/r_1$ was assumed caused by reducing the gas pressure. The assumed conditions therefore correspond closely to the experiments of Wagner and Kerker [14] using aerosols in helium. These experimental data seem to be the best available for significant Kn_D and significant Kn. Both the present theory for β and Fuchs' theory agree with the experimental data within the experimental error (of less than ten percent) in the range of Kn up to 12.

9. AEROSOL DEPOSITION IN A FINE CAPILLARY.

We wish to determine in this section the rate of deposition of aerosol particles onto the wall of a fine capillary. Since the local rate of particle deposition depends on the distance z from the capillary inlet plane, an equivalent and

Table 2. Comparison of the correction factor β calculated with the present theory and Fuchs' theory.

Kn	C_s	β Present Theory	β Fuchs' Theory	% difference
0.1	1.123	1.096	1.087	0.82
0.2	1.248	1.214	1.204	0.82
0.5	1.653	1.594	1.576	1.13
0.7	1.947	1.865	1.841	1.29
1.0	2.406	2.282	2.247	1.53
2.0	4.002	3.661	3.579	2.24
3.0	5.629	4.961	4.824	2.76
5.0	8.907	7.281	7.038	3.34
7.0	12.20	9.250	8.926	3.50
8.0	13.84	10.12	9.765	3.51
10.0	17.13	11.65	11.26	3.35
12.0	20.43	12.96	12.56	3.10
15.0	25.37	14.56	14.17	2.68
20.0	33.61	16.54	16.22	1.93
30.0	50.08	18.94	18.76	0.95
50.0	83.04	21.05	21.04	0.05
70.0	116.0	21.90	21.93	-0.14
100.0	165.4	22.46	22.49	-0.13
200.0	330.2	22.95	22.95	0
500.0	824.6	23.11	23.10	0.04
10^3	1.649×10^3	23.14	23.12	0.09
10^4	1.648×10^4	23.15	23.12	0.13

more convenient quantity for describing deposition in the capillary is the penetration function $\phi(a,z)$. This function is the fraction of particles of radius a which penetrate the distance z into the capillary without adsorbing on the wall. For an aerosol with $N(a)da$ particles/cm^3 of radius between a and $a + da$ flowing into a capillary of length L, the fraction of the total number of aerosol particles penetrating the capillary is

$$\Phi(L) = \int_{a=0}^{\infty} \phi(a,L)N(a)\,da \Bigg/ \int_{a=0}^{\infty} N(a)\,da \ .$$

while the fraction adsorbed on the capillary wall is $1 - \Phi$. Once $\phi(a,z)$, or simply $\phi(z)$, is known, similar expressions can be used to obtain the penetration of other quantities such as aerosol mass, surface or activity.

The steady convective diffusion of an aerosol in a circular tube in which axial diffusion is negligible compared to convection is described by

$$\frac{\partial^2 n}{\partial r^2} + \frac{1}{r}\frac{\partial n}{\partial r} = \frac{1}{D}\frac{\partial (nv)}{\partial z} \tag{9.1}$$

where n is the concentration of particles of radius a, r the radial position in the tube, D the diffusion coefficient of the particles and $v = v(r)$ is the axial fluid velocity. Fluid velocity components in other than axial directions are assumed negligible. Later we shall also make the assumption of fully developed flow over the entire capillary length. The boundary conditions are

$$\begin{aligned} n(r,0) &= n_o \\ n(0,z) &\text{ is regular} \\ n(R[1-Kn_D],z) &= -(2R\,Kn_D/\delta)\left.\frac{\partial n}{\partial r}\right|_{r=R} \end{aligned} \tag{9.2}$$

where R is the capillary radius and $Kn_D = 2kT/(\bar{c}fR)$. The last boundary condition comes from (7.3). It tacitly assumes $a/R << 1$. Although this assumption is not necessary, it is made here for simplicity. The present solution also applies for $a \lesssim R$ if R is replaced by $R - a$ in (9.2) and in Kn_D and at later points in the analysis.

Since $Kn = \lambda/R$ is not generally negligible in fine capillaries, the Poiseuille velocity distribution cannot be used. Rather, we assume the fluid velocity profile for fully developed flow

$$v(r) = \sigma\bar{v}[1 - \gamma(r/R)^2]$$

where $\bar{v}$ is $Q/\pi R^2$, Q being the volumetric flow rate through the tube. This velocity profile satisfies both the laws of hydrodynamics and the slip velocity condition at the wall when

Kn is significant. The values of the dimensionless coefficients σ and γ have not yet been established from first principles so that flow rate through a capillary cannot presently be predicted in the general case (see, for example, Suetin, *et al.* [13]). But using the volumetric flow rate Q as an experimental parameter, values of the coefficients σ and γ are easily established.

The velocity profile in the tube must satisfy the simple force balance

$$-2\pi RL\eta \left.\frac{dv}{dr}\right)_{r=R} = \pi R^2 \Delta p$$

where Δp is the pressure drop across the tube length L. This condition requires

$$\sigma\gamma = R^2\Delta p/(4\eta L\bar{v}) \ . \qquad (9.3)$$

The second condition is

$$Q = \pi R^2 \bar{v} = \int_{r=0}^{R} v(r)\ 2\pi r\ dr$$

which requires

$$\gamma = 2(1 - 1/\sigma) \ . \qquad (9.4)$$

Combination of (9.3) and (9.4) gives

$$\sigma = 1 + R^2\Delta p/(8\eta L\bar{v}) \ . \qquad (9.5)$$

This result gives for the slip velocity at the tube wall

$$v(r{=}R) = \bar{v}\left\{1 - \frac{R^2\Delta p}{8\eta L\bar{v}}\right\}$$

in agreement with the limiting conditions $Kn \to 0$ for which $\bar{v} = R^2\Delta p/(8\eta L)$ and $v(R) \to 0$, and $Kn \to \infty$ for which $\Delta p \to 0$ and $v(R) \to \bar{v}$. The coefficients thus span the range between continuum flow: $\sigma = 2$, $\gamma = 1$, and free molecule flow: $\sigma = 1$, $\gamma = 0$. The coefficients cannot be presently expressed in terms of Kn, but they are exactly specified by (9.4) and (9.5) in terms of the experimental quantities Q (or $\bar{v}$) and Δp.

Expressions (9.1) and (9.2) are non-dimensionalized by the use of the dimensionles variables $r' = r/R$, $n' = n/n_o$ and $z' = Dz/(\sigma\bar{v}R^2)$. Expression (9.1) thus becomes

$$\frac{\partial^2 n'}{\partial r'^2} + \frac{1}{r'}\frac{\partial n'}{\partial r'} = (1 - \gamma r'^2)\frac{\partial n'}{\partial z'} \tag{9.6}$$

with boundary conditions

$$n'(r',0) = 1$$

$$n'(0,z') \text{ is regular} \tag{9.7}$$

$$n'(r'=[1-Kn_D],z') = -(2Kn_D/\delta)\left.\frac{\partial n'}{\partial r'}\right)_{r'=1} .$$

In all subsequent expressions containing n, r and z the primes will be understood.

When Kn_D is sufficiently small, the last boundary condition can be replaced by its truncated Taylor's series expansion, as in (7.5), so that

$$n(r=1,z) = -(2/\delta - 1)\, Kn_D \left.\frac{\partial n}{\partial r}\right)_{r=1} . \tag{9.7a}$$

With this boundary condition the system (9.6) and (9.7) can be solved by separation of variables and Sturm-Liouville theory. The solution is a "slip flow" solution since it is restricted to sufficiently small Kn_D.

Assume a solution of the form

$$n(r,z) = \rho_i(r)\, \zeta_i(z) .$$

Substitution into (9.6) and separation of variables gives $\rho_i(r)$ as the solution of

$$\rho_i'' + \frac{1}{r}\rho_i' + \omega_i^2 (1 - \gamma r^2)\, \rho_i(r) = 0 \tag{9.8}$$

with boundary conditions

$$\rho_i(0) \text{ is regular}$$

$$\rho_i(1) = -(2/\delta - 1)\, Kn_D\, \rho_i'(r=1) \tag{9.9}$$

and $\zeta_i(z)$ as

$$\zeta_i(z) = A_i \exp(-\omega_i^2 z)$$

where the coefficients A_i and eigenvalues ω_i^2 must be determined.

The solution

$$\rho_i(r) = \sum_{n=0}^{\infty} a_n(\omega_i^2)\, r^n$$

is regular at $r = 0$ and satisfies (9.8) provided

$$a_o = 1$$

$$a_1 = a_3 = a_5 = a_7 = \ldots . = a_{2n+1} = 0$$

$$a_2 = -\omega_i^2/4$$

$$a_4 = +\omega_i^2(\omega_i^2/4 + \gamma)/16$$

$$\vdots$$

$$a_{2n+2} = -\omega_i^2 \frac{a_{2n} - \gamma a_{2n-2}}{(2n+2)^2} .$$

The general solution thus takes the form

$$n(r,z) = \sum_{i=1}^{\infty} A_i e^{-\omega_i^2 z} \sum_{n=0}^{\infty} a_{2n}(\omega_i^2) r^{2n} .$$

Before determining the coefficients and eigenvalues we convert to the penetration $\phi(z)$ defined earlier. Accordingly,

$$\phi(z) = 2\sigma \int_{r=0}^{1} n(r,z)(1 - \gamma r^2) r \, dr$$

$$= \frac{4}{2-\gamma} \sum_{i=1}^{\infty} A_i(\gamma,\omega_i^2) e^{-\omega_i^2 z} \sum_{n=0}^{\infty} a_{2n}(\omega_i^2) \left\{ \frac{2n+4-\gamma(2n+2)}{(2n+4)(2n+2)} \right\}$$

$$= \sum_{i=1}^{\infty} B_i(\gamma,\omega_i^2) e^{-\omega_i^2 z} .$$

To determine the eignevalues ω_i^2 we use the boundary condition (9.7a). This requires

$$\sum_{n=0}^{\infty} \{1 + 2n(2/\delta - 1) Kn_D\} a_{2n}(\omega_i^2) = 0 .$$

Thus, the eigenvalues depend on the coefficient of the fluid velocity profile γ, on the sticking probability δ and on the particle Knudsen number Kn_D. Tables 3, 4 and 5 show the first six eigenvalues for various Kn_D values when $\delta = 1$ and $\gamma = 0$, 0.5 and 1.0 .

Table 3. Eigenvalues and coefficients for $\gamma = 0$ (free molecule flow through a capillary).

Kn_D	$i =$	1	2	3	4	5	6
0	$\omega_i^2 =$	5.78319	30.4713	74.8870	139.040	222.932	326.563
	$B_i =$	0.69166	0.13127	0.05341	0.02877	0.01794	0.01125
0.1	$\omega_i^2 =$	4.75021	25.3332	63.3120	119.603	194.827	289.336
	$B_i =$	0.80388	0.12598	0.03869	0.01523	0.00696	0.00353
0.2	$\omega_i^2 =$	3.95936	22.2137	58.0295	112.833	187.103	280.998
	$B_i =$	0.87214	0.09535	0.02075	0.00642	0.00252	0.00116
0.3	$\omega_i^2 =$	3.36390	20.3669	55.4762	109.951	184.057	277.861
	$B_i =$	0.91276	0.06932	0.01203	0.00334	0.00124	0.00055
0.4	$\omega_i^2 =$	2.91051	19.2003	54.0341	108.406	182.464	276.244
	$B_i =$	0.93767	0.05116	0.00767	0.00201	0.00073	0.00032
0.5	$\omega_i^2 =$	2.55824	18.4123	53.1206	107.450	181.492	275.262
	$B_i =$	0.95366	0.03877	0.00527	0.00134	0.00048	0.00021

Table 4. Eigenvalues and coefficients for $\gamma = 0.5$ (transition or slip flow through a capillary).

Kn_D	i =	1	2	3	4	5	6
0	ω_i^2 =	6.47641	36.1924	89.8982	167.527	269.062	394.496
	B_i =	0.72680	0.11940	0.04708	0.02505	0.01536	0.00020
0.1	ω_i^2 =	5.45150	31.1557	78.3677	147.702	239.731	354.873
	B_i =	0.81527	0.11352	0.03671	0.01558	0.00758	0.00098
0.2	ω_i^2 =	4.64778	27.8208	72.2133	139.209	229.500	343.394
	B_i =	0.87221	0.09105	0.02230	0.00757	0.00314	0.00039
0.3	ω_i^2 =	4.02261	25.6742	68.9119	135.226	225.124	338.782
	B_i =	0.90850	0.07000	0.01392	0.00418	0.00161	0.00020
0.4	ω_i^2 =	3.53186	24.2373	66.9437	133.002	222.769	336.354
	B_i =	0.93215	0.05397	0.00927	0.00260	0.00097	0.00013
0.5	ω_i^2 =	3.14060	23.2277	65.6588	131.600	221.310	334.864
	B_i =	0.94809	0.04230	0.00655	0.00176	0.00064	0.00008

Table 5. Eigenvalues and coefficients for γ = 1.0 (continuum flow through a capillary).

Kn_D	i =	1	2	3	4	5	6
0	ω_i^2 =	7.31359	44.6095	113.921	215.241	348.564	513.890
	B_i =	0.81905	0.09753	0.03250	0.01543	0.00168	0.00000
0.1	ω_i^2 =	6.33404	40.5081	105.487	201.646	329.192	488.261
	B_i =	0.86923	0.08320	0.02358	0.00977	0.00379	0.00000
0.2	ω_i^2 =	5.55381	37.6387	100.270	194.094	319.401	476.348
	B_i =	0.90293	0.06760	0.01633	0.00602	0.00217	0.00000
0.3	ω_i^2 =	4.92768	35.6031	96.9237	189.606	313.917	469.977
	B_i =	0.92594	0.05441	0.01156	0.00393	0.00135	0.00000
0.4	ω_i^2 =	4.41898	34.1151	94.6517	186.705	310.492	466.107
	B_i =	0.94205	0.04405	0.00847	0.00273	0.00091	0.00000
0.5	ω_i^2 =	4.00000	32.9926	93.0272	184.697	308.171	463.521
	B_i =	0.95362	0.03608	0.00642	0.00199	0.00066	0.00000

Finally, to complete the solution for $\phi(z)$, we determine the coefficients A_i and B_i. By use of Sturm-Liouville theory

$$A_i = \int_{r=o}^{1} \rho_i(r)(1-\gamma r^2)\, rdr \Big/ \int_{r=o}^{1} \rho_i^2(r)(1-\gamma r^2)\, rdr$$

so that, by carrying out the integration and using the definition of B_i,

$$B_i = \frac{4}{2-\gamma}\left\{\sum_{n=0}^{\infty} \frac{[2+(2n+2)(1-\gamma)]a_{2n}}{(2n+4)(2n+2)}\right\}^2$$

$$\Big/ \sum_{n=0}^{\infty}\left\{\frac{[2+(4n+2)(1-\gamma)]a_{2n}^2}{(4n+4)(4n+2)} + 2a_{2n}\sum_{m=n+1}^{\infty}\frac{[2+(2n+2m+2)(1-\gamma)]a_{2m}}{(2n+2m+4)(2n+2m+2)}\right\}.$$

Thus, the penetration function $\phi(z)$ is completely specified. Values of the coefficients B_i are listed in Tables 3, 4 and 5 for various Kn_D values when $\delta = 1$ and $\gamma = 0$, 0.5 and 1.0.

Figures 2, 3 and 4 show the function $\phi(z)$ versus $z = DL/(\sigma R^2 \bar{v})$ for the sticking probability $\delta = 1$. The curves show the penetration is substantially increased by significant Kn_D, as may occur in fine capillaries.

Measurement of aerosol deposition in a fine capillary or onto a thin wire would be easier than measurement of the rate of coagulation of an aerosol. Since all these processes are governed by the same diffusion law and deposition boundary condition, experimental verification of one theory will be very suggestive of the validity of another. Thus, some of the fundamental properties of aerosol coagulation and deposition, such as the influence of particle bounce which may cause important attenuation of the coagulation or deposition rate according to the calculations presented here and the earlier analysis of Dahneke [3] of the mechanics of particle-particle and particle-surface collisions, could be most conveniently studied in, say, a capillary penetration experiment.

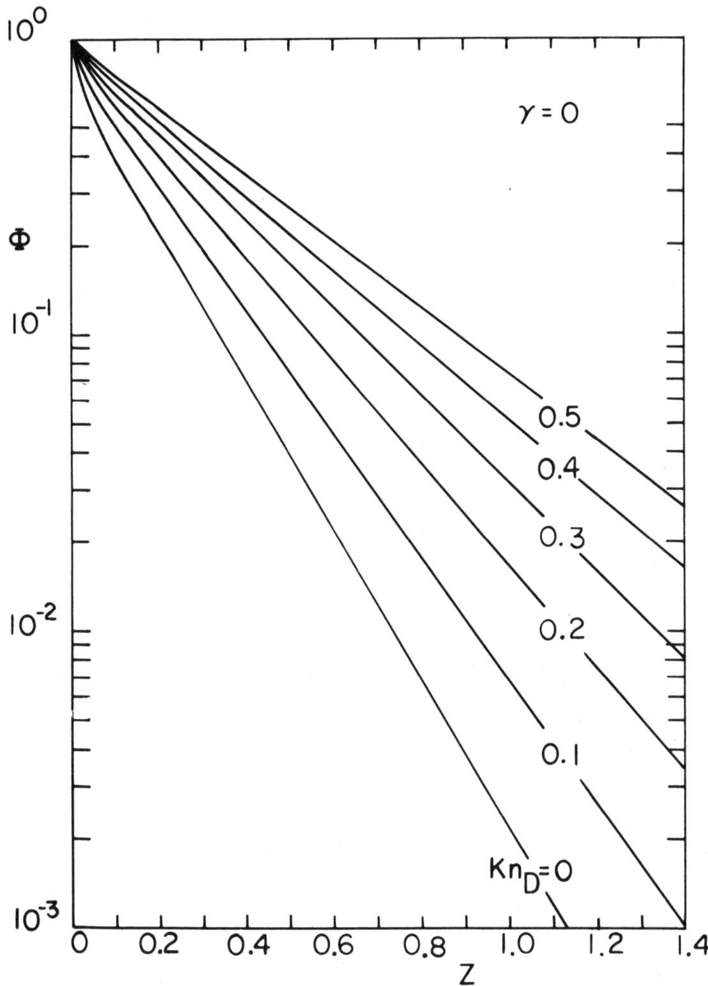

Fig. 2. Penetration versus dimensionless length for various Kn_D assuming $\delta = 1$ and $\gamma = 0$ (free molecule flow). These results also apply for sedimentation of particles at velocity $\bar{v}$ through a vertical tube, provided $\bar{v}$ is sufficiently large so that sedimentation dominates vertical diffusion. They also apply to the problem of diffusional deposition on the wall of a long tube from a motionless aerosol at uniform concentration n_0 when time t=0. In this case the variable $z = DL/(\sigma\bar{v}R^2)$ is replaced by $z = Dt/R^2$ and Φ is the fraction of particles remaining in the fluid-borne state.

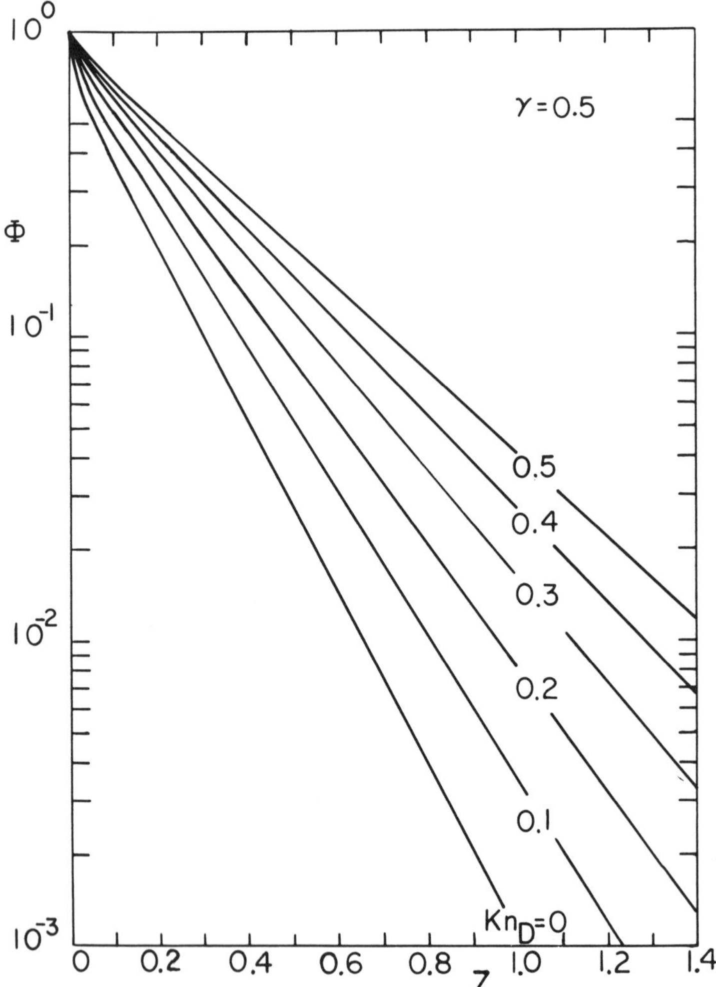

Fig. 3. Penetration versus dimensionless length for various Kn_D assuming $\delta = 1$ and $\gamma = 0.5$ (transition flow).

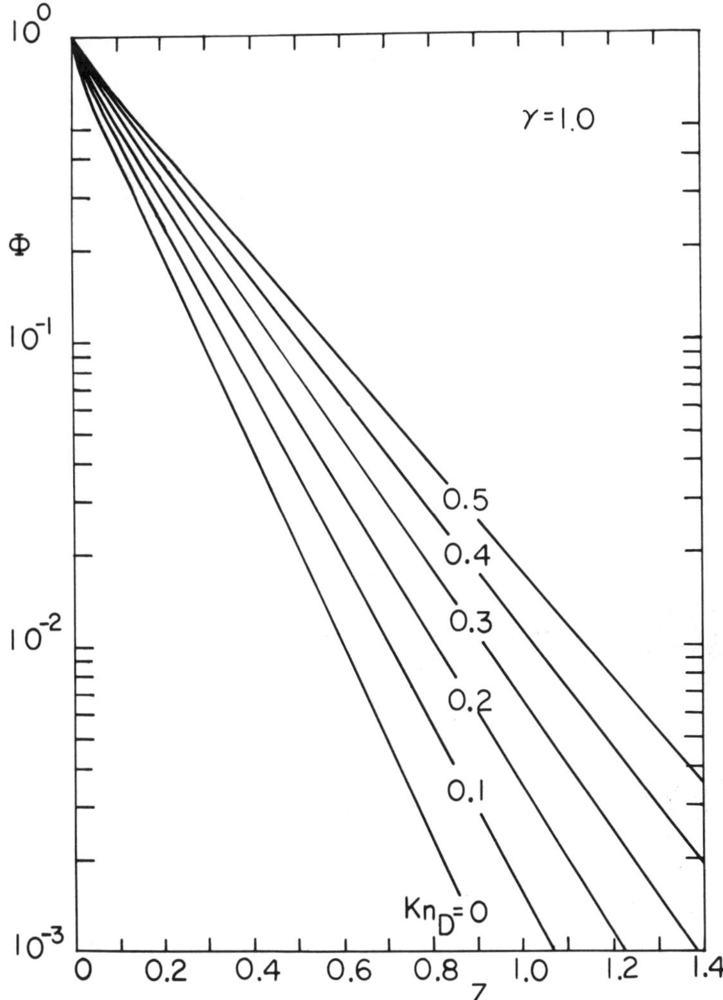

Fig. 4. Penetration versus dimensionless length for various Kn_D assuming $\delta = 1$ and $\gamma = 1.0$ (continuum flow).

REFERENCES

1. Chapman, S. and F. W. Dootson, Thermal diffusion, Phil. Mag. 33 (1917), 248-253.
2. ___________ and T. G. Cowling, The Mathematical Theory of Non-uniform Gases, Cambridge University Press, 1960.
3. Dahneke, B., The capture of aerosol particles by surfaces, J. Colloid and Interface Sci. 37 (1971), 342-353.
4. Davis, E. J. and A. K. Ray, Submicron droplet evaporation in the continuum and non-continuum regimes, J. Aerosol Sci. 9 (1978), 411-422.
5. Einstein, A., Investigations on the Theory of the Brownian Movement, Dover Publications, 1956.
6. Fuchs, N. A., Concerning the evaporation rate of small droplets in gases, Phys. Z. Sowjetunion 6 (1934), 224-243.
7. ___________, On the theory of coagulation, Z. f. Physik. Chemie A-171 (1934), 199-209.
8. ___________, The Mechanics of Aerosols, Pergamon Press, 1964.
9. Furth, R., An elementary theory of thermal diffusion, Roy. Soc. London, Proc. A 179 (1942), 461-469.
10. Maxwell, J. C., Collected Scientific Papers, Cambridge University Press, 1890.
11. Monchick, L. and H. Reiss, Studies of evaporation of small drops, J. Chem. Phys. 22 (1954), 831-836.
12. Schmoluchowski, M. v., Mathematical theory of the kinetics of coagulation of colloidal solutions, Z. f. Physik. Chemie 92 (1917), 129-168.
13. Suetin, P. E., B. T. Porodnov, V. G. Chernjak and S. F. Borisov, Poiseuille flow at arbitrary Knudsen numbers and tangential momentum accommodation, J. Fluid Mech. 60 (1973), 581-592.
14. Wagner, P. E. and M. Kerker, Brownian coagulation of aerosols in rarefied gases, J. Chem. Phys. 66 (1977), 638-646.
15. Wright, P. G., On the discontinuity involved in diffusion across an interface (the delta of Fuchs), Disc. Faraday Soc. 30 (1960), 100-112.

16. Wylie, C. R., Advanced Engineering Mathematics, 2nd Edition, McGraw-Hill, 1960.

The author acknowledges the assistance of Amiram Rasolt for his computer calculations of the data presented in Tables 3,4, 5 and Figures 2,3,4 and the support of National Science Foundation Grant CPE-8019543. The contents of Sections 2-5 were presented at the 160th National Colloid Symposium of the ACS, Grand Island, New York, June 1977.

Barton Dahneke
Department of Chemical Engineering
University of Rochester
Rochester, New York 14627

Simulation of Aerosol Dynamics

J. Brock

1. Introduction.

The term aerosol is generally applied to a dilute suspension of particles in a host gas. Atmospheric suspended particulate matter and water clouds are the most obvious examples of such disperse systems. Particle sizes may range from those associated with molecular clusters, to some indefinite upper limit which depends on the particular system. Although the host gas is usually a dilute gas, dense gases may be included in the definition, but not liquids.

The suspected adverse effects of atmospheric aerosols and the role of aerosols in meteorological and climatological processes have led to increased interest in the processes of formation, growth and transport of aerosols. More recently, industrial and military applications have added to this interest. Apart from such applications, the study of aerosols involves one with fundamental problems of interest in fluid dynamics, statistical mechanics, kinetic theory, probability theory, quantum chemistry, etc.

A number of books and reviews (e.g. 1,2,3,4,5,) are available dealing with various aspects of aerosol science. The reader is referred to these sources for a complete development of this subject.

Aerosol particles in nature and many technical applications are very complex in terms of their morphology, internal structure, chemical composition, etc. Unlike colloidal

THEORY OF DISPERSED MULTIPHASE FLOW

ISBN 0-12-493120-0

dispersions in liquids which may be stabilized, there are no known stable aerosols and the process of coagulation (sticking, interparticle collisions) of aerosol particles is always present. In addition, aerosol particles are in general subject to the processes of molecular condensation and evaporation at rates much larger than those in the corresponding liquid phase dispersions.

In the discussion which follows, the aerosol consists of a dilute host gas and spherical aerosol particles which in sticking collisions always coalesce. Even with these restrictions a large number of parameters are necessary to characterize such a system. Some of the important dimensionless groups appearing in the description of aerosol dynamics for the system described are:

Knudsen number : $Kn_k = L_h / R_k$ (1.1)

Mach number: $Ma_k = | \vec{V}_{kh} | / \bar{V}_h$ (1.2)

Brown number: $Br_k = \bar{V}_k / \bar{V}_h$ (1.3)

Schmidt number: $Sc_k = R_k^2 n_h L_h$ (1.4)

Kelvin number: $Ke_k = 2 \sigma\gamma / R_k \kappa T$ (1.5)

where subscript k represents the property of the particle and subscript h that of the host gas. R_k is the particle radius, L_h the host gas molecular mean free path, $\vec{V}_{kh}$ the relative particle-gas velocity, $\bar{V}$ the mean thermal speed, n_h the number density, σ the particle's surface tension, γ the particle's molecular volume, and κT the thermal energy.

In the simulation of aerosol dynamics, the Knudsen number plays an essential role. At atmospheric host gas pressure when $Kn_k \to \infty$, the aerosol particles may be regarded as large molecules, although unstable ones. For many purposes in this limit, therefore, the dynamics of the aerosol is described in part by evolution equations such as the Fokker-Planck and Boltzmann equations (3). As $Kn_k \to 0$, the particles move in a continuum and hydrodynamic theory is invoked (e.g. 3). For intermediate values of Kn_k, a complete theory of aerosol dynamics is lacking and various approximations are used (3,5).

In the discussion which follows, $Ma_k << 1$, which is equivalent to a requirement of very small Reynolds numbers, Re_k, since $Re_k \sim 4Ma_k / Kn_k$. Br_k is a measure of the transition to Brownian motion from the molecular which occurs for $Br_k << 1$. Sc_k signifies the relative importance of momentum transfer compared to diffusive transfer of particles.

This paper is divided into two parts. In the first, the generation and initial stages of growth of condensation aerosols are considered. For uniform systems (systems without macroscopic gradients), aerosol evolution by homogeneous nucleation, coagulation and condensation is described. The second part discusses the simulation of aerosol dynamics in non-uniform systems. Problems in description of aerosol dynamics in this case are outlined. Several examples are cited including aerosol evolution in axisymmetric laminar jets and dispersion of aerosol plumes in the atmosphere.

2. Aerosol Formation and Growth in Uniform Systems.

An aerosol of coalescing particles formed from a single chemical species is usually characterized by its particle size distribution. For such aerosols, the goal of aerosol dynamics simulation is to describe the evolution of this distribution. For uniform systems this goal, at least in principle, can be achieved. In non-uniform systems (turbulent, for example) much more remains to be done.

This section begins with a discussion of the rate equations commonly used to describe formation of ultrafine particles by homogeneous nucleation. The roles of coagulation and condensation in development of the particle size distribution of the stable particles are then examined.

2.1 Homogeneous nucleation and growth.

The principal mechanism for formation of ultrafine aerosol particles is homogeneous nucleation. This involves the aggregation of monomeric molecular species at sufficiently high supersaturations. After formation, the stable aggregates or particles may continue to grow by the processes of coagulation and condensation. If the host gas is non-uniform, the characteristics of the aerosol may also be modified by the non-uniformity.

The physical system to be considered consists of a monomeric vapor species undergoing homogeneous nucleation in an inert, dilute host gas. This system corresponds to that frequently realized in practice. The monomeric vapor molecules have an initial number concentration n_{10}, molecular mass m_1, and molecular radius R_1. The corresponding properties of the host gas are given by n_h, m_h, and R_h. It will be supposed that $n_{10} << n_h$ and that the system is uniform. The specification of uniformity is introduced to delineate clearly the particle formation and growth processes. The system temperature and pressure are T and p respectively. With the condition $n_{10} << n_h$ the average T and p will remain sensibly constant during the evolution of the aerosol.

The appearance of a liquid phase in this system remains as one of the unsolved classical problems of physics. No extensive discussion of current theory and experiment for this problem will be given; there exist many reviews of vapor phase nucleation (e.g. 6,7). A complete theory should proceed from a description of the dynamics of molecular motion in a suitable phase space; this apparently has not yet been done. Currently, except for molecular dynamics studies (e.g. 8,9) the theories are phenomenological and have proceeded from master equations (e.g. 10) and, much earlier, from intuitive rate expressions (e.g. 7).

In the currently accepted picture of homogeneous nucleation, stable clusters capable of continued growth will form from monomeric vapor molecules when the supersaturation ratio, $S = n_1 / n_{v1}$ is greater than some suitably defined critical value, S_c. In the definition of S, n_{v1} refers to the saturation vapor concentration of monomer vapor at the system temperature T.

For the homogeneous system considered, suppose that the initial supersaturation ratio $S_o = n_{10} / n_{v1}$, is: $S_o > S_c$ at $t = 0$. Then the intuitive rate expression for the homogeneous nucleation and growth process is:

$$\frac{dn_k}{dt} = (1/2) \sum_{j=1}^{k-1} a_{j,k-j}\, b_{j,k-j}\, n_j\, n_{k-j}$$

$$- n_k \sum_{j=1}^{\infty} a_{k,j}\, b_{k,j}\, n_j - \sum_{j=1}^{k/2} e_{j,k}\, n_k$$

$$+ \sum_{j=k+1}^{\infty} e_{j-k,j}\, n_j; \; k = 1, 2, \ldots \qquad (2.1)$$

where t is time and n_k represents the number concentration of a cluster with k monomers. $a_{k,j}$ and $b_{k,j}$ are respectively the collision efficiency and collision rate between two clusters with k and j monomers. $e_{j,k}$ is the total rate at which a cluster of size j evaporates from a cluster of size k.

For the system specified, eq. (2.1) describes the complete evolution of the aerosol formed by homogeneous nucleation. However, it is usual practice (e.g. 7) to proceed in a somewhat different manner in studying aerosol evolution. One recognizes that particles (clusters) smaller than a critical size k^* are unstable while those larger than k^* are stable and may grow indefinitely so long as the monomer supersaturation ratio, S, which decreases from the initial S_o, remains greater than one. For the period when $S \geq S_c$, and $k > k^*$:

$$\frac{dn_k}{dt} = (1/2) \sum_{j=1}^{k-k^*-1} a_{j,k-j}\, b_{j,k-j}\, n_j\, n_{k-j}$$

$$- n_k \sum_{j=1}^{\infty} a_{k,j}\, b_{k,j}\, n_j + c_{k-1}\, n_1\, n_{k-1}$$

$$- c_k\, n_1\, n_k - d_k\, n_k + d_{k+1}\, n_{k+1}, \qquad (2.2)$$

As is the usual practice, eq. (2.2) separates coagulation, the first two terms on the right hand side, from condensation and evaporation, the third through the sixth terms. As is consistent also with current practice, collisions between stable, $k \geq k^*$, and unstable, $k < k^*$, clusters are neglected. The same assumption is made in homogeneous nucleation theory, where collisions between unstable clusters are not accounted. The coefficients c_k and d_k represent respectively the condensation and evaporation rates for a stable cluster of size k. $b_{k,j}$ represents the coagulation rate and $a_{k,j}$ the sticking efficiency, usually assumed to be unity.

For $k = k^*$

$$\frac{dn_{k^*}}{dt} = I(k,^* t) - n_k^* \sum_{k = k^*}^{\infty} a_{k^*,k}\, b_{k^*,k}\, n_k - c_{k^*}\, n_1\, n_{k^*} + d_{k^*+1}\, n_{k^*+1} \tag{2.3}$$

where $I(k^*,t)$ is the rate at which critical clusters are supplied by homogeneous nucleation. The initial conditions for (2.2) and (2.3) are:

$$n_k\ (t = 0) = 0;\ k = k^*,\ k^* + 1, \ldots \tag{2.4}$$

The corresponding equations for the unstable clusters are:

$$\frac{dn_k}{dt} = a_{k-1}\, b_{k-1}\, n_1\, n_{k-1} - a_k\, b_k\, n_1\, n_k - e_k\, n_k + e_{k+1}\, n_{k+1}\ ;\ 2 \le k \le k^* - 1 \tag{2.5}$$

$$\frac{dn_{k^*-1}}{dt} = - a_{k^*-1}\, b_{k^*-1}\, n_1\, n_{k^*-1} - e_{k^*-1}\, n_{k^*-1} \tag{2.6}$$

and

$$I(k^*,t) = a_{k^*-1}\, b_{k^*-1}\, n_1\, n_{k^*-1} \tag{2.7}$$

Eqs. (2.2) - (2.6) are coupled to the monomer equation:

$$\frac{dn_1}{dt} = - n_1 \sum_{k=1}^{\infty} a_k\, b_k\, n_k + \sum_{k=2}^{\infty} e_k\, n_k \tag{2.7}$$

It is assumed that so long as $S \geq S_c$, there is no net flux of particles to k^*-1 from the stable particles with $k \geq k^*$. The initial conditions for eqs. (2.5) - (2.7) are:

$$n_k\ (t = 0) = 0\ ;\ k = 2,\ 3,\ \ldots\ k^*-1$$

$$n_1\ (t = 0) = n_{10} \tag{2.8}$$

It should be noted that the neglect of collisions between unstable clusters has not been verified and doubts have been raised about this assumption (e.g. 9).

The development of these rate equations is purely formal. Little is known in detail about the nature of the coefficients a_k and e_k for the unstable cluster. In the classical theory (e.g. 7) these coefficients are evaluated by essentially equilibrium arguments. In spite of these limitations, these equations provide a useful vehicle for discussion and aid in understanding experimental results from aerosol formation studies.

It is physically unrealistic to suppose that the initial supersaturation S_o is obtained without a time delay which is at the least the order of the relaxation time for the gas, $t_r \sim L_{1h} / \bar{v}_1$ where L_{1h} is the molecular mean free path for monomer - host gas collisions and v_1 is the mean molecular speed of the monomer. The shortest nucleation times would correspond to small critical nucleus size k^*, small values of e_k and a_k near unity. In this case the nucleation time, t_n, would approach the value, $t_n \sim (m_1 / \pi\kappa T)^{1/2} / 4n_1 R_1^2$, the inverse monomer collision frequency. Since it is supposed that $n_{10} << n_h$, $t_r << t_n$ for "fast" nucleation and the initial conditions posed are physically realistic.

For short times after initiation of nucleation, $Kn_k = L / R_k >> 1$ for all the particles, which therefore behave as large molecules in a dilute gas (3). In this "free molecule regime" the rate coefficients for the stable clusters are (3):

$$b_{k,j} = [8\pi\kappa T (m_k + m_j) / m_k m_j]^{1/2} (R_k + R_j)^2 \quad (2.9)$$

$$c_k = \alpha\pi R_k^2 \bar{v}_1 \quad (2.10)$$

$$d_k = \alpha\pi R_k^2 \bar{v}_1 (P_{v1} / \kappa T) \exp (Ke_k) \quad (2.11)$$

where P_{v1} is the vapor pressure of bulk monomer and α is the condensation/evaporation coefficient.

In nucleation experiments, the total number concentration, $M_o(t)$, of stable clusters is frequently measured:

$$M_o(t) = \sum_{k=k^*}^{\infty} n_k \tag{2.12}$$

Taking this moment of (2.2) and (2.3):

$$\frac{dM_o}{dt} = (1/2) \sum_{k=k^*}^{\infty} \sum_{j=k^*}^{k-k^*} a_{j,k-j}\, b_{j,k-j}\, n_j\, n_{k-j}$$

$$- M_o \sum_{j=k}^{\infty} a_{j,k}\, b_{j,k}\, n_j + I(k^*,t) \tag{2.13}$$

since M_o is conserved by the condensation and evaporation processes, so long as $S \geq S_c$. For small times, the approximation $b_{k,j} \sim b$, a constant since $b_{k,k} \sim m_k^{1/6}$ from (2.9). In this case:

$$\frac{dM_o}{dt} = -(1/2)bM_o^2 + I \tag{2.14}$$

This has the solution for constant I

$$M_o = (2I/b)^{1/2} \tanh\left[(Ib/2)^{1/2}\, t\right] \tag{2.15}$$

From this equation one sees the familiar experimental result (e.g. 11) that the higher the nucleation rate, which is determined by S_o, the higher is the initial number concentration of stable particles. For high nucleation rates and small times for the stable particles, the condensation/evaporation rates are small compared to the coagulation rate, so that to the same approximation:

$$\frac{dn_{k^*}}{dt} = I(k^*) - bn_{k^*} \sum_{k=k^*}^{\infty} n_k \tag{2.16}$$

$$\frac{dn_k}{dt} = (1/2)\, b \sum_{j=k^*}^{k-k^*-1} n_j\, n_{k-j} - b\, n_k \sum_{j=k^*}^{\infty} n_j \tag{2.17}$$

and with the transform:

$$\phi = \sum_{k=k^*}^{\infty} n_k\, z^k$$

the transform solution of (2.16) and (2.17) for I = constant:

$$\phi = \left[2(z^{k^*} - 1)\, I/b\right]^{1/2} \tan\left(\left[(z^{k^*} - 1)\, Ib/2\right]^{1/2} t\right)$$

$$+ (2I/b)^{1/2} \tanh\left[(Ib/2)^{1/2}\, t\right] \tag{2.18}$$

so that the first few moments, $M_i = \sum_{k=k^*}^{\infty} k^i n_k$, are:

$M_1 = k^* It$; $M_2 = k^*(k^*-1)\ It + (1/3) b\ k^{*2}\ I^2\ t^3$, which shows the effect of coagulation in increasing the polydispersity of the distribution.

It is possible to express the growth equations (2.2) and (2.3) in continuous approximation. This same continuous approximation has been commonly used in deriving the classical nucleation rate from the rate equations (e.g. 7). The continuous version of (2.2) is expressed in terms of the number density function $n(x,t)$ where $n(x,t)dx$ is the number of particles having masses in the range x, dx at time t. For uniform system described above, the evolution equation for $n(x,t)$ is:

$$\frac{\partial n(x,t)}{\partial t} = (1/2) \int_{x^*}^{x} dx'\ b(x-x',x')\ n(x-x',t)\ n(x',t) - n(x,t) \int_{x^*}^{\infty} dx'\ b(x',x)\ n(x',t) - \frac{\partial}{\partial x} [\Psi(x,t) n(x,t)] + \dot{\nu}_N \quad (2.19)$$

x^* is the mass of the smallest stable particle in the aerosol: $x^* = m_1 k^*$. $\Psi(x,t)$ is the condensation/evaporation rate of a particle, dx/dt. $\dot{\nu}_N$ represents the source strength of critical nuclei introduced by homogeneous nucleation. With the assumption, as above, that a single critical nucleus size is produced:

$$\dot{\nu}_N = x\ I(k^*,t)\ \delta(x-x^*) \quad (2.20)$$

Eq. (2.19) is coupled to the monomer equation for the mass concentration $c = m_1\ n_1$ which for $S > S_c$ has the form:

$$\frac{\partial c(t)}{\partial t} = - \int_{x^*}^{\infty} dx\ \Psi(x,c)\ n(x,t) - x^*\ I(k^*,t) + \overset{\circ}{R} \quad (2.21)$$

$\overset{\circ}{R}$ is the source function for monomer produced, for example by chemical reaction. Eqs. (2.19) and (2.21) have been studied numerically (e.g. 12) with the rate coefficients:

$$b(x',x) = 4\pi(R + R')(D + D')\beta$$

$$\beta = 1/(1 + 4(D + D')/(x^{1/3} + x'^{1/3})(\bar{V}^2 + \bar{V}'^2)^{1/2})$$

$$x = (4/3)\pi R^3 \rho;\ \bar{V} = (8\kappa T/\pi x)^{1/2};$$

$$D = (kT/6\pi\mu R)(1 + Kn(A + Be^{-CKn^{-1}})) \tag{2.22}$$

$$\Psi(x) = \pi R^2 \bar{V}_1 c_v (S - e^{Ke}) \pi^{1/2} Kn \cdot \left[\frac{1 + Kn\left(\frac{4\pi Kn}{3}\right) + 1.016}{\frac{4\ Kn}{3} + 1} \right]^{-1} \tag{2.23}$$

where:

$$\bar{V}_1 = \left(\frac{8\kappa T}{\pi m_1}\right)^{1/2};\ Kn = L/R;\ L = 2D_{1h}(m_1/2\kappa T)^{1/2};$$

$$S = c/c_v$$

$$I(k^*) = a_{1,k^*}(P_{v1}/\kappa T)^2 (2\sigma/\pi m_1)^{1/2}(m_1/\rho) \exp(-16\pi(m_1/\rho)^2(\sigma/\kappa T)^3/3(\ln S)^2), \tag{2.24}$$

$b(x,x')$ is the Brownian coagulation coefficient for coalescing, spherical particles proposed by Fuchs (2), derived by semi-empirical methods but agreeing with many of the best experimental data (13,14) over $0 < Kn < \infty$.

$\Psi(x,t)$ also correlates available experimental data satisfactorily (15) over the range of $0 < Kn < \infty$. The expression for $I(k^*)$ is due to Volmer, Becker and Doring and is known to be inexact (7). However, it agrees with experiment to within 5 percent for more than a dozen chemical compounds; this agreement refers to the prediction of the critical supersaturation, S_c, as a function of temperature. (Eq. (2.24), however, is not a reliable equation for predicting the actual rate of nucleation (e.g. (7)). Also, it should be pointed out that the applicability of eq. (2.24) has been questioned when nucleation occurs in the presence of pre-existing particles (16).

Figs. 2.1 and 2.2 show some of the rate processes and moments of $n(x,t)$ in a numerical study (12) of eqs. (2.20)

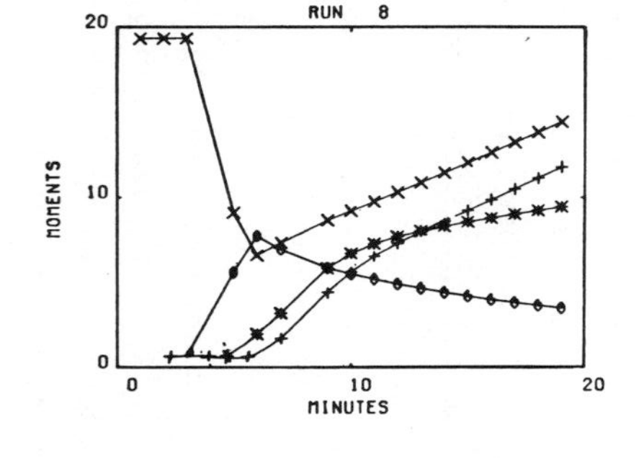

Figure 2.1. Moments of particle size distribution for nucleation and growth of sulfuric acid aerosol from sulfuric acid vapor at low humidity. Number concentration, cm^{-3}, x 2E5, 0; mass concentration g/cc x 1E-13, + : surface, cm^{-2} x 1E-7, *; 1/number, cm^{3} x 1E-5,X.

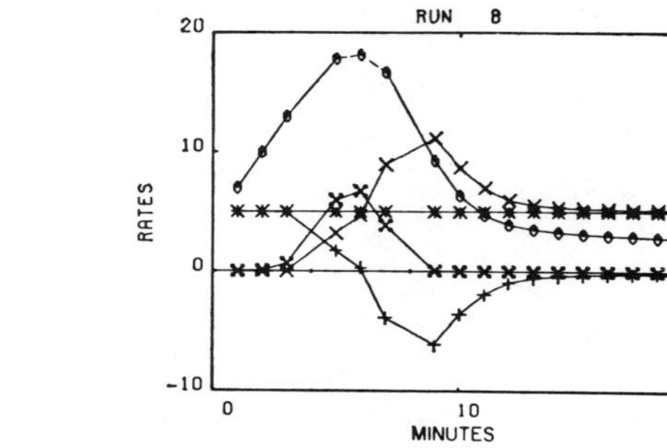

Figure 2.2. Rate processes in evolution of sulfuric acid aerosol nucleated from sulfuric acid vapor at low humidity. Monomer concentration g/ccx2E-14, 0; total rate g/cc-s x 2E-16 ;+; input rate g/cc-s x 2E-16 * ; nucleation rate g/cc-s x 2E-18, X.

and (2.21). These results represent the creation of a supersaturated sulfuric acid vapor in dry air at 25°C, latm., produced by a constant source of sulfuric acid vapor. The rate coefficients eqs. (2.22) - (2.24) were used in eq.(2.19). As is evident from the figures, owing to the very low source strength $\mathring{R}$ of sulfuric acid vapor, nucleation occurs slowly leading to a particle number concentration $M_O(t)$ which peaks after about 5 minutes from the time $\mathring{R}$ is "switched on." After this time, sufficiently high concentrations of particles exist to suppress homogeneous nucleation by condensing the monomer and lowering the supersaturation below the critical value, S_c, for this example. Subsequently, growth of particles occurs by condensation and coagulation as shown in Figs. 2.1 and 2.2. These results illustrate qualitatively the operations of the various processes involved in the formation and growth of ultrafine particles.

2.2 Coagulation and Condensation.

As illustrated by the example in the previous section, the stable particle concentration increases with time as homogeneous nucleation proceeds. Condensation on these particles and nucleation both lead to a decrease in supersaturation ratio, S, so that eventually $S < S_c$, the critical supersaturation ratio, and production of new stable clusters stops. In general, for most applications involving generation of ultrafine particles (jets, shocks, flames, etc.) this picture is valid. Subsequent to nucleation the stable particles grow by the coagulation and condensation/evaporation processes. In this section these are examined in more detail.

For the same uniform system discussed previously, after nucleation the evolution of the ultrafine aerosol can be represented by the conservation equations in the continuous representation for the number density function n(x,t) and the vapor mass concentration, c. For a uniform system with no sources of particles:

$$\frac{\partial n(x,t)}{\partial t} = (1/2) \int_{x*}^{x} dx' b(x-x',x') n(x-x',t) n(x,t)$$

$$- \; n(x,t) \int_{x*}^{\infty} dx \; b(x',x) \; n(x,t)$$

$$- \frac{\partial}{\partial x} \; [\Psi(x,t) \; n(x,t)] \qquad (2.25)$$

where $n(x,t)dx$ is the number of particles with masses in the range x, dx at time t. Eq. (2.25) is coupled to the vapor mass conservation equation:

$$\frac{\partial c}{\partial t} = - \int_{x*}^{\infty} dx \Psi(x,c) \; n(x,t) \; - \; x^* \; \Psi \; (x^*,c) \; n(x^*,t) \qquad (2.26)$$

where x^* represents the mass of the critical nucleus size so that evaporation from this size leads to unstable particles and an addition to the monomer concentration.

Some of the properties of these processes can be examined by considering the first two moments of (2.25) - the total number concentration, M_o and total mass concentration M_1:

$$M_o = \int_{x*}^{\infty} n(x,t)dx \qquad (2.27)$$

$$M_1 = \int_{x*}^{\infty} x \; n(x,t)dx \qquad (2.28)$$

It is simple to show that the moment equations for (2.25) and (2.26) are (3):

$$\frac{\partial M_o}{\partial t} = -(1/2) \int_{x*}^{\infty} \int_{x*}^{\infty} dx'dx \; b(x',x) \; n(x,t) \; n(x,t)$$

$$+ \; \Psi \; (x^*,c) \; n(x^*,t) \qquad (2.29)$$

$$\frac{\partial M_1}{\partial t} = \int_{x*}^{\infty} dx \; \Psi \; (x,c,t) n(x,t) \; + \; x^* \; \Psi \; (x^*,c) \; n(x^*,t) \qquad (2.30)$$

As can be demonstrated easily (e.g. 3) the coagulation process conserves particle mass and this term vanishes in obtaining eq. (2.30). Also, adding (2.26) and (2.30) gives:

$$\frac{\partial}{\partial t} \; (M_1 + c) = 0 \qquad (2.31)$$

so that, as obvious from the specifications of the physical system, the total mass concentration of particles plus monomer is invariant. The number concentration, $M_o(t)$ decreases not only by coagulation but also by the instability below x^*, which is determined by the critical supersaturation S_c for the system temperature, T. According to the classical theory, the radius of the critical nucleus, $R_k{}^*$, is:

$$R_k{}^* = 2\sigma\gamma \; / \; KT \ln S_c \tag{2.32}$$

and therefore: $x^* = (4/3)\ \pi\ R_k^3{}^* \rho$

As particle growth proceeds at $S < S_c$, the point separating the regions of evaporation and condensation is given by:

$$R_o = 2\sigma\gamma \; / \; \kappa T \ln S \tag{2.33}$$

since $\Psi(R_o) = 0$. As the stable particles grow by coagulation and condensation, R_o increases as S decreases. Clearly, as $S \to 1$, $R_o \to \infty$. In the initial stages of growth, particle sizes are small and those with radii $R < R_o$ will evaporate rapidly when $Ke >> 1$. This evaporation supplies monomeric vapor which permits condensation to proceed for particles with $R > R_o$.

In many applications monomeric vapor is not conserved as in the system described. Instead mixing with external streams may rapidly lower S, perhaps to the extentthat $S < 1$. In this case the particles will, of course, only evaporate. If the vapor pressure of the substance composing the particles is very small relative to the time scale of interest, then evaporation may be neglected and coagulation will remain as the only growth process.

The work of Sutugin (e.g. 17) has shown that under some conditions formation and growth of ultrafine particles will proceed entirely through the coagulation mechanism. Regardless of the process suppressing the condensation/evaporation process, it is of interest to consider briefly the characteristic features of the coagulation process.

In practice, one is often interested not in the details of $n_k(t)$ or $n(x,t)$ during the nucleation regime or during early stages of growth but in the form of $n(x,t)$ for long times - the asymptotic distribution. Several questions can be posed: given an initial density function $n(x,o)$ in a system in which only coagulation occurs, will the form of this density function be changed by coagulation, and if so what will be the form of the new function?

These questions can be studied in the case of coagulation in a uniform system in which the aerosol evolves according to eq. (2.25) without the condensation/evaporation term:

$$\frac{\partial n(x,t)}{\partial t} = (1/2) \int_{x^*}^{x} dx\ b(x-x',x')n(x-x',t)n(x',t) - n(x,t) \int_{x^*}^{\infty} dx' b(x',x)n(x',t) \qquad (2.34)$$

subject to the initial condition $n(x,0)$. The questions posed have been investigated analytically by a number of investigators (e.g. 18, 19, 20) for restricted forms of $n(x,0)$ and restricted forms of $b(x',x)$. For Brownian coagulation with $b(x',x)$ given by eq. (2.22) and more complex density functions, it is only possible to study these questions by numerical analysis.

We have recently carried out (21) numerical investigations of eq. (2.34) with $b(x',x)$ given by eq. (2.22) with the view of determining the existence of an asymptotic limit distribution $n(x,t\to\infty)$ and the empirical distribution providing the best fit to $n(x,t\to\infty)$. This investigation used the five initial number density functions shown in Fig. (2.3) in non-dimensional form for convenience: log normal, exponential, first-order gamma, gamma, and log gamma. Each of the five density functions has two adjustable parameters which were set by requiring that the mass concentration and geometric mean particle size be the same for each of the functions. In this example, after times in excess of 10^5 sec. the initial density functions are "forgotten" and all five are merged essentially into the same density function. The rate at which the five density functions approach each other can be seen from Figs. 2.4, 2.5 and

Figure 2.3. Reduced initial density functions used in study of asymptotic limit distributions for Brownian coagulation. Initial mass concentration= 5E-11 g/cc. Initial mean diameter = 8E-7 cm.

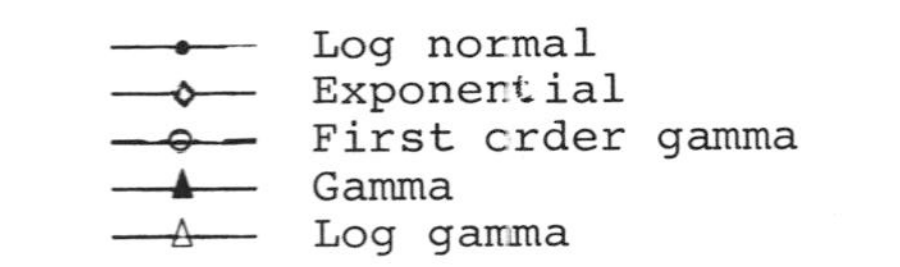

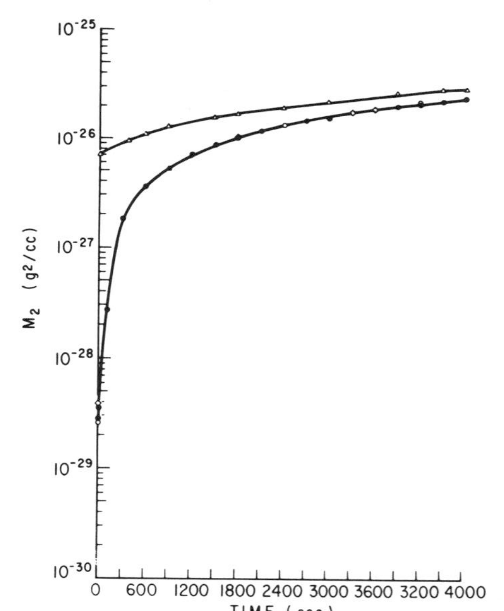

Figure 2.4. Effect of coagulation on second moments of five initial density functions as a function of time: approach to asymptotic limit distribution. See Fig. 2.3 for details.

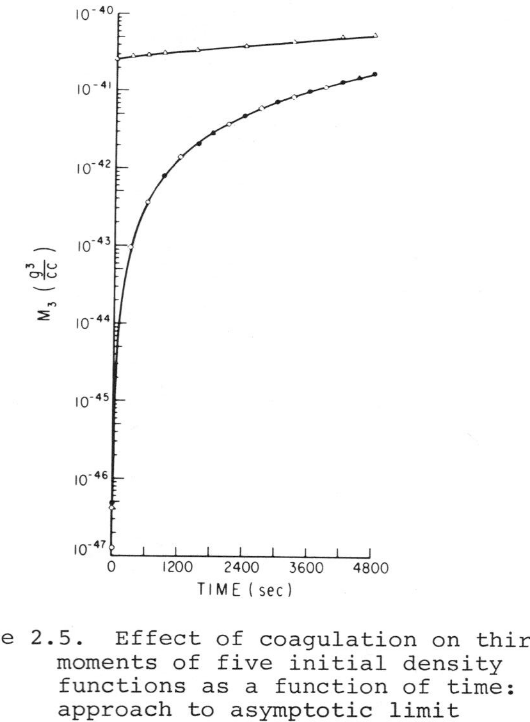

Figure 2.5. Effect of coagulation on third moments of five initial density functions as a function of time: approach to asymptotic limit distribution. See Fig. 2.3 for details.

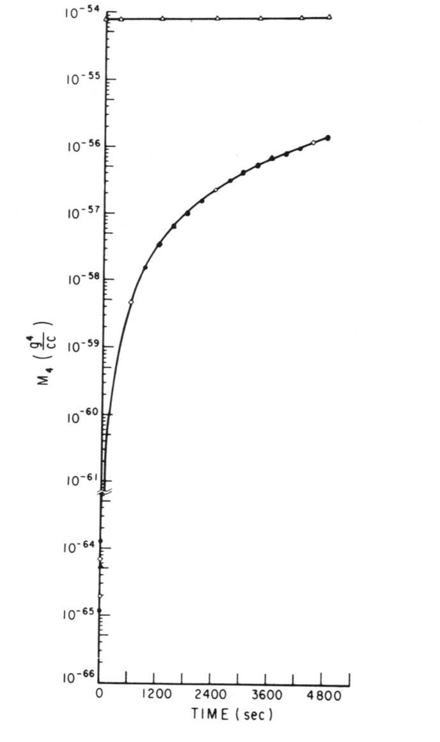

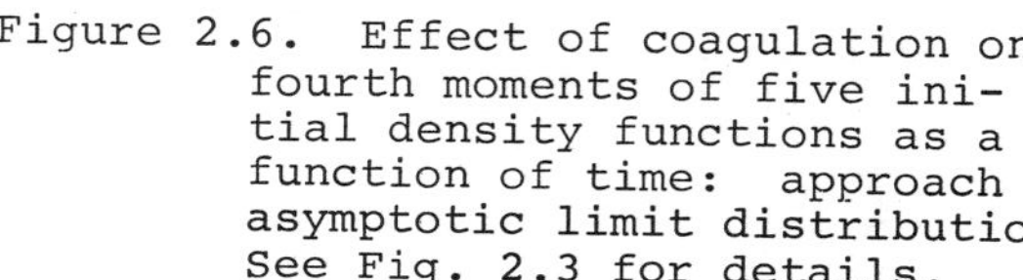

Figure 2.6. Effect of coagulation on fourth moments of five initial density functions as a function of time: approach to asymptotic limit distribution. See Fig. 2.3 for details.

2.6 for the moments M_2, M_3 and M_4. The log gamma function merges with the other four at a noticeably slower rate. This is caused by the long tail of this initial function.

The best fit of the asymptotic function approached by the initial functions is an important question because it is intuitively supposed in the literature that the log normal density function should describe aged condensation aerosols. Two tests have been used in determining the best empirical density function: closeness of fit and randomness of fit. Five candidate density functions were chosen for these tests: log normal, first-order gamma, gamma, log gamma, and beta of the second kind. These were chosen because they have been used in the literature by various investigators as representative of the density function of condensation aerosols. By a variety of fitting procedures, it was determined that the log gamma density function was marginally superior to the log normal density function in providing an empirical fit to the numerical asymptotic limit function. Power spectral analysis of residuals revealed that the three other functions (first order gamma, gamma, beta of second kind) were consistently biased and were therefore unacceptable.

The various studies of asymptotic limit distributions produced by coagulation have all been initiated with unimodal or truncated density functions. The question arises as to the limit distribution for multimodal density functions. It has been pointed out (22) that the diagonal matrix elements of the Brownian coagulation coefficients are small in magnitude compared to the off-diagonal elements whose magnitudes are greatest between the smallest and largest particles. Therefore, growth in particle size within a mode is a slower process than the rate of attachment of particles of this mode by another mode of larger particles, if the number concentrations of the two modes are the same. This qualitative picture has been confirmed by numerical calculations (22), although the existence of an asymptotic limit was not investigated. The previous study (21) certainly suggests that a multimodal function will eventually approach a unimodal asymptotic limit function.

As coagulation of an aerosol with a unimodal density function proceeds, the characteristic time of coagulation $t_{coag.}$ increases. Approximately, $t_{coag.} \sim 1/bM_o$. Coagulation can therefore be neglected when the interest is in times small in comparison with t_{coag}. For this situation, the evolution equation (2.25) reduces to:

$$\frac{\partial n(x,t)}{\partial t} = - \frac{\partial}{\partial x} [\Psi(x,S)\ n(x,t)] \tag{2.35}$$

which is coupled to the monomer conservation equation. In terms of the supersaturation ratio, S:

$$\frac{dS}{dt} = -\int_{x^*}^{\infty} dx\ \Psi(x,S,t)n(x,t) - x^*\ \Psi(x^*,S)\ n(x^*,t) \tag{2.36}$$

The condition $n_{10} << n_h$ introduced at the beginning insures that the condensation/evaporation process is isothermal. Without this condition, eqs. (2.35) and (2.36) would be coupled to the energy conservation equation, as indeed is the case in consideration of rapid growth of aerosols.

If the supersaturation, S, is controlled externally eq. (2.35) is decoupled from (2.36). In this case, the solution of (2.35) is very simple and can be effected by the method of characteristics. For two cases, similarity is easily demonstrated.

When $S = S_o$, a constant, Ψ given by eq. (2.23) is a function only of x. With the transformations:

$y = \int^x \frac{dx}{\Psi(x)}$, $F = \Psi n$, eq. (2.35) becomes:

$$\frac{\partial F}{\partial t} + \frac{\partial F}{\partial x} = 0 \tag{2.37}$$

which has the solution $F = F_o(y-t)$ where F_o is some arbitrary initial function. Consequently:

$$n(x,t) = \frac{1}{\Psi(x)}\ F_o\ (\int^x \frac{dx}{\Psi(x)} - t) \tag{2.38}$$

If $S = S(t)$ where $S(t)$ is determined by external variation, then similarity is also easily shown for the special case: $\Psi = G(x)H(t)$. This separation would be valid, for example, for Ψ given by eq. (2.23) when $S >> Ke$. With the transformations: $y = \int^x \frac{dx}{G(x)}$, $u = \int^t H(t)dt$, $F = G(x)n$, eq. (2.35) becomes:

$$\frac{\partial F}{\partial u} + \frac{\partial F}{\partial y} = 0 \tag{2.39}$$

and $F = F(y-u)$ so that:

$$n(x,t) = \frac{1}{G(x)} F_o \left(\int^x \frac{dx}{G(x)} - \int^t H(t)dt \right) \tag{2.40}$$

Other special cases of similarity can be demonstrated. More generally, the solution of (2.35) is possible by standard procedures.

When S is not externally controlled, eqs. (2.35) and (2.36) are coupled and no simple solution appears to exist. Eqs. (2.35) and (2.36) with (2.23) are non-linear integro-differential equations and only numerical solution appears to be feasible. In the next part of this section, some approaches to this numerical problem will be considered briefly.

From the previous discussions of the separate processes of coagulation and condensation/evaporation, it follows that the intermediate and asymptotic behavior of the coupled equations (2.25) and (2.26) with realistic initial conditions and realistic rate coefficients (2.22) and (2.23) can only be studied numerically. The following discussion describes methods that have been used to date with some success.

2.3 Numerical simulation of coagulation and condensation.

A variety of methods have been proposed for numerical simulation of coagulation and condensation. Work before 1970 on coagulation has been reviewed by Drake (23). Since this time, a large number of numerical studies have been carried out (e.g. 12, 22, 24, 25, 26, 27, 18) in uniform systems. In view of the many numerical approximation schemes which have been proposed, it would be desirable to develop test problems for evaluating these schemes. For very large scale atmospheric aerosol dynamics simulations, approximate methods may be suitable, given the existing uncertainties in input data. The emphasis in this discussion will be on methods of high accuracy which we have employed in simulation studies.

Simulation of condensation aerosol dynamics involves particle radii covering around four orders of magnitude $\sim 10^{-8} - 10^{-4}$ cm. The logarithmic transformation suggested by Berry (29) has been employed by us in a number of studies (e.g., 12, 21, 22, 30, 31):

$$x(J) = x(J_o)\exp(q(J-J_o)) \quad (2.41)$$

J is a positive number greater than or equal to J_o. $x(J_o)$ is the mass of a particle starting at J_o. q is a numerical parameter which can be selected to give equally spaced integer J values. From the definition of the density function,

$$n(x(J)) = g(J)/qx(J) \quad (2.42)$$

With (2.41) and (2.42), eq. (2.25) becomes:

$$\frac{\partial g(J,t)}{\partial t} = \int_{J_o}^{J_u} dJ'\, b_1\,(\tilde{J}, J')g(\tilde{J},t)g(J',t)$$

$$- g(J,t) \int_{J_o}^{\infty} dJ' b(J,J')g(J',t)$$

$$- \frac{\partial}{\partial J}\,[\Psi(J)g(J)/qx(J)] \quad (2.43)$$

and: $J_u = J - \ln 2/q,\ J \geq 2;$

$\tilde{J} = J + (1/q)\ \ln\,[1-\exp(q(J'-J))];$

$b_1(\tilde{J},J') = x(J)/x(\tilde{J})\ b(x-x',x'),$

with b(x-x',x') given by eq. (2.22). The adjustable parameter q has a great advantage in "fine tuning" for mass or diameter spacings to increase numerical accuracy.

The coagulation process can be simulated numerically with high accuracy. We have chosen methods which optimize both accuracy and efficiency. Accuracy has been studied (12,22) by comparison with analytical solutions for restricted forms of b(x,x') and with Brownian coagulation, eq. (2.22), by testing for conservation of mass. We use cubic spline for numerical quadrature and interpolation of the coagulation terms. Gear's method is used for time integration. Comparisons noted above show that simulation by these methods is accurate and reliable and that errors can be reduced to any desired level.

In comparison with the coagulation term, the condensation/evaporation term in eq. (2.43) is deceptively simple in appearance. However, it is a first-order hyperbolic equation whose numerical solution is difficult, as evidenced by the numerous published attempts at numerical solution of similar hyperbolic equations (simulation of advection, for example). The difficulty lies in the fact that most numerical schemes for hyperbolic equations give rise to numerical dispersion and numerical diffusion. The numerical dispersion, due to the combination of large phase errors and insufficient damping of short waves, manifests itself by the unphysical wakes behind and ahead of the simulated regions of high concentration. Numerical diffusion lowers the peak values of the concentration distribution but increases the values around the peak. Numerical methods with "upwinding" can remove numerical dispersion but create unrealistic numerical diffusion (32). Numerical methods without upwinding, such as the finite element method with linear basis functions (32, 33, 34, 35), do not introduce much numerical diffusion but create numerical dispersion. Such dispersion is undesirable for simulation of condensation/evaporation because the dispersion level increases with time and for prolonged simulations dispersion finally dominates and creates numerical instability. Also, numerical dispersion may erase the real "signatures" of the condensation/evaporation processes.

Many methods have been tried to reduce dispersion, such as by introduction of a dissipative term in the Galerkin finite element formulation (36,37) or use of filtering schemes (38,39). These have not been found to be suitable for condensation/evaporation simulation.

A robust numerical scheme for condensation/evaporation should be free of numerical dispersion while numerical diffusion is minimized. Eulerian numerical schemes create numerical dispersion whereas Lagrangian schemes do not suffer from this difficulty. Using Bonnerot and Jamet's approach (40), Varoglu and Finn (41) derived a finite ele-

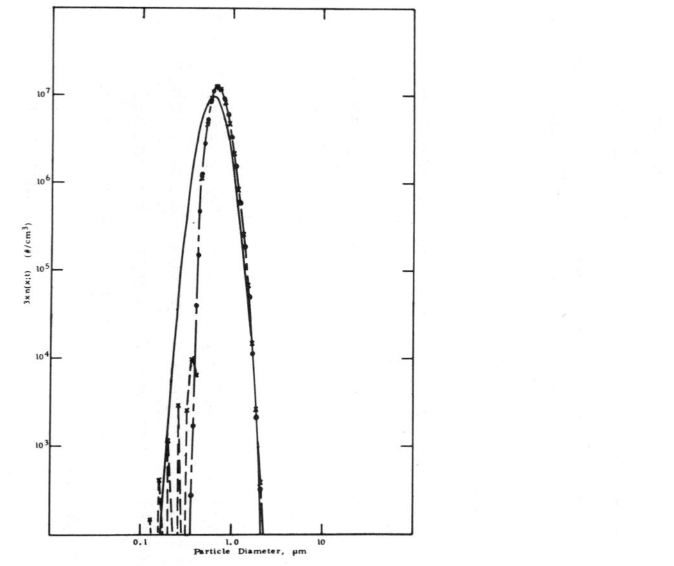

Figure 2.7. Comparison between linear finite element method and Tsang's method for numerical solution of condensation equation for continuum growth.

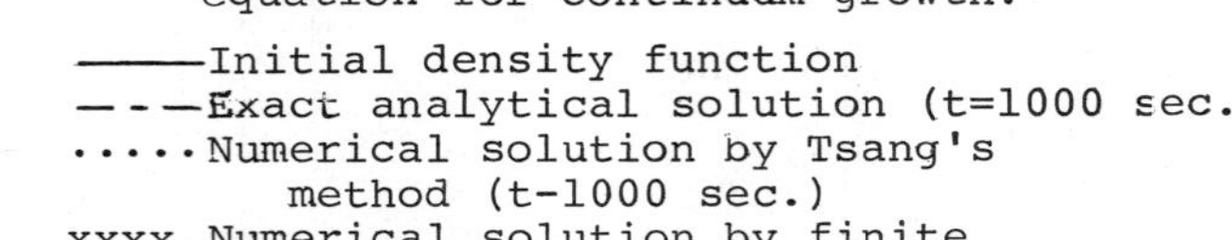

——— Initial density function
– – – Exact analytical solution (t=1000 sec.)
····· Numerical solution by Tsang's method (t-1000 sec.)
xxxx Numerical solution by finite element method (t=1000 sec.)

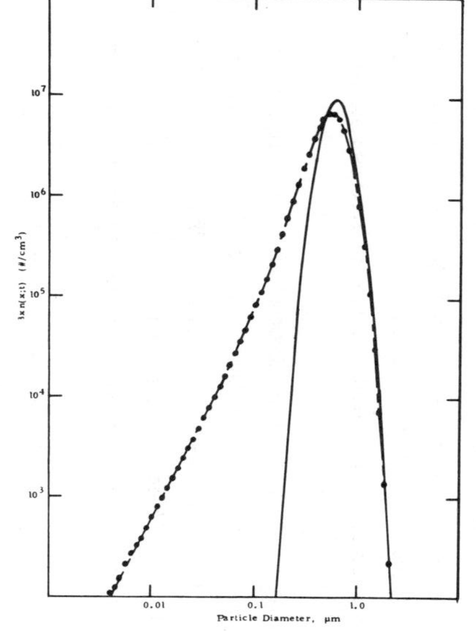

Figure 2.8. Application of Tsang's method for numerical solution of evaporation equation for continuum evaporation.

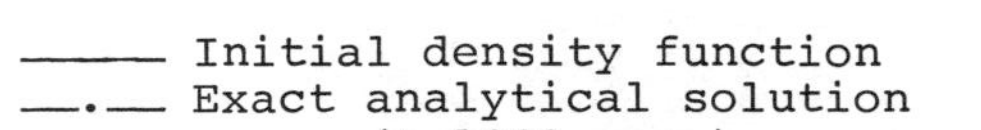
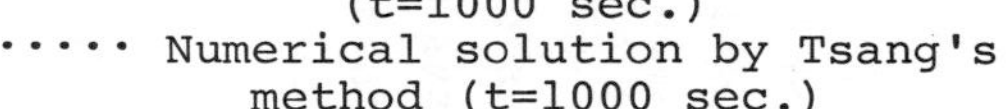

——— Initial density function
—·— Exact analytical solution (t=1000 sec.)
····· Numerical solution by Tsang's method (t=1000 sec.)

ment method incorporating characteristics for the diffusion-convection equation. This scheme is free of numerical dispersion. Recently, Neumann (42) derived an Eulerian-Langrangian numerical scheme for the diffusion-convection equation, which is also free from numerical dispersion and controls numerical diffusion.

In our work, we have modified and combined Varoglu and Finn's method with Neumann's method because of the requirements of the condensation/evaporation term. This new method, Tsang's method, has been successful. Fig. 2.7 shows a comparison between the linear finite element and our method for continuum ($Kn \to o$) condensation according to the law $\Psi = \varepsilon x^{1/3}$ with the analytical solution included for comparison. For the linear finite element method, the magnitude of dispersion increases with time. Tsang's method is more accurate than the linear finite element method over the whole computational domain. For continuum evaporation, $\Psi = -\varepsilon x^{1/3}$, Tsang's method gives also good agreement with the analytical solution as shown in Fig. 2.8.

These preliminary results are encouraging. Work is now underway on the numerical simulation of eqs. (2.25) and (2.26) using Tsang's method.

3. Aerosol Formation and Growth in Non-Uniform Systems.

Aerosol formation and growth in uniform systems is an abstraction realized in practice only approximately as in, for example, large systems subjected to adiabatic cooling so that throughout much of the system volume there are only very small spatial gradients. The evolution of the droplet spectra in atmospheric water clouds generated by rising air masses is one important example which approximates such a uniform system (e.g. (44)). However, in many technical applications, aerosols form and grow in non-uniform systems as in flames, jets and plumes.

In this section, aerosol evolution in non-uniform systems is examined. The incomplete nature of theory in this case is discussed. The complexity of evolution in non-uniform systems makes numerical simulation essential. Two examples of such simulation are given for coagulation in a laminar coaxial jet and for coagulation in an atmospheric plume.

3.1 Particle formation and growth.

Conventional nucleation theory proceeds from the assumption of a uniform system (e.g. 6,7). This classical theory has been modified to allow for diffusional migration of clusters during nucleation (45), but there is no comprehensive theory. The situation is analogous to the theory of kinetics of gas phase chemical reactions which currently is phenomenologically based. There is, therefore, no kinetic theory of nucleation in non-uniform systems.

The theory of growth of stable (in the sense of nucleation) particles is, for the most part, also phenomenological. For aerosol evolution in a non-uniform system one writes for example the extension of eq. (2.19):

$$\frac{\partial n(x,\vec{r},t)}{\partial t} + \nabla \cdot n(x,\vec{r},t)\vec{V}_o(\vec{r},t) + \nabla \cdot n(x,\vec{r},t)\vec{V}_x(x,\vec{r},t)$$

$$=(1/2)\int_{x^*}^{x} dx' b(x-x',x',\vec{r},t)n(x-x',\vec{r},t)n(x',\vec{r},t)$$

$$- n(x,\vec{r},t) \int_{x^*}^{\infty} dx' b(x,x',\vec{r},t)n(x',\vec{r},t)$$

$$- \frac{\partial}{\partial x} [\Psi(x,\vec{r},t)n(x,\vec{r},t)] + \dot{\nu}_N(\vec{r},t) \qquad (3.1)$$

where now $n(x,\vec{r},t)dx$ is the number of particles with masses in the range x,dx at a point in space $\vec{r}$ at time t. For $M_o \ll n_h$, $\vec{V}_h$ is the mass average velocity of the host gas and $\vec{V}_x$ is the average velocity of particles of mass x relative to the mass average velocity. The various rate coefficients on the right-hand side of eq. (3.1) are functions of $\vec{r}$ and t and may be modified in form by the non-uniformities.

Eq. (3.1) is coupled to the conservation equations of mass, momentum and energy for the host gas (46). The description of aerosol dynamics becomes more difficult when $M_o \sim n_h$ and coagulation with condensation/evaporation processes occur at rates of the order of the molecular relaxation times.

Eq. (3.1) is a purely formal statment in that there is no general theory from which we can extract $\vec{V}_x$ and the rate coefficients. For $M_o << n_h$ (a commonly realized constraint in practice) the various processes contributing to $\vec{V}_x$ - diffusion, thermophoresis, diffusiophoresis, stressphoresis, photophoresis, etc. (3) - are studied separately as a function of Kn and introduced in eq. (3.1). However, the phenomena giving rise to $\vec{V}_x$ are complex and there is no reason to expect that such linear superposition will be generally correct (5). This is even more the case for an aerosol undergoing at the same time rapid coagulation and condensation/evaporation.

Of course, for turbulent host gas, eq. (3.1) is stochastic. An example will be given below of the evolution of an aerosol plume in atmospheric turbulence. Even in the absence of turbulence, a stochastic description of aerosol evolution is necessary when particle concentrations are small. In fact, it has been recognized for some time that the development of droplet spectra in rain clouds by coagulation and condensation should be treated as a stochastic process (e.g. 23). Therefore in the case eq. (3.1) is certainly incorrect as is its uniform version eq. (2.19).

In principle, generalized kinetic equations derived from the Liouville equation for the particle-host gas system could be derived. However, such general treatment of an aerosol appears to be unlikely in the near future.

The theoretical base for general deterministic or stochastic population balance models for aerosol dynamics is at present unsatisfactory. However, in one special case a satisfactory theory exists. As noted previously, when Kn >> 1 with the host gas at standard conditions, the particles are sufficiently small that they may be regarded as large molecules. Under certain additional conditions the evolution of this system can be studied with the well-developed kinetic theory of dilute gases. In order to apply the kinetic theory, it is necessary that the coagulation and condensation/evaporation processes be restricted. It is realistic to ignore condensation/evaporation if one limits the consideration to involatile particles. However, as pointed out earlier, coagulation is always present in an aerosol. In order to apply kinetic theory, therefore, it is necessary that the evolution of the velocity distribution functions be independent of particle-particle collisions. Therefore, the requirement is made that the particle number concentration in size class, k, n_k, be much less than the host gas molecule concentration, n_h - that is, $n_k << n_h$. Also, the assumption of collisional delocalization must hold (47) so that gradients in the system are negligible over distances of the order of the particle diameter. With these assumptions, the aerosol-host gas system can be derived approximately by the appropriate set of Boltzmann equations for the mixture for model particles that are rigid spheres:

$$\frac{\partial f_k}{\partial t} + \vec{w}_k \cdot \nabla f_k = K(f_k, f_h), \; k = 1, 2, \ldots \tag{3.2}$$

$$\frac{\partial f_h}{\partial t} + \vec{w}_h \cdot \nabla f_h = K(f_h, f_h) \tag{3.3}$$

where $f_k(w_k, \vec{r}, t)$ is the velocity distribution function of particles of size k, $\vec{w}_k$ is the molecular velocity and K the binary collision operator (47). With the additional restriction $(m_h/m_k)^{1/2} << 1$, (3.2) and (3.3) describe the "quasi-Lorentz" gas. Mason and Chapman (48) have shown that "normal" solutions of (3.2) and (3.3) give results for the mean particle motion $\vec{V}_x$ which agree with experimental data for aerosol particles when allowance is made for diffuse and inelastic scattering for the gas-particle collisions.

In this restricted case, the conservation equation (3.1) can be applied in the form:

$$\frac{\partial n_k(\vec{r}, t)}{\partial t} + \nabla \cdot n_k(\vec{r}, t) \; \vec{V}_o + \nabla \cdot n_k(\vec{r}, t) \; \vec{V}_x = 0 \tag{3.4}$$

or its continuous version:

$$\frac{\partial n(x, \vec{r}, t)}{\partial t} + \nabla \cdot n(x, \vec{r}, t) \; \vec{V}_o + \nabla \cdot n(x, \vec{r}, t) \; \vec{V}_x = 0 \tag{3.5}$$

with $\vec{V}_x$ given by kinetic theory. Coagulation can be reintroduced to eqs. (3.4) and (3.5) if it is assumed that coagulation does not alter the state of the gas as specified through eqs. (3.2) and (3.3).

An application of these ideas is given below for the case of the evolution of a coagulating aerosol in a laminar coaxial jet.

3.2. Coagulation in a laminar coaxial jet.

In view of the uncertainties in the theory of aerosol dynamics, it is desirable to study the evolution of an aerosol under conditions permitting the application of the complete theory embodied in eqs. (3.2) - (3.5) and the coupled mass, momentum and energy equations. A system permitting this is an aerosol evolving in a laminar coaxial jet. The aerosol and its coagulation process must meet the requirements discussed above.

Calculations have been carried out for an aerosol coagulating in a laminar coaxial jet (51). In this arrangement, aerosol in a host gas at temperature T_1, is introduced through a nozzle into a co-flowing stream at T_2 of the same gas free of aerosol. In the specific example studied, the nozzle of the jet is a thin-walled circular tube with a radius, RN, of 0.7 mm, and coaxial to it is a circular tube with a radius of 11.55 mm. For the isothermal jet, initially at a temperature of 293°K, the average axial velocity at the nozzle exit (fully developed laminar flow) is 91 cm/sec, and $Re_{nozzle} = 85.2$; the same exit velocity is used for the nonisothermal jet, initially at 563°K, but $Re_{nozzle} = 27.6$ For both jets, the interior jet flow is coaxial with an annular flow of air at 293°K with an initial average axial velocity (fully developed flow) of 11.04 cm/sec and $Re_{annulus} = 170$.

A series of initial particle size distributions were studied. We give here a single example: an aerosol with a spatially uniform initial log normal size distribution (mean particle mass, $x_o = 1.94 \times 10^{-20}$ g, mean particle diameter, $D_{po} = 1.56 \times 10^{-3}$ μm; geometric standard deviation, $\sigma_g = 1.51$) with a concentration of 6.30×10^{-7} g/cm^3.

The flow system is modeled in a cylindrical coordinate system as an axisymmetric, two-dimensional, steady, non-isothermal, laminar flow.

With the conditions stated, the particles are passive contaminants and the mass, momentum and energy equations (46) for this problem may be solved independently of the aerosol conservation equation. The numerical calculation of the velocity and temperature fields was carried out by a difference method and a variable mesh size to accommodate the rapid changes taking place in the entrance region.

For this problem, the aerosol evolution equation is:

$$\begin{aligned} \nabla \cdot \vec{V}_o\, n(x,\vec{r}) &= \nabla \cdot D(x)\nabla n(x,\vec{r}) + \nabla \cdot \vec{V}_{xt} n(x,\vec{r}) \\ &+ \int_0^{x/2} b(x',x-x')\, n(x-x',\vec{r})\, n(x',\vec{r})\, dx \\ &- n(x,\vec{r}) \int_0^{\infty} b(x,x')\, n(x',\vec{r})\, dx' \end{aligned} \qquad (3.6)$$

where: $D(x) = (\kappa T/6\pi\mu R)\left[1 + (L/R)(A + B\, e^{-CR/L})\right]$

and: $$\vec{V}_{xt} = \frac{0.1132\, \lambda_{tr}\ R\exp(-0.41\ R/L)\ \nabla \ln T}{\bar{V}_h \mu/[1 + (L/R)(A + B\, e^{-CR/L})]}$$

A, B and C are constants with values 1.246, 0.42 and 0.87 respectively (3) and μ is the viscosity coefficient of the host gas. The other contribution to $\vec{V}_x$ besides diffusion is the thermophoresis velocity, $\vec{V}_{xt}$ with λ_{tr} the translational part of the thermal conductivity of the host gas (3).

The calculation of coagulation was carried out by the methods described earlier. Figs. 3.1 and 3.2 show the evolution of the particle number density function (= 3 x n(x)) as a function of downstream distance in a region around the periphery of the jet for isothermal ($T_1 = T_2 = 293°K$) and non-sothermal ($T_1 = 563°K$, $T_2 = 293°K$) cases. At the periphery of the heated and unheated jets, near the nozzle, the particle number and mass density functions are bimodal. In the heated jet, near the nozzle, the pronounced modes at smaller

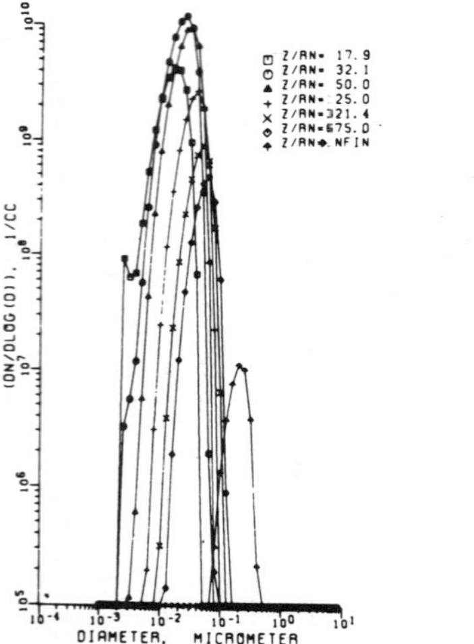

Figure 3.1. Aerosol number density functions for isothermal jet in a region around periphery of jet for various nondimensional axial distances, Z/RN. RN is the nozzle radius, R is radial distance from the axis, Z is the axial distance from nozzle exit.

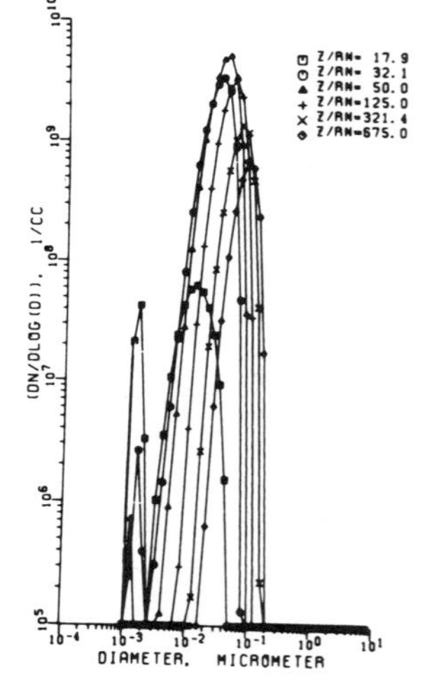

Figure 3.2. Aerosol number density function for nonisothermal jet in a region around periphery of jet for various nondimensional axial distances, Z/RN. RN is nozzle radius, R is radial distance from the axis, Z is the axial distance from nozzle exit.

particle diameters arise from radial transfer by Brownian diffusion plus thermophoresis and convection from the interior of the jet; the modes at larger particle diameters are due principally to thermophoresis and radial convection.

Of course, the assumption that the aerosol was a passive contaminant is equivalent in this case to the assumptions leading to eqs. (3.2) and (3.3). Without this assumption, there would be no acceptable theoretical basis from which to proceed.

3.3 Coagulation in an atmospheric plume.

As noted, when the host gas is turbulent, eq. (3.1) becomes a stochastic equation. Under the stated assumptions it may still be possible to find $\vec{V}_x$. One then can proceed in the usual way for treating turbulent systems and invoke various closure assumptions for the correlation terms (e.g. 46).

As an example of the application of eq. (3.1) in a turbulent system some results are presented for coagulation of an aerosol in a plume from a cross-wind line source. For application to larger aerosol particles a term $\vec{G}(x)\cdot\nabla n(x,\vec{r},t)$ must be added to the right hand side of eq. (3.1), $\vec{G}$ is the gravitational sedimentation velocity of a particle of mass x.

With the customary Reynolds averaging of the modified eq. (3.1), one gets an infinite set of equations for the correlation terms. In the analysis presented, first-order closure is used and the eddy diffusivity $D_t(z)$ is introduced. The correlation $\overline{n'n'}$ in the coagulation equation can be neglected when the characteristic coagulation time is much greater than the Lagrangian time scale of turbulence (50,51); this condition is met in the downwind field of the plume considered here. Neutral atmospheric stability is assumed. In this case, D_t is well-known (52,53) from both experiment and theory. Then, the modified eq. (3.1) for the crosswind line source can be written:

$$\frac{\partial \bar{n}(x,X,Z,t)}{\partial t} + U(Z) \frac{\partial \bar{n}(x,X,Z,t)}{\partial X}$$

$$= \frac{\partial}{\partial Z} \left(D_t(Z) \left[\frac{\partial \bar{n}(x,X,Z,t)}{\partial Z} \right] \right)$$

$$+(1/2) \int_{x^*}^{x} dx' \; b(x,x')\bar{n}(x-x',X,Z,t)$$

$$\cdot \bar{n}(x',X,Z,t)$$

$$- \bar{n}(x,X,Z,t) \int_{x^*}^{\infty} b(x',x)\bar{n}(x',X,Z,t)dx'$$

$$+ G_Z(x) \frac{\partial \bar{n}(x,X,Z,t)}{\partial Z} \qquad (3.7)$$

where X is the downwind distance, Z is the height above ground level, and U(Z) is the mean wind speed. Eq. (3.7) is solved with initial and boundary conditions:

$$\bar{n}(x,X,Z,0) = 0; \; \bar{n}(x,Z,\zeta,t) = \frac{Q_o \; \delta(Z-\zeta) n_s(x)}{U(\zeta)} ;$$

$\frac{\partial \bar{n}}{\partial Z} = 0$, $Z = Z_o$; $\bar{n}V_d(x) = -D_t \frac{\partial \bar{n}}{\partial Z}$, $Z = 0$. Q_o is the source strength, ζ its height above ground, n_s the source density function, Z_o the mixing height. V_d is the particle deposition velocity, representeing irreversible loss of particles to the ground surface. Eq. (3.7) has been studied numerically (30, 31) using procedures described in part in Section 2 and more fully in (30,31). An example will be given here of the effect of coagulation and deposition on the mean extinction coefficient, $\bar{\sigma}_{ext}$, for propagation of infrared radiation (3.5 μm wavelength) through the plume. $\bar{\sigma}_{ext}$ is given by:

$$\bar{\sigma}_{ext} = \pi \left(\frac{3}{4\pi\rho} \right)^{2/3} \int_0^{\infty} Q_{ext} \; x^{2/3} \; \bar{n}(x)\,dx \qquad (3.8)$$

where Q_{ext} is the normalized extinction efficiency for radiation of a particular wavelength.

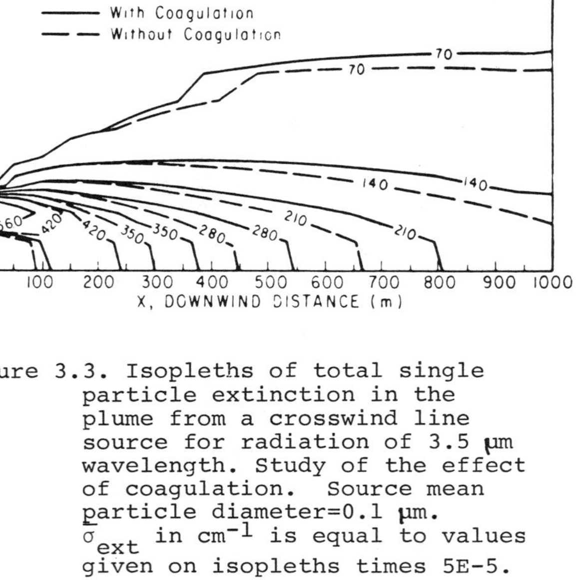

Figure 3.3. Isopleths of total single particle extinction in the plume from a crosswind line source for radiation of 3.5 μm wavelength. Study of the effect of coagulation. Source mean particle diameter=0.1 μm. $\overline{\sigma}_{ext}$ in cm^{-1} is equal to values given on isopleths times 5E-5.

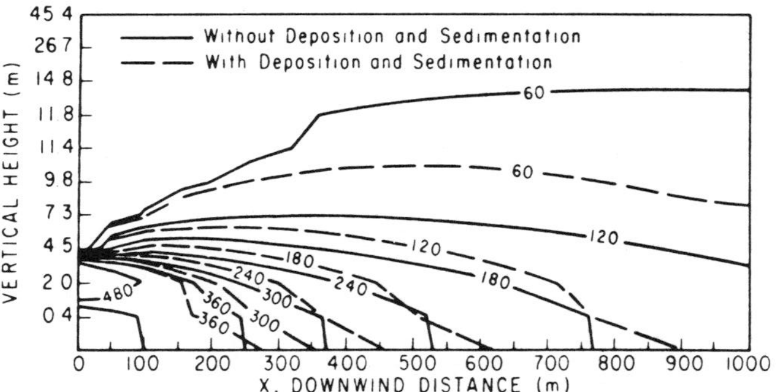

Figure 3.4. Isopleths of total single particle extinction in the plume from a crosswind line source for radiation of 3.5 μm wavelength. Study of the effect of sedimentation and deposition. Source mean particle size=3.0 μm. $\overline{\sigma}_{ext}$ in cm^{-1} is equal to values given on isopleths times 1E-4.

Figures 3.3 and 3.4 show the effects of coagulation and deposition on $\bar{\sigma}_{ext}$ calculated from (3.7) and (3.8). Fig. 3.3 represents an aerosol with a source mean particle diameter of 0.1 μm. For Fig. 3.4 the source mean particle diameter is 3.0 μm. Reference (31) can be consulted for more detail. In a coagulating plume, an interesting coupling arises between the processes of coagulation, sedimentation and deposition. Coagulation increases the sizes of particles; this increase leads to increased rates of sedimentation and, in the size ranges studied, increased rates of dry deposition. Therefore, although the coagulation process does not directly alter the mass concentration, it does result indirectly in decreases in mass concentration through its coupling to sedimentation and deposition. From Fig. 3.3, it is clear that coagulation appreciably alters extinction in this case by increasing particle size in the sensitive region of Q_{ext} (below the geometric limit), even though sedimentation and deposition rates are increased. The results of Fig. 3.4 are easy to understand simply from the fact that deposition decreases $\bar{n}(x)$ in the plume and therefore decreases $\bar{\sigma}_{ext}$.

4. Summary.

The formation and growth of condensation aerosols is a complex process, even in the uniform system described in Section 2. In the "real world" of shocks, flames, and jets, highly non-uniform systems, the theoretical basis for studying aerosol dynamics is lacking, except for special cases, some of which are recounted in Section 3. The description of the dynamics of high density condensation aerosols in non-uniform systems is a challenging problem for the future for workers in non-equilibrium statistical mechanics as well as for fluid dynamicists interested in multiphase flow.

REFERENCES

1. Green, H., and Lane, W., Particulate Clouds: Dusts, Smokes and Mists, 2nd ed., Van Nostrand Co., Inc., Princeton, N.J., 1964.
2. Fuchs, N., Mechanics of Aerosols, Pergamon Press, Oxford, 1964.
3. Hidy, G.M. and Brock, J.R., The Dynamics of Aero-Colloidal Systems, Pergamon Press, Oxford, 1970.
4. Friedlander, S.K., Smoke, Dust and Haze, Wiley, N.Y., 1977.
5. Marlow, W.H. (Ed.), Aerosol Microphysics I, II, Springer-Verlag, Berlin, 1980.
6. Zettlemoyer, A. (Ed.), Nucleation Phenomena, Elsevier, N.Y., 1977.
7. Abraham, F., Homogeneous Nucleation Theory, Academic Press, N.Y., 1974.
8. Berne, B., in W.H. Marlow (Ed.), Aerosol Microphysics II, Springer-Verlag, Berlin, 1982.
9. Zurek, W., and Schieve, W., J. Chem. Phys. 68 840 (1978).
10. Kitahara, K., Metin, H., and Ross, J., J. Chem. Phys. 63 3156-3160 (1975).
11. Higuchi, W.I. and O'Konski, C.T., J. Colloid Sci. 15 14-49 (1960).
12. Middleton, P.B. and Brock, J.R., J. Colloid Interface Sci. 54 249 (1976).
13. Wagner, P.E. and Kerker, M., J. Chem. Phys. 66 638 (1977).
14. Shon, S-N, Kasper, G., and Shaw, D.T., J. Colloid Interface Sci. 73 233-243 (1980).
15. Davis, E.J., Ravindran, P. and Ray, A.K., Chem. Eng. Commun. 5 251-268 (1980).
16. Friedlander, S.K., J. Colloid Interface Sci. 67, 387-388 (1978).
17. Sutugin, A.G., Lushnikov, A.A., and Chernyaeva, G. A., J. Aerosol Sci. 4 295-305 (1973).
18. Schumann, T., J. Roy. Met. Soc. 66 195 (1940).

19. Friedlander, S.K. and Wang, C.I., J. Colloid Interface Sci. 22 126 (1966).
20. Mulholland, G.W. and Baum, H.R., Phys. Rev. Letters 45 761-763 (1980).
21. Yom, K. and Brock, J.R., To be published.
22. Suck, S.H. and Brock, J.R., J. Aerosol Sci. 10 581-590 (1979).
23. Drake, R. in G.M. Hidy and J.R. Brock, Eds., Topics in Aerosol Research Vol. III, Pergamon, Oxford, 1971.
24. Sutugin, A. and Fuchs, N.A., J. Aerosol Sci. 1 287-293 (1970).
25. Bleck, R., J. Geophys. Res. 65 5165-5171 (1970).
26. Gelbard, F. and Seinfeld, J.H., J. Comp. Phys. 28 357-375 (1978).
27. Gelbard, F., Tambour, Y., and Seinfeld, J.H., J. Colloid Interface Sci. 76 541-556 (1980).
28. Whitby, K.T., EPA Report, 1980.
29. Berry, E.X., J. Atmos. Sci. 24 688 (1967).
30. Tsang, T.H. and Brock, J.R., Atmos. Environ., in Press, 1982.
31. Tsang, T.H. and Brock, J.R., J. Appl. Optics, in Press, 1982.
32. Long, P.E. and Pepper, D.W., J. Appl. Meterol. 20 146-156 (1981).
33. Pepper, D.W. and Baker, A.J., Num. Heat Transfer 2 81-95 (1979).
34. Baker, A.J. and Solimon, M.O., J. Comp. Phys. 32 289-324 (1979).
35. Baker, A.J. and Solimon, M.O., Comp. Meth. Appl. Mech. Engr. 27 215-237 (1981).
36. Dendy, J.E., SIAM J. Num. Meth. 13 233-247 (1974).
37. Raymond, W.H. and Garder, A., Mon. Weath. Rev. 104 1583-1590 (1976).
38. Pepper, D.W., Kerr, C.D., and Long, P.E., Atmos. Environ. 13 223-237 (1979).
39. Orszag, S.A. and Gottlieb, D., Lecture Notes in Mathematics, #771, 381-389 (1981).

40. Bonnerot, R. and Jamet, P., Int. J. Num. Meth. Eng. 8 811-820 (1974).
41. Varoglu, E. and Finn, W.D.L., J. Comp. Phys. 34 371-389 (1980).
42. Neumann, S.P., J. Comp. Phys. 41 270-294 (1981).
43. Tsang, T.H. and Brock, J.R., To be published.
44. Sedunov, Y.S., Physics of Drop Formation in the Atmosphere, Wiley, N.Y., 1974.
45. Becker, C. and Reiss, H., J. Chem. Phys. 65 2066-2070 (1976).
46. Bird, R.B., Stewart, W.E. and Lightfoot, E., Transport Phenomena, Wiley, N.Y., 1960.
47. Chapman, S. and Cowling, T.G., The Mathematical Theory of Non-Uniform Gases, Cambridge Univ. Press, Cambridge, 1958.
48. Mason, E. and Chapman, S., J. Chem. Phys. 36 627 (1962).
49. Kajiuchi, T., and Brock, J.R., J. Colloid Interface Sci. 83 66-76 (1981).
50. Lamb, R.G., Atmos. Environ. 7 257 (1973).
51. Brock, J.R., Faraday Symposium No. 7, "Fogs and Smokes", Chemical Society, London, 1973.
52. Wyngaard, J.C., Cote, O.R. and Rao, K.S., Adv. Geophys. 18A 192 (1974).
53. Robins, A.G., Atmos. Environ. 7 257 (1973).

This work was supported by the Chemical Systems Laboratory U.S. Army, Aberdeen Proving Ground and Aerosol Research Branch, U.S. Environmental Protection Agency.

Department of Chemical Engineering
University of Texas
Austin, Texas 78712

Continuum Modeling of Two-Phase Flows

D. A. Drew

Introduction

The flow of two materials, one dispersed throughout the other, has received much attention in recent times. Unfortunately, at this time, there seems to be no set of equations which is regarded as fundamental, from which other models can be derived as approximations. (Consider the analogy with fluid mechanics, where the incompressible, inviscid equations are thought to be valid approximations outside of shear layers and boundary layers, when thermal and sonic effects are unimportant.)

Many researchers derive equations of motion by applying an averaging process to the microscopic equations of motion. The choice of averaging process is dictated by the taste of the researcher as well as the particular problem studied. In this paper, we give a derivation of the averaged equations by applying a generic average. The relation of the generic average to time- and space-averaging is discussed.

Once believable equations of motions have been formulated, it is natural to study their predictions in relatively simple flow situations. Often the constitutive assumptions used in the model are derived and/or tested on uniform flow situations. We give a discussion of transition layers in two-phase flows. A transition layer is a thin

THEORY OF DISPERSED MULTIPHASE FLOW

ISBN 0-12-493120-0

region where the concentration of one material changes rapidly in space. An example is the "interface" between a carbonated liquid and its foamy "head". An adequate description of these transitions provides a harsh test of the constitutive assumptions used in the model.

EQUATIONS OF MOTION

Each material is assumed to be a continuum, governed by the partial differential equations of continuum mechanics. The materials are separated by an interface, which is a surface. At the interface, jump conditions express the conditions of conservation of mass and momemtum.

The equations of motion for each phase are (Truesdell and Toupin 1960)

(1) conservation of mass

$$\frac{\partial \rho}{\partial t} + \nabla \cdot \rho \mathbf{v} = 0 \tag{1}$$

(2) conservation of linear momentum

$$\frac{\partial \rho \mathbf{v}}{\partial t} + \nabla \cdot \rho \mathbf{v}\mathbf{v} = \nabla \cdot \mathbf{T} + \rho \mathbf{f} \tag{2}$$

valid in the interior of each phase. Here ρ denotes the density, $\mathbf{v}$ the velocity, $\mathbf{T}$ the stress tensor, and $\mathbf{f}$ the body force density. Conservation of angular momentum becomes $\mathbf{T} = \mathbf{T}^t$, where t denotes the transposed. At the interface, the jump conditions are

(1) jump condition for mass

$$[\rho(\mathbf{v} - \mathbf{v}_i)\cdot\mathbf{n}] = 0 \tag{3}$$

(2) jump condition for momentum

$$[\rho\mathbf{v}(\mathbf{v} - \mathbf{v}_i)\cdot\mathbf{n} - \mathbf{T}\cdot\mathbf{n}] = 0 \ . \tag{4}$$

Here $[\ ]$ denotes the jump across the interface, $\mathbf{v}_i$ is the velocity of the interface, and $\mathbf{n}$ is the unit normal (Aris 1962). We shall assume that $\mathbf{n}$ points out of phase k, and that the jump between f in phase k and f in phase ℓ defined by $[f] = f^\ell - f^k$, where a superscript k denotes the limiting value from the phase k side. As a sign convention for the curvature, we assume that κ is positive (concave) toward $-\mathbf{n}$. The mass of the interface and surface stresses have been neglected. We do not discuss any thermodynamic relations in this paper.

Constitutive equations must be supplied to describe the behavior of each material involved. For example, if one material is an incompressible liquid, then specifying the value of ρ, and assuming $\mathbf{T} = -p\mathbf{I} + \mu(\nabla\mathbf{v} + (\nabla\mathbf{v})^t)$ determines the nature of the behavior of the fluid in that phase. Similar considerations are possible for solid particles or a gas. The resulting differential equations, along with the jump conditions, provide a fundamental description of the detailed or exact flow.

Usually, however, the details of the flow are not required. For most purposes of equipment or process design, averaged, or macroscopic flow information is sufficient. Fluctuations, or details in the flow must be resolved only to the extent that they affect the mean flow (like the Reynolds stresses affect the mean flow in a turbulent flow).

Averaging

In order to obtain equations which do not contain the details of the flow, it has become customary to apply some sort of averaging process. We present a generic averaging method, and its results.

Let $\langle\ \rangle$ denote an averaging process so that if $f(\mathbf{x},t)$ is an exact microscopic field, then $\langle f\rangle(\mathbf{x},t)$ is the corresponding averaged field.

An averaging process assigns average values to certain variables. The ensemble, or set of possible outcomes, can be taken to be the possible flows in some apparatus where the initial and boundary conditions which are prescribed are equivalent in some sense. For example, for spherical particles, it may be necessary to give the statistical distribution of the positions and velocities of the centers of the particles at time $t = 0$ such that the average number density and average particle velocity is the same for all equivalent flows. We shall assume that there is some ensemble Ω, with some appropriate weighting $\mu(\omega)d\omega$ so that the average of f is given by

$$\langle f\rangle(\mathbf{x},t) = \int_\Omega f(\mathbf{x},t,\omega)\mu(\omega)d\omega \ .$$

Two cases can be discussed. If the flow is nearly steady, so that a time translation τ makes no essential difference in the ensemble, it may be enough to consider the subset of the entire ensemble which consists of translations in time of amount τ. We assign a weight $\mu(\tau)$ to the likelihood of the flow whose outcome at time t is $f(\mathbf{x},t-\tau)$, where $f(\mathbf{x},t)$ is the outcome at time t in some flow. The average of f is then taken to be

$$\langle f\rangle_t(\mathbf{x},t) = \int_{-\infty}^{\infty} f(\mathbf{x},t-\tau)\mu(\tau)d\tau \ .$$

This is classical time averaging; it is often used with

$$\mu(\tau) = \begin{cases} \frac{1}{T} & \text{if } 0 \leq \tau \leq T \\ 0 & \text{otherwise ,} \end{cases}$$

although other averages are possible.

If there are no boundaries in the flow (that is, boundary effects are unimportant), then small spatial translations should make no difference in the ensemble. In analogy to the above, the average

$$\langle f\rangle_s(\mathbf{x},t) = \int_{R^3} f(\mathbf{x}+\mathbf{s},t)\mu(\mathbf{s})d\mathbf{s}$$

can be defined. This is the classical space average; it is often used with

$$\mu(\mathbf{s}) = \begin{cases} \frac{1}{V} & \text{if } \mathbf{s} \in V \\ 0 & \text{otherwise} \end{cases}$$

where V is some volume (for example, a sphere). Again, other averages are possible.

Thus, in some sense, ensemble averaging contains space and time averages as special cases. The averaging process is assumed to satisfy

$$\langle f+g\rangle = \langle f\rangle + \langle g\rangle \tag{5}$$

$$\langle\langle f\rangle g\rangle = \langle f\rangle\langle g\rangle \tag{6}$$

$$\langle c\rangle = c \tag{7}$$

$$\left\langle\frac{\partial f}{\partial t}\right\rangle = \frac{\partial}{\partial t}\langle f\rangle \tag{8}$$

$$\left\langle\frac{\partial f}{\partial x_i}\right\rangle = \frac{\partial}{\partial x_i}\langle f\rangle \ . \tag{9}$$

The first three of these relations are called Reynolds rules, the fourth is called Leibnitz' rule, and the fifth is called Gauss' rule.

In order to apply the average to the equations of motion for each phase, we introduce the phase function $X_k(\mathbf{x},t)$ which is defined to be

$$X_k(\mathbf{x},t) = \begin{cases} 1 & \text{if } x \text{ is in phase } k \text{ at time } t \\ 0 & \text{otherwise .} \end{cases} \tag{10}$$

We shall deal with X_k as a generalized function, in particular in regard to differentiating it. Recall that a derivative of a generalized function can be defined in terms of a set of "test functions" ϕ, which are "sufficiently smooth" and have compact support. Then $\frac{\partial X_k}{\partial t}$ and $\frac{\partial X_k}{\partial x_i}$ are defined by

$$\int_{R^3\times R} \frac{\partial X_k}{\partial t} (\mathbf{x},t)\phi(\mathbf{x},t)d\mathbf{x}dt = $$

$$= - \int_{R^3\times R} X_k(\mathbf{x},t) \frac{\partial \phi}{\partial t} (\mathbf{x},t)d\mathbf{x}dt \ , \tag{11}$$

$$\int_{R^3\times R} \frac{\partial X_k}{\partial x_i} (\mathbf{x},t)\phi(\mathbf{x},t)d\mathbf{x}dt = $$

$$= - \int_{R^3\times R} X_k(\mathbf{x},t) \frac{\partial \phi}{\partial x_i} (\mathbf{x},t)d\mathbf{x}dt \ . \tag{12}$$

It can be shown that

$$\frac{\partial X_k}{\partial t} + \mathbf{v}_i \cdot \nabla X_k = 0 \tag{13}$$

in the sense of generalized functions.

If f is smooth except at S, then $f\nabla X_k$ is defined via

$$\int_{R^3\times R} f\nabla X_k \phi dxdt = \int_{-\infty}^{\infty} \int_S \mathbf{n}_k f^k \phi \ dSdt \ , \tag{14}$$

where $\mathbf{n}_k$ is the unit normal exterior to phase k, and f^k denotes the limiting value of f on the phase-k side of S.

It is also clear that ∇X_k is zero, except at the interface. Equation (14) describes the behavior of ∇X_k at the interface. Note that it behaves as a "delta-function", picking out the interface S, and has the direction of the normal interior to phase k.

Averaged Equations

In order to derive averaged equations for the motion of each phase, we multiply the equation of conservation of mass valid in phase k (1) by X_k and average. Noting that

$$X_k \frac{\partial \rho}{\partial t} = \frac{\partial}{\partial t} X_k \rho - \rho \frac{\partial X_k}{\partial t} = \frac{\partial}{\partial t} X_k \rho + \rho \mathbf{v}_i \cdot \nabla X_k \tag{15}$$

and

$$X_k \nabla \cdot \rho \mathbf{v} = \nabla \cdot X_k \rho \mathbf{v} - \rho \mathbf{v} \cdot \nabla X_k , \tag{16}$$

we have

$$\frac{\partial}{\partial t} \langle X_k \rho \rangle + \nabla \cdot \langle X_k \rho \mathbf{v} \rangle = \langle [\rho(\mathbf{v}-\mathbf{v}_i)]^k \cdot \nabla X_k \rangle . \tag{17}$$

Similar considerations for the momentum equations yield

$$\frac{\partial}{\partial t} \langle X_k \rho \mathbf{v} \rangle + \nabla \cdot \langle X_k \rho \mathbf{v}\mathbf{v} \rangle = \nabla \cdot \langle X_k T \rangle + \langle X_k \rho f \rangle + \langle [\rho \mathbf{v}(\mathbf{v}-\mathbf{v}_i) - \mathbf{T}]^k \cdot \nabla X_k \rangle . \tag{18}$$

The terms

$$\langle [\rho(\mathbf{v}-\mathbf{v}_i)]^k \cdot \nabla X_k \rangle = \Gamma_k \tag{19}$$

and

$$\langle [\rho \mathbf{v}(\mathbf{v}-\mathbf{v}_i) - \mathbf{T}]^k \cdot \nabla X_k \rangle = \mathbf{M}_k \tag{20}$$

are the interfacial source terms. As noted, ∇X_k picks out the interface, and causes discontinuous quantities multiplying it to be evalued on the phase-k side of the interface.

The jump conditions come from equations (3) and (4). We have

$$\sum_{k=1}^{2} \langle [\rho(\mathbf{v}-\mathbf{v}_i)]^k \cdot \nabla X_k \rangle = \sum_{k=1}^{2} \Gamma_k = 0 \tag{21}$$

$$\sum_{k=1}^{2} \langle [\rho \mathbf{v}(\mathbf{v}-\mathbf{v}_i) - \mathbf{T}]^k \cdot \nabla X_k \rangle = 0 . \tag{22}$$

Applying a more specific averaging process (time averaging, for example) requires a different set of manipulations regarding the interfacial source terms (Anderson & Jackson 1967, Drew 1971, Ishii 1975, Delhaye and Achard 1976). Almost all of the derivations for specific averaging processes seem to be more complicated than the above; however, the trade-off for the simple derivation is that all manipulations now involve generalized functions.

The volumetric concentration (or volume fraction, or relative residence time) of phase k is defined by

$$\alpha_k = \langle X_k \rangle \ . \tag{23}$$

We note that

$$\frac{\partial \alpha_k}{\partial t} = \langle \frac{\partial X_k}{\partial t} \rangle \tag{24}$$

and

$$\nabla \alpha_k = \langle \nabla X_k \rangle \ . \tag{25}$$

There are two types of averaged variables which are useful in two-phase mechanics, namely the phasic, or X_k-weighted average, and the mass-weighted average. Which is appropriate is suggested by the appearance of the quantity in the equation of motion. The phasic average of the variable ϕ is defined by

$$\tilde{\phi}_k = \langle X_k \phi \rangle / \alpha_k \tag{26}$$

and the mass weighted average of the variable ψ is defined by

$$\hat{\psi}_k = \langle X_k \rho \psi \rangle / \alpha_k \tilde{\rho}_k \ . \tag{27}$$

It is convenient to write the stresses $\tilde{\mathbf{T}}_k$ in terms of pressures plus extra stresses. Thus,

$$\tilde{\mathbf{T}}_k = -\tilde{p}_k \mathbf{I} + \tilde{\tau}_k \ . \tag{28}$$

It is expected that readers familiar with fluid dynamical concepts are familiar with the concept of pressure in fluids; in this case, $\tilde{p}_k$ can be thought of as the average of the microscopic pressure. If one of the phases consists of solid particles, the concept is less familiar. In this case, the microscopic stress (involving small elastic deformations, for example) is thought of being made up of a spherical part (acting equally in all directions) plus an extra stress. The spherical part, when averaged, yields the pressure $\hat{p}_k$ in equation (28).

It has further become customary to separte various parts of the interfacial momentum transfer term. This is done by defining the interfacial velocity of the kth phase by

$$\Gamma_k \mathbf{v}_{k,i} = \langle [\rho \mathbf{v}(\mathbf{v}-\mathbf{v}_i)]^k \cdot \nabla X_k \rangle , \tag{29}$$

and the interfacial pressure on the kth phase by

$$p_{k,i} |\nabla \alpha_k|^2 = \langle p^k \nabla X_k \rangle \cdot \nabla \alpha_k . \tag{30}$$

Equation (30) is the dot product of $\nabla \alpha_k$ of the "standard" definition (Ishii 1975) of the interfacial pressure. The standard definition uses three equations to define one scalar quantity, and cannot be a generally valid definition. Here the remaining part of the contribution of the pressure at the interface is lumped with the viscous stress contribution at the interface, and is treated through the use of a constitutive equation. Thus, we write

$$\mathbf{M}^k = \Gamma_k \mathbf{v}_{k,i} - p_{k,i} \nabla \alpha_k + \mathbf{M}_k^d , \tag{31}$$

where $\mathbf{M}_k^d = \langle (p-p_{k,i})^k \nabla X_k - \boldsymbol{\tau}_k^k \cdot \nabla X_k \rangle$ is referred to as the interfacial force density, although it does not contain the effect of the average force on the interface due to the average interfacial pressure. The term $-p_{k,i} \nabla \alpha_k$, which does contain the force due to the average interfacial pressure, is sometimes referred to as the bouyant force. The reason for this terminology is, of course, that the buoyant force on an object is due to the distribution of the pressure of the surrounding fluid on its boundary.

With equations (23) and (26) - (31), the equations of motion (17) and (18) become

$$\frac{\partial \alpha_k \tilde{\rho}_k}{\partial t} + \nabla \cdot \alpha_k \tilde{\rho}_k \hat{\mathbf{v}}_k = \Gamma_k \tag{32}$$

$$\frac{\partial \alpha_k \tilde{\rho}_k \hat{\mathbf{v}}_k}{\partial t} + \nabla \cdot \alpha_k \tilde{\rho}_k \hat{\mathbf{v}}_k \hat{\mathbf{v}}_k = -\alpha_k \nabla \hat{p}_k + \nabla \cdot \alpha_k (\hat{\boldsymbol{\tau}}_k + \tilde{\boldsymbol{\sigma}}_k) + \Gamma_k \mathbf{v}_{k,i} + (p_{k,i} - \hat{p}_k) \nabla \alpha_k + \mathbf{M}_k^d . \tag{33}$$

The jump conditions (21) and (22) are

$$\sum_{k=1}^{2} \Gamma_k = 0 \tag{34}$$

$$\sum_{k=1}^{2} [\Gamma_k \mathbf{v}_{k,i} + p_{k,i} \nabla \alpha_k + \mathbf{M}_k^d] = 0 \quad . \tag{35}$$

Adequate models for compressibility and phase change require consideration of thermodynamic processes. These are beyond the scope of this paper; therefore we shall restrict our attention to incompressible materials where no phase change occurs. Thus we assume that

$$\hat{\rho}_k = \text{constant} \tag{36}$$

and

$$\Gamma_k = 0 \quad . \tag{37}$$

In order to simplify the notation, we shall drop all symbols denoting averaging.

CONSTITUTIVE EQUATIONS

In order to have a useable model, relations must be given which specify the stresses $(\boldsymbol{\tau} + \boldsymbol{\sigma}_k)$, the interfacial force density $\mathbf{M}_k^d$, and the pressure differences $p_k - p_{k,i}$, consistent with the equations of motion and the jump conditions.

The fundamental process consists of proposing forms for the necessary terms within the framework of the principles of constitutive equations, finding solutions of the resulting equations, and verifying against experiments. The ideal end result of the process is a set of equations which could be used to predict the behavior of the two-phase flow, for example with a computer code. With the equations should come a set of conditions for the validity of the values of the constants and other functions used in the constitutive equations.

The stresses $\boldsymbol{\tau}_k + \boldsymbol{\sigma}_k$, the interfacial force density $\mathbf{M}_k^d$ and the pressure differences $p_k - p_{k,i}$ are assumed to be functions of α_k, $\partial\alpha_k/\partial t$, $\nabla\alpha_k$, $\mathbf{v}_k$, $\nabla\mathbf{v}_k$, $\partial\mathbf{v}_k/\partial t$, ... , where ... represents the material properties, such as the viscosities and densities of the two materials, and other geometric parameters such as the average particle size, or the interfacial area density.

For concreteness, we shall refer to phase one as the particulate, or dispersed phase and include in that description solid particles, droplets, or bubbles. Phase two is then the continuous, or carrier phase, and can be liquid or gas. We shall denote

$$\alpha = \alpha_1 \tag{38}$$

so that

$$1-\alpha = \alpha_2 . \tag{39}$$

It is evident that both α and $1-\alpha$ need not be included as independent variables in forming constitutive equations.

Drew and Lahey (1979) consider the general process of constructing constitutive equations. The simplest reasonable set of assumptions leads to models for the motions of the two materials which may be ill-posed. Drew (1982) gives a review of the state of affairs.

Essentially, the problem and the reason for its importance can be summarized in the following manner. The simplest model assumes that the interaction forces are due to viscous drag, and that pressure forces equilibrate across particles instantaneously. Thus,

$$\mathbf{M}_1^d = \mathbf{M}_2^d = \frac{3}{8} \alpha\rho_1 \frac{C_D}{r} |\mathbf{v}_1-\mathbf{v}_2|(\mathbf{v}_2-\mathbf{v}_1) \tag{40}$$

and

$$p_1 = p_2 = p_{1,i} = p_{2,i} = p . \tag{41}$$

If the stresses $\boldsymbol{\tau}_k + \boldsymbol{\sigma}_k$ are ignored, the model has complex characteristics, and hence is ill-posed. See Ramshaw and Trapp (1978). The ill-posed nature of the model leads (theoretically) to solutions which grow rapidly on a small length scale. This leads some to conjecture that the model is trying to resolve events on the microscale. The application of an average, however, is supposed to lose details of the flow; indeed, both (40) and (41) specifically ignore flow details on the scale of the particle size and smaller. The obvious conclusion is that equations (32 - 37), with assumptions (40), (41) and neglect of viscous forces lack some mechanism which is important on the particle scale. It is noteworthy that no difficulties arise in numerical simulations with the ill-posed system. The reason for this is that the instability inherent in the ill-posed system appears on a scale comparable to the particle radius to a power which depends on the exact form of the drag law used for C_D. Thus, if the mesh is not refined to the particle size, no instabilities will be seen. This suggests that computation with the ill-posed system with a reasonable mesh will most likely give good results.

The above argument indicates some difficulties for stability calculations. When is an instability not an instability? Presumably it is unobserved if its wavelength is too small. It would be satisfying to find the missing effect in the model and show how the model reduces to the isobaric, inviscid model mentioned above. There is a long list of candidates for the forgotten mechanism (Drew 1982) but no clear winner has emerged. The sensible approach seems to be to examine various models on a mesoscale, that is, on a scale which is small compared to the usual experimental verification flows (such as laminar settling in a still fluid), but large compared to the particle scale. These mesoscale calculations may indicate what sorts of terms are needed to give a valid description on a smaller scale.

There is another reason for seeking the more complete model. If the ill-posed model is the limit of some more complete model with some effects neglected, if the neglect can be done by a set of formal manipulations on the more complete model, the result may indicate whether some well-posed model might do as well as the above ill-posed model. Moreover, it is always of interest to find reduced models which contain the same features as the original model.

We shall examine the effects of the viscous and Reynolds stresses on a mesoscale motion, namely transition layers in vertical flow. If the flow is vertical, and we denote the vertical velocities by

$$\mathbf{v}_1 = u(z,t)\mathbf{k} \qquad (42)$$

$$\mathbf{v}_2 = v(z,t)\mathbf{k} \ , \tag{43}$$

we have

$$\frac{\partial \alpha}{\partial t} + \frac{\partial \alpha u}{\partial z} = 0 \tag{44}$$

$$\alpha u + (1-\alpha)v = j(t) \tag{45}$$

$$\alpha\rho_1\left(\frac{\partial u}{\partial t} + u\frac{\partial u}{\partial z}\right) = -\alpha\frac{\partial p}{\partial z} + \alpha b(v-u) - \alpha\rho_1 g + \frac{\partial}{\partial z}\alpha\sigma_1 \tag{46}$$

$$(1-\alpha)\rho_2\left(\frac{\partial v}{\partial t} + v\frac{\partial v}{\partial z}\right) = -(1-\alpha)\frac{\partial p}{\partial z} + \alpha b(u-v) - (1-\alpha)\rho_2 g + \frac{\partial}{\partial z}(1-\alpha)\sigma_2 + \frac{\partial}{\partial z}(1-\alpha)\mu\frac{\partial v}{\partial z} \ . \tag{47}$$

The function $j(t)$ is the volumetric flux, and will be taken to be constant. The stress models are

$$\tau_1 = 0 \tag{48}$$

$$\tau_2 = \mu\frac{\partial v}{\partial z} \tag{49}$$

to represent the viscosity of the fluid. The particles are assumed to be inviscid. In addition, the Reynolds stress terms σ_k will be taken to be constants.

For problems of sedimentation and transitions in fluidized beds, we shall assume that $j(t)$ = constant for $t > 0$.

Kinematic Waves

If we ignore inertia in equations (46) and (47) and set $\sigma_1 = \sigma_2 = 0$ and $\mu = 0$, we have a model which reduces to a one-dimensional scalar conservation law. If the pressure is eliminated from the resulting momentum equations, we have

$$v-u = \frac{(1-\alpha)(\rho_1-\rho_2)g}{b} \ . \tag{50}$$

Using (45) gives

$$u = j - (1-\alpha)(v-u) \ .$$

Hence

$$\alpha u = \alpha j - \frac{\alpha(1-\alpha)^2(\rho_1-\rho_2)g}{b} = f(\alpha) \ . \tag{51}$$

Equation (44) with αu given by (51) is a scalar conservation law for $\alpha(z,t)$.

Solutions can be found by the method of characteristics; on the characteristics,

$$\frac{d\alpha}{dt} = 0 \tag{52}$$

$$\frac{dz}{dt} = f'(\alpha) = \text{constant} \ . \tag{53}$$

As long as no discontinuities occur, α is constant on characteristics.

If discontinuities occur, they must propagate at a speed given by

$$S = \frac{[f]}{[\alpha]} \ , \tag{54}$$

where $[\phi]$ denotes the jump in the quantity ϕ across the discontinuity. A discontinuity, or shock will persist if characteristics tend to come into the shock. If characteristics diverge from the shock, then it will smooth out. In Figure 1, a shock from $(\alpha_-, f(\alpha_-))$ to $(\alpha_+, f(\alpha_+))$ is not stable if the region where α_+ occurs is above the shock. This is because the characteristics at α_+ are moving faster than the shock. Note that contact discontinuities are possible with this model.

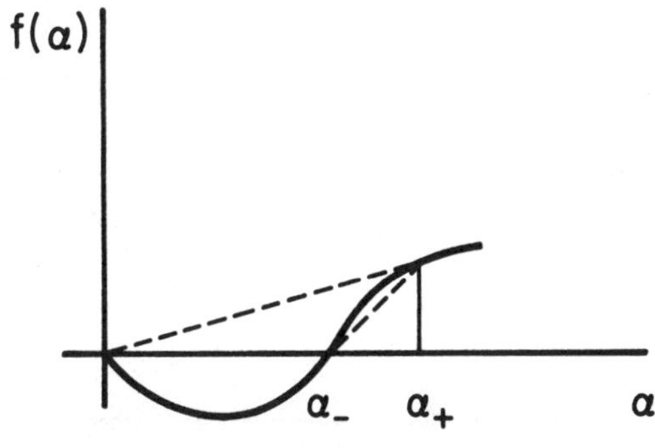

Figure 1

For fluidization, $j > 0$. In order to describe a transition in fluidization, we assume that the bed is fluidized at $t = 0-$ with a concentration α_+ which corresponds to some value of the volumetric flow rate j_+. At time $t = 0$, the volumetric flow rate is instantaneously changed to j. The flow-concentration diagram is given in

Figure 1. The simplest situation is depicted, consisting of an upward traveling bottom shock (transition) and an upward traveling top shock.

Figure 2 shows the consentration at some time t_1, and Figure 3 shows the solution in various regions of the $t - z$ plane.

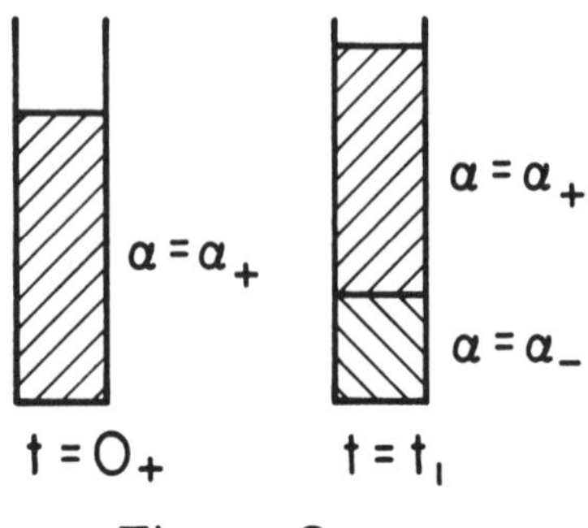

Figure 2

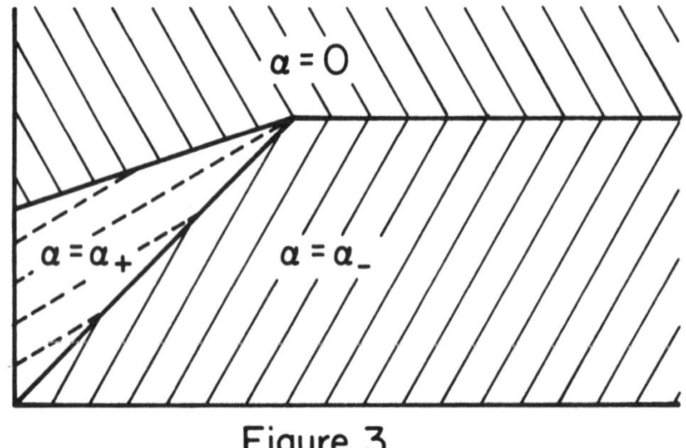

Figure 3

Diffusional Regularization

One way to understand shocks is to include in the model some means of smearing them out. This can be done in the present problem by including the diffusion terms. We continue to ignore inertia and viscosity. Repeating the procedure outlined above yields

$$\alpha u = f(\alpha) - D \frac{\partial \alpha}{\partial z} , \tag{55}$$

where

$$D = \frac{(1-\alpha)[(1-\alpha)\sigma_1 + \alpha\sigma_2]}{b} . \tag{56}$$

The equation for α is a nonlinear diffusion equation

$$\frac{\partial \alpha}{\partial t} + \frac{\partial f(\alpha)}{\partial z} = \frac{\partial}{\partial t} \left(D \frac{\partial \alpha}{\partial z}\right) . \tag{57}$$

In order to examine the transition, we let

$$\alpha = \alpha(z - st) \tag{58}$$

where s is the speed of propagation of the traveling wave representing the transition. We obtain

$$\frac{d}{d\xi} [f(\alpha) - s\alpha] = \frac{d}{d\xi} \left(D \frac{d\alpha}{d\xi}\right) . \tag{59}$$

Integrating from $-\infty$ to ξ gives

$$f(\alpha) - [f(\alpha_-) + s(\alpha - \alpha_-)] = D \frac{d\alpha}{d\xi} . \tag{60}$$

The quantity in the bracket is the equation for the chord (in Figure 1 for example), and $f(\alpha)$ is the flow-concentration curve there. If the curve lies above the chord, then $d\alpha/d\xi$ is positive. If the curve lies below the chord, then $d\alpha/d\xi$ is negative. The transitions go from α_- to α_+ where α_+ is the value of α where the left hand side of (60) is zero, that is, where the curve and chord intersect. If D is small, the transition region is thin. The diffusional regularization corresponds to the results obtained from shock stability considerations.

Inclusion of the inertia of both phases complicates the situation immensely. It can be shown that the thing which corresponds to the shock in the kinematic wave model is not a shock in the model which includes inertia.

In order to study the transition, assume that α, u, v and p are functions of $\xi = z - st$, where s is the speed of the transition wave, and $\alpha \to \alpha_-$, $u \to u_-$, $v \to v_-$ as $\xi \to \infty$, $\alpha \to \alpha_+$, $u \to u_+$, $v \to v_+$ as $\xi \to \infty$. $\alpha_\pm$, $u_\pm$ and $v_\pm$ are related to s by (45) and (54). Substituting in the equations (44 - 47) and eliminating p, u, and v gives

$$\mu(1-\alpha_-)(v_- - s)\frac{\alpha}{(1-\alpha)^2}\alpha'' + \hat{\beta}(\alpha)\alpha' + \hat{g}(\alpha) = 0 \tag{61}$$

where

$$\hat{\beta}(\alpha) = \sigma_1 + \sigma_2 \frac{\alpha}{1-\alpha} - \frac{\alpha_-^2 \rho_1 (u_- - s)^2}{\alpha^2} - \frac{\alpha}{(1-\alpha)^3}\rho_2 (1-\alpha_-)^2 (v_- - s)^2 \tag{62a}$$

and

$$\hat{g}(\alpha) = -\frac{b}{(1-\alpha)^2}[f(\alpha) - \alpha_- u_- - s(\alpha - \alpha_-)] . \tag{62b}$$

Note that if $\mu = 0$, and we wish to retain the picture given by the characteristics of the <u>scalar</u> conservation law, we can do so only if

$$\hat{\beta}(\alpha) > 0 \tag{63}$$

for α between α_- and α_+. In a sense, equation (63) suggests that inertia will be unimportant for transitions if diffusion is sufficiently large; sufficiently large means so that equation (63) is satisfied.

Note from equation (61) with $\mu = 0$ and $\sigma_1 = \sigma_2 = 0$, that inertia without diffusion gives results <u>opposite</u> to the results from the scalar conservation law.

<u>The Effect of Viscosity</u>

In equation (61), let $\alpha' = w(\alpha)$. Then $\alpha'' = w\frac{dw}{d\alpha}$, and equation (61) becomes

$$\mu(1-\alpha_-)(v_- - s)\frac{\alpha}{(1-\alpha)^2}\frac{d(w^2/2)}{d\alpha} + \mu(1-\alpha_-)(v_- - s)\frac{\alpha}{(1-\alpha)^3}w^2 + \hat{\beta}(\alpha)w + \hat{g}(\alpha) = 0 . \tag{64}$$

Let us now define $G(\alpha)$ by

$$G'(\alpha) = g(\alpha) \tag{65}$$

and

$$\begin{aligned} G(\alpha_-) &= 0 \quad \text{if} \quad g(\alpha) > 0 \quad \text{for} \quad \alpha_- < \alpha < \alpha_+ \\ G(\alpha_+) &= 0 \quad \text{if} \quad g(\alpha) < 0 \quad \text{for} \quad \alpha_- < \alpha < \alpha_+ . \end{aligned} \tag{66}$$

(We also assume that $(1-\alpha)(v-s) > 0$.) Equation (64) becomes

$$\frac{d}{d\alpha}\left[\frac{1}{2}\left(\frac{w}{1-\alpha}\right)^2 + G(\alpha)\right] = -\beta(\alpha)w \; . \tag{67}$$

The curves $H(\alpha,w) = \frac{1}{2}\left(\frac{w}{1-\alpha}\right)^2 + G(\alpha)$ are closed curves centered at α_- if $g(\alpha) > 0$, and α_+ if $g(\alpha) < 0$. See Figure 4. On a trajectory leaving α_+ (if $g(\alpha) > 0$) or α_- (if $g(\alpha) < 0$), the function $\alpha(\xi)$ satisfies

$$\frac{d}{d\xi} H(\alpha,w) = -\beta(\alpha)w^2 \tag{68}$$

which is negative if $\beta(\alpha) > 0$. Thus H decreases on a trajectory, giving the transition from α_+ to α_- (if $g(\alpha) > 0$) or from α_- to α_+ (if $g(\alpha) < 0$). Since

$$g(\alpha) = -\frac{b}{\alpha(1-\alpha)^2, \mu(1-\alpha_-)(v_- - s)}\,(f(\alpha) - \alpha_- u_- - s(\alpha-\alpha_-)) \; ,$$

we see that if the curve f is above the chord, $s(\alpha-\alpha_-) + \alpha_- u_-$, then $g(\alpha) < 0$, and this type of transition can occur from α_- to α_+. On the other hand, if the curve $f(\alpha)$ is below the chord $s(\alpha-\alpha_-) + \alpha_- u_-$, then $g(\alpha) > 0$, and the transition can occur from α_+ to α_-. This again agrees with the picture given by the characteristics for the one-dimensional conservation law.

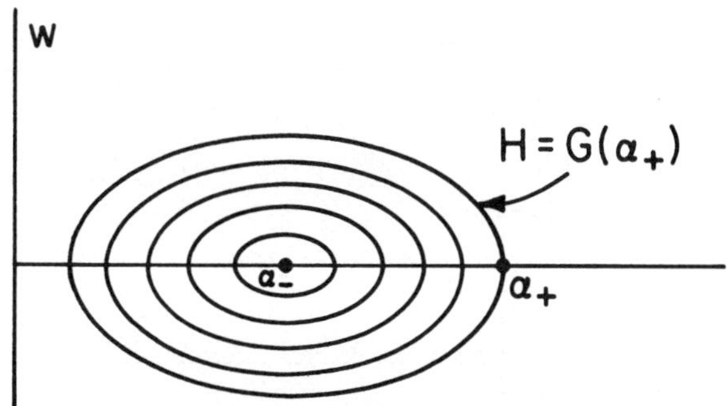

Figure 4

Conclusion

There are two distinct features of two-phase flow modeling addressed in this paper, namely the averaging process, and transition layers.

The generic averaging process includes certain classical averages as special cases and gives the same results as these classical averages when applied to the equations of motion.

Transition layers are often observed in two-phase flows, and various models for them have been used. The simplest model, that of the kinematic wave, describes the situation when inertia, viscosity and diffusivity are negligible, but diffusivity dominates over inertia.

REFERENCES

Anderson, T. B. and Jackson, R. 1967. A fluid mechanical description of fluidized beds, Ind. Eng. Chem. Fund. 6, 527-539.

Aris, R. 1962. Vectors, Tensors and the Basic Equations of Fluid Mechanics, Prentice Hall, Englewood Cliffs, NJ.

Delhaye, J. M. and Achard, J. L. 1976. On the averaging operators introduced in two-phase flow modeling, CSNI Specialists Meeting on Transient Two-Phase Flow, Toronto, Canada.

Drew, D. A. 1971. Average field equations for two-phase media, Studies in Appl. Math. L:133-166.

Drew, D. A. and Lahey, R. T. 1979. Application of general constitutive principles to the derivation of multi-dimensional two-phase flow equations, Int. J. Multi-phase Flow, 5:243-264.

Drew, D. A. 1982. Mathematical Modeling of Two-Phase Flow, to appear in Annual Review of Fluid Mechanics.

Ishii, M. 1975. Thermo-fluid Dynamic Theory of Two-Phase Flow, Eyrolles, France.

Ramshaw, J. D. and Trapp, J. A. 1978. Characteristics, stability and short-wavelength phenomena in two-phase flow equation systems, Nuclear Sci. and Eng. 66:93-102.

Donald A. Drew
Department of Mathematical Sciences
Rensselaer Polytechnic Institute
Troy, NY 12181

A Multiphase Mixture Theory for Fluid-Particle Flows

J. W. Nunziato

1. INTRODUCTION

There are a number of problems of technological interest arising in hydrology, suspension rheology, sedimentation, and bed fluidization, which require a model for a two-phase mixture of suspended particles in a viscous fluid. As a result, several different approaches have been developed to deal with the different transport processes which occur. In flow problems where only the forces exerted on boundaries by a dilute suspension are required, a single fluid model suffices in which the effect of the particulate phase is accounted for in terms of an effective viscosity [1-3]. If information about the flow field in the vicinity of particles is desired, then a more detailed analysis of the fluid motion around a single particle is necessary. Problems of this nature have been studied extensively for particles in shear flows and for gravitational settling in fluids at rest; with and without boundaries. The analytical approach is described by Happel and Brenner [1] and a more recent review is given by Leal [4]. One of the interesting results of this type of analysis is the phenomenon of lateral particle migration observed experimentally by Segre and Silberberg [5].

However, for many applications, what is desired is knowledge of the collective motion of the particles and the motion of the particles relative to the flow field.

THEORY OF DISPERSED MULTIPHASE FLOW

ISBN 0-12-493120-0

For example, in a given flow, segregation of the particles may be possible and knowledge of the location of high particle concentration could be useful. In this case, it is necessary to recognize that one is dealing with a two-phase mixture, each phase having its own density, velocity, and volume fraction. To analyze these problems, the approach has been to model the mixture as a continuum and account for the interaction between phases through the exchange of momentum and energy. This general theoretical approach is detailed by Ishii [6], Drew and Segel [7], Drew and Lahey [8], and Nunziato, Passman, and Walsh [9-12]. More specialized models have been developed in the context of bed fluidization by Murray [13] and Anderson and Jackson [14], and in the context of particulate sedimentation by Bedford, Hill, and Drumheller [15,16] and Thacker and Lavelle [17]. One advantage of the continuum approach is that it provides the theoretical structure by which models can be extended to high concentrations.

Although the above approaches to two-phase flow modelling are separated for discussion purposes, they are not entirely independent. Almost all the continuum theories rely on the small-scale hydrodynamic models of single particle motion to infer constitutive relations for the pressure forces and for the drag and lift forces on particles. In addition, in the absence of slip between the fluid and the particles, the models for the fluid and the solid particles are chosen to insure that the continuum theories yield the correct limit for the viscosity of very dilute mixtures, i.e., as the solid volume fraction ϕ_s tends to zero.

In this paper, we discuss the application of a continuum mixture theory to the description of fluid-particle flows. The general setting of the theory lies in the context of thermodynamics and utilizes the approach of Nunziato, Passman, and Walsh [9-12]. In Section 2, we review the kinematics and the equations of motion for a two-phase mixture in which the continuous phase is an incompressible viscous fluid and the dispersed phase consists of incompressible solid particles which are spherical. The entropy inequality for two-phase mixtures is recorded in Section 3 and modified to account for the constraints of incompressibility of the phases and of saturation of the mixture. Using the method of Lagrange multipliers, distinct hydrostatic pressures are introduced for each phase and the notion of an interface pressure arises naturally. In Section 4, we discuss the general form of constitutive equations for the mixture consistent with invariance requirements, establish the thermodynamic restrictions, and discuss the well-known closure problem. This closure problem arises because of more unknowns than field equations and is usually solved by setting all pressures equal. This assumption is not always satisfactory, however, since it implies aphysical behavior and can lead to unstable solutions of the flow equations. Thus, an alternative approach is suggested in which dynamic pressure effects are utilized to describe pressure differences. In Section 5, we specialize the theory to model isothermal flows of dilute suspensions. This model includes a viscosity associated with the particles to account for the increased dissipation arising from the mere presence of the particles in the suspending fluid medium. In addition, we utilize results for the forces on single particles to develop specific models for the pressures arising in the theory as well as for the momentum exchange due to lift and drag forces. Section 6 discusses the application of the theory to describe translational Brownian motion. In particular, we show that the present model reduces to a diffusive theory similar to that considered by Drew [18] where the diffusivity is related to the derivative of a thermodynamic potential. This result is consistent with the ideas developed by Batchelor [19]. Section 7 presents an analysis of steady Poiseuille flow. The effect of viscous, drag, lift, dynamic pressure and diffusive forces (Brownian motion) on the velocity and concentration profiles are discussed. Our results show that in general the particles lag the fluid and that the concentration of particles has a maximum off the axis suggestive of the Segre-Silberberg effect. In Section 8, we assume that the resulting solutions apply to a capillary viscometer and we evaluate the rheological properties of the suspension including the effective viscosity and normal stresses. Consistent with previous studies, the suspension is found to behave macroscopically as a non-Newtonian fluid.

2. KINEMATICS AND THE EQUATIONS OF MOTION

In this paper, we consider the behavior of a two-phase mixture of small solid particles in a viscous fluid. At every spatial position $\underset{\sim}{x}$ in the region R occupied by the mixture and at every time t, each constituent is assigned a _material density_ $\gamma_a(\underset{\sim}{x},t)$ (a = s (solid), f (fluid)) which represents the mass of the ath constituent per unit volume occupied by the constituent. Both the dispersed solid phase and the continuous fluid phase are assumed to be incompressible and thus the densities γ_s, γ_f are constants.

In considering multiphase mixtures, it is also useful to define the _partial density_ $\rho_a(\underset{\sim}{x},t)$ which is related to the material density by

$$\rho_a = \phi_a \gamma_a \quad . \tag{2.1}$$

Here ϕ_a, $0 < \phi_a < 1$, is the _volume fraction_ of the ath constituent and indicates the proportion of the total volume occupied by the constituent. Clearly ϕ_a is a measure of the local volume changes which result from the relative motion of the phases and we require that the mixture always be saturated:

$$\phi_s + \phi_f = 1. \tag{2.2}$$

It should be evident that the volume fraction of either phase (and hence the partial density) may vanish at any position $\underset{\sim}{x}$ and t as long as they both do not vanish simultaneously. Thus, particle-free zones in the flow may exist in which case $\phi_f = 1$. The mixture is said to be _dilute_ if the solid volume fraction is small compared to unity ($\phi_s << 1$) everywhere in the flow.

The _density of the mixture_ ρ is defined by

$$\rho = \phi_s \gamma_s + \phi_f \gamma_f \tag{2.3}$$

$$= \gamma_f + \phi_s (\gamma_s - \gamma_f). \tag{2.4}$$

Notice that the mixture is incompressible only if the particles are neutrally buoyant; $\gamma_s = \gamma_f$. In situations where this is not the case, where $\gamma_s \neq \gamma_f$, the compressibility of the mixture can have a significant effect on its rheological behavior [4].

In considering the flow of the mixture, we assume that the motion of each phase is as smooth as required and assign to each phase a *velocity* field $\underset{\sim}{v}_a(\underset{\sim}{x},t)$. The *velocity gradient* with respect to the spatial positions $\underset{\sim}{x}$ is defined by

$$\underset{\sim}{L}_a = \nabla \underset{\sim}{v}_a, \tag{2.5}$$

and the symmetric part of $\underset{\sim}{L}_a$ is called the *rate of deformation tensor*, $\underset{\sim}{D}_a$;†

$$\underset{\sim}{D}_a = (\underset{\sim}{L}_a + \underset{\sim}{L}_a^T)/2 . \tag{2.6}$$

The *velocity of the mixture* $\underset{\sim}{v}$ is defined in terms of the total momentum:

$$\rho \underset{\sim}{v} = \rho_s \underset{\sim}{v}_s + \rho_f \underset{\sim}{v}_f \tag{2.7}$$

and determines the barycentric motion of the mixture. Then the *diffusion velocity* $\underset{\sim}{u}_a$ for the ath constituent describes the motion of the constituents relative to the barycentric motion and is given as

$$\underset{\sim}{u}_a = \underset{\sim}{v}_a - \underset{\sim}{v}. \tag{2.8}$$

We also note that the velocity gradient for the mixture at each $(\underset{\sim}{x},t)$ is

$$\underset{\sim}{L} = \nabla \underset{\sim}{v} \tag{2.9}$$

and the corresponding rate of deformation is

$$\underset{\sim}{D} = (\underset{\sim}{L} + \underset{\sim}{L}^T)/2. \tag{2.10}$$

Next, we wish to consider the general equations of motion for each of the phases and for the mixture. In writing these equations, we make two important assumptions.

(i) Each phase of the mixture behaves as if it were a single material except when it is interacting and hence exchanging momentum and energy with the other phase; and

(ii) The equations of motion for the mixture are also the same as those for a single material and result from

---

†$(\ )^T$ designates the transpose of a tensor.

summing the equations of motion for the individual phases over all phases.

Assumption (i) asserts that for chemically inert constituents, balance of mass, momentum, and energy for each constituent (a = f,s) take the form [20]

$$\grave{\rho}_a = - \rho_a \nabla \cdot \underset{\sim}{v}_a, \tag{2.11}$$

$$\rho_a \grave{\underset{\sim}{v}}_a = \nabla \cdot \underset{\sim}{\sigma}_a + \rho_a \underset{\sim}{b}_a + \overset{+}{\underset{\sim}{m}}_a, \tag{2.12}$$

$$\rho_a \grave{e}_a = \underset{\sim}{\sigma}_a \cdot \underset{\sim}{D}_a - \nabla \cdot \underset{\sim}{q}_a + \rho_a r_a + \overset{+}{e}_a - \overset{+}{\underset{\sim}{m}}_a \cdot \underset{\sim}{v}_a . \tag{2.13}$$

where the backward prime denotes the material time derivative following the motion of a particular phase; for example,

$$\grave{\rho}_a = \partial \rho_a / \partial t + \nabla \rho_a \cdot \underset{\sim}{v}_a \quad . \tag{2.14}$$

In (2.11) - (2.13), $\underset{\sim}{\sigma}_a$ is the symmetric stress tensor, $\underset{\sim}{\sigma}_a = \underset{\sim}{\sigma}_a{}^T$ (particle rotation is neglected), $\underset{\sim}{b}_a$ is the external body force (e.g., gravity), $\overset{+}{\underset{\sim}{m}}_a$ is the momentum exchange resulting from the forces acting on the dispersed phase due to the presence and motion of the continuous fluid phase, e_a is the internal energy (per unit mass), $\underset{\sim}{q}_a$ is the local heat flux, r_a is the external heat supply, and $\overset{+}{e}_a$ is the energy exchange due to the local heat transfer between phases.

It is important to note that the equations of motion (2.11) - (2.13) are identical in form to those developed by Ishii [6] and Drew and Segel [7] using averaging methods and subsequently employed by Drew and Lahey [8, 18]. Along with appropriate constitutive equations for the stresses $\underset{\sim}{\sigma}_a$, the momentum exchange $\overset{+}{\underset{\sim}{m}}_a$, the internal energies e_a, the heat fluxes $\underset{\sim}{q}_a$, and the energy exchange $\overset{+}{e}_a$, these equations of motion form the basis of our analysis of fluid-particle flows.

Our second assumption asserts that if we add together the equations of motion for each phase, then we should recover the usual equations of motion for a single substance. Indeed, carrying out this rather tedious calculation [9-12], we obtain the mixture equations

$$\dot{\rho} = - \rho \nabla \cdot \underset{\sim}{v} \ , \tag{2.15}$$

$$\rho \dot{\underset{\sim}{v}} = \nabla \cdot \underset{\sim}{\sigma} + \rho \underset{\sim}{b} \ , \tag{2.16}$$

$$\rho \dot{e} = \underset{\sim}{\sigma} \cdot \underset{\sim}{D} - \nabla \cdot \underset{\sim}{q} + \rho r \ , \tag{2.17}$$

where†

$$\underset{\sim}{\sigma} - \rho\underset{\sim}{v}\otimes\underset{\sim}{v} = \Sigma(\underset{\sim}{\sigma}_a - \rho_a\underset{\sim}{v}_a\otimes\underset{\sim}{v}_a) \;, \tag{2.18}$$

$$\rho\underset{\sim}{b} = \Sigma\,\rho_a\underset{\sim}{b}_a \;, \tag{2.19}$$

$$\rho(e + \underset{\sim}{v}\cdot\underset{\sim}{v}/2) = \Sigma\,\rho_a(e_a + \underset{\sim}{v}_a\cdot\underset{\sim}{v}_a/2) \;, \tag{2.20}$$

$$\underset{\sim}{q} - \underset{\sim}{\sigma}\underset{\sim}{v} - \rho(e + \underset{\sim}{v}\cdot\underset{\sim}{v}/2)\underset{\sim}{v} = \Sigma\underset{\sim}{q}_a - \underset{\sim}{\sigma}_a\underset{\sim}{v}_a - \rho_a(e_a + \underset{\sim}{v}_a\cdot\underset{\sim}{v}_a/2)\underset{\sim}{v}_a \;, \tag{2.21}$$

$$\rho(r + \underset{\sim}{b}\cdot\underset{\sim}{v}) = \Sigma\,\rho_a(r_a + \underset{\sim}{b}_a\cdot\underset{\sim}{v}_a) \;, \tag{2.22}$$

and the superimposed dot is the material time derivative following the barycentric motion of the mixture; e.g.,

$$\dot{\rho} = \partial\rho/\partial t + \nabla\rho\cdot\underset{\sim}{v} \;. \tag{2.23}$$

The expressions (2.18) - (2.22), as well as (2.3) and (2.7), are often referred to as the <u>summing rules</u> for the mixture. It is interesting to note that these definitions are consistent with the classical kinetic theory of gases. It is also worth pointing out that the relation for the total stress $\underset{\sim}{\sigma}$ includes the effects of diffusion. This result implies that it is not necessary to include fictitious "inertial coupling" terms in the momentum equation for each phase as has been suggested in other works [21].

In deriving (2.15) - (2.17) it is also necessary to require that the interactions $\underset{\sim}{m}_a^+$, e_a^+ satisfy the summing rules

$$\underset{\sim}{m}_f^+ = -\underset{\sim}{m}_s^+ \;, \; e_f^+ = -e_s^+ \;. \tag{2.24}$$

These results impose important constraints on the motions possible in a two-phase mixture. For example, $(2.24)_1$ implies that any momentum lost by the dispersed particles due to lift and drag forces acting on the particles will be gained by the fluid.

3. THE ENTROPY INEQUALITY

Central to our discussion is the use of the Second Law of thermodynamics to establish restrictions on the forms of the constitutive equations for a fluid-particle mixture. To state the appropriate form of the Second Law, we assign to each phase an entropy η_a. Furthermore, for most flows of interest, we can assume that the dispersed and

---

† For brevity, we use ΣJ_a to represent the sum $J_f + J_s$. Also we use $\otimes$ to denote a tensor product.

continuous phases are locally in thermal equilibrium and assign to them a common absolute temperature θ. Then, the Second Law asserts that for all flows the mixture must satisfy the entropy inequality

$$\Sigma \left[\rho_a \dot{\eta}_a + \nabla \cdot (\underset{\sim}{q}_a/\theta) - \rho_a r_a/\theta\right] \geq 0 \tag{3.1}$$

For the mixtures considered here, it is convenient to express the constitutive equations in terms of the Helmholtz free energy ψ_a defined in the usual manner:

$$\psi_a = e_a - \theta\eta_a . \tag{3.2}$$

With this in mind, we can use (3.2), along with (2.16) and (2.17), to write (3.1) as the dissipation inequality

$$\begin{aligned}\Sigma \ \Big[\rho_a(\dot{\psi}_a + \eta_a\dot{\theta}) &- \underset{\sim}{\sigma}_a \cdot \underset{\sim}{D}_a + \overset{+}{\underset{\sim}{m}}_a \cdot \underset{\sim}{u}_a \\ &+ \rho_a\eta_a\underset{\sim}{u}_a \cdot \nabla\theta + \underset{\sim}{q}_a \cdot \nabla\theta/\theta\Big]/\theta \leq 0\end{aligned} \tag{3.3}$$

It is important note, however, that this inequality is subject to some constraints on the motion of the phases. As a consequence of the requirement of incompressibility of each phase, balance of mass (2.14) implies that all changes in volume fractions must be related to the velocity field:

$$\dot{\phi}_a + \phi_a \nabla \cdot \underset{\sim}{v}_a = 0 \quad , \tag{3.4}$$

Furthermore, the mixture is saturated, and thus (2.2) implies that

$$\Sigma \ (\dot{\phi}_a - \nabla\phi_a \cdot \underset{\sim}{u}_a) = 0 \tag{3.5}$$

To account for these three constraints in (3.3), we utilize the method of Lagrange multipliers. That is, we define the Lagrange multipliers p_f, p_s, and π which with (3.4) and (3.5) permit the dissipation inequality to be expressed as

$$\begin{aligned}\Sigma \ \Big[\rho_a(\dot{\psi}_a + \eta_a\dot{\theta}) &- \underset{\sim}{\sigma}_a \cdot \underset{\sim}{D}_a + \overset{+}{\underset{\sim}{m}}_a \cdot \underset{\sim}{u}_s \\ &+ \rho_a\eta_a\underset{\sim}{u}_a \cdot \nabla\theta + \underset{\sim}{q}_a \cdot \nabla\theta/\theta \\ &- p_a \ (\dot{\phi}_a + \phi_a \nabla \cdot \underset{\sim}{v}_a) + \pi(\dot{\phi}_a - \nabla\phi_a \cdot \underset{\sim}{u}_a)\Big]/\theta \leq 0\end{aligned}$$

Noting (2.5) and (2.6), this inequality can be rearranged to yield the desired result:

$$\begin{aligned}\Sigma \ \Big[\rho_a(\dot{\psi}_a + \eta_a\dot{\theta}) &- \underset{\sim}{\tau}_a \cdot \underset{\sim}{D}_a + \overset{+}{\underset{\sim}{n}}_a \cdot \underset{\sim}{u}_a \\ &+ \rho_a\eta_a\underset{\sim}{u}_a \cdot \nabla\theta + \underset{\sim}{q}_a \cdot \nabla\theta/\theta \\ &+ (\pi - p_a) \ \dot{\phi}_a\Big]/\theta \leq 0 \ ,\end{aligned} \tag{3.6}$$

where $\underset{\sim}{\tau}_a$ is the <u>extra stress</u> and $\overset{+}{\underset{\sim}{n}}_a$ is the <u>extra momentum transfer</u>,

$$\underset{\sim}{\tau}_a = \underset{\sim}{\sigma}_a + \phi_a p_a \underset{\sim}{1} \quad , \tag{3.7}$$

$$\overset{+}{\underset{\sim}{n}}_a = \overset{+}{\underset{\sim}{m}}_a - \pi\nabla\phi_a \quad , \tag{3.8}$$

respectively. The Lagrange multipliers p_f, p_s, and π have specific physical interpretations; that is, they represent the reaction forces required to maintain the constraints on the motion. For example, the quantities p_f and p_s are the hydrostatic pressures associated with the incompressibility of each phase. The quantity π is also a pressure which is required to insure that the mixture remains saturated for all time. Since this pressure must act at the interface between the dispersed and continuous phases and affect the momentum exchange as is suggested by (3.8), π is called the <u>interface pressure</u> [6]. In a sense, π reflects the pressure distribution over the surface of a particle as it moves through the flow field (cf. Figure 1).

The results (3.7), along with (3.4), indicate that the equations of motion (2.11) and (2.12) can be replaced by

$$\grave{\phi}_a = - \phi_a \nabla \cdot \underset{\sim}{v}_a \quad , \tag{3.9}$$

$$\rho_a \grave{\underset{\sim}{v}}_a = - \phi_a \nabla p_a + \nabla \cdot \underset{\sim}{\tau}_a + \rho_a \underset{\sim}{b}_a + \underset{\sim}{n}_a^+ + (\pi - p_a) \nabla \phi_a \ . \tag{3.10}$$

An immediate consequence of (3.8), (2.24), and (2.2) is

$$\underset{\sim}{n}_f^+ = - \underset{\sim}{n}_s^+ \ . \tag{3.11}$$

Finally, since the constituents have a common temperature, (2.16) can be replaced with the energy equation for the mixture (2.17).

Two observations should be made with regard to the revised form of the momentum equations (3.10). First, it is apparent that in utilizing the saturation constraint in our thermodynamic arguments, we have neatly resolved any further debate on whether the volume fraction should be inside or outside the gradient in the pressure gradient term. The form shown is the one desired on physical grounds. Second, the equations delineate the principle sources of momentum; that is, due to pressure gradients, body (i.e., gravitational) forces, viscous forces represented by $\underset{\sim}{\tau}_a$, lift and drag forces represented by $\underset{\sim}{n}_a^+$, and diffusive forces represented by $(\pi - p_a)\nabla\phi_a$. This latter term provides the mechanism by which we can account for the contact pressure at high concentrations (cf. Anderson and Jackson [14]) and the diffusive pressure associated with the Brownian motion of small particles at dilute concentrations (cf. Drew [18]). It should be noted that Ishii [6] also recorded a result of the form (3.10).

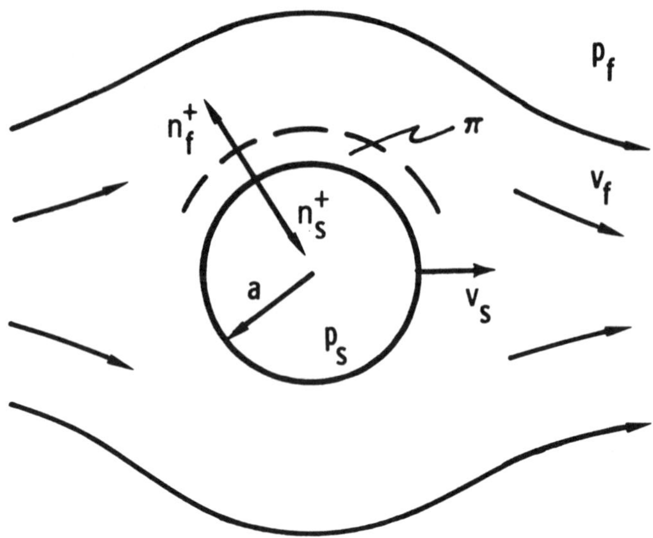

Figure 1. The pressure fields and forces affecting the motion of a single particle, with radius a, in a flow field.

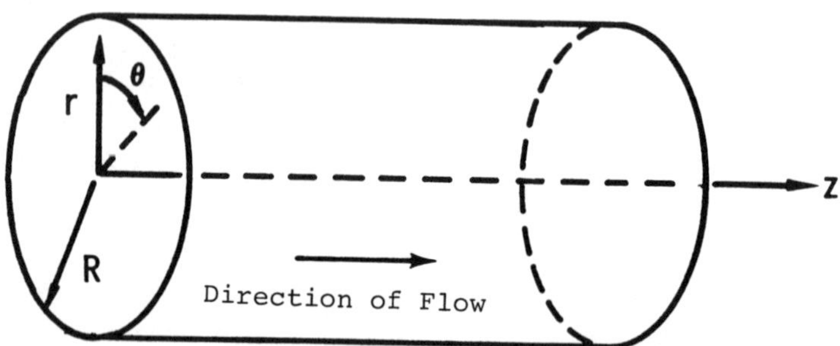

Figure 2. The geometry for steady Poiseuille flow.

4. CONSTITUTIVE EQUATIONS FOR FLUID-PARTICLE FLOWS

In this section, we consider the general features of constitutive equations for fluid-particle mixtures which appear appropriate for arbitrary concentration. In addition to the assumption of incompressibility of each phase, we suppose that both the fluid and the dispersed particles are homogeneous and isotropic. We further require the particles to be spherical and of uniform size with radius a.

In formulating these general expressions, we appeal to two basic principles; material frame-indifference [22] and phase separation [7]. The first states that any dependence of a response function on the velocity $\underset{\sim}{v}_a$ and its gradient $\underset{\sim}{L}_a$ must be expressed in terms of the diffusion velocity $\underset{\sim}{u}_a$ and the rate of deformation $\underset{\sim}{D}_a$. The principle of phase separation expresses the idea that in a mixture of discrete phases, the free energy ϕ_a, the entropy η_a, the extra stress $\underset{\sim}{\tau}_a$, and the heat flux $\underset{\sim}{q}_a$ of a given phase only depends on the properties and motion of that phase. However, the extra momentum and energy exchange, $\overset{+}{\underset{\sim}{n}}_a$ and $\overset{+}{e}_a$, occurring between the two phases may depend on the properties and motions of both phases.

With this in mind, let us first consider the constitutive equations for the fluid. Physically, we would expect the extra stress in a viscous fluid, $\underset{\sim}{\tau}_f$, to be determined by the rate of deformation $\underset{\sim}{D}_f$ and the temperature θ. In addition, it is well-known that the local heat flux $\underset{\sim}{q}_f$ is determined by the temperature gradient $\nabla\theta$ and that the free energy ϕ_f is a function only of the temperature. Thus, we assume that for the continuous fluid phase,

$$\phi_f = \phi_f(\theta) \quad , \qquad \eta_f = \eta_f(\theta) \ , \tag{4.1}$$

$$\underset{\sim}{\tau}_f = \underset{\sim}{\tau}_f(\theta,\ \phi_f,\ \underset{\sim}{D}_f) \quad , \tag{4.2}$$

$$\underset{\sim}{q}_f = \underset{\sim}{q}_f(\theta,\ \phi_f,\ \underset{\sim}{D}_f,\ \nabla\theta) \quad . \tag{4.3}$$

In the case of the solid particles, results on granular material flows [23,24] clearly suggest that at high concentrations, the extra stress $\underset{\sim}{\tau}_s$ can be viscous dominated due to particle collisions. Hence, we would again expect a dependence on the rate of deformation $\underset{\sim}{D}_s$. In fact, we would expect the constitutive equations of the solid to to mirror those of the fluid with perhaps one exception.

The existence of a static pressure due to interparticle contact forces at high concentrations suggests letting the free energy ψ_s also depend on the solid volume fraction ϕ_s. Thus, we assume for the dispersed solid phase,

$$\psi_s = \psi_s(\theta,\phi_s),\ \eta_s = \eta_s(\theta,\phi_s), \tag{4.4}$$

$$\underset{\sim}{\tau}_s = \underset{\sim}{\tau}_s\ (\theta,\phi_s,\underset{\sim}{D}_s), \tag{4.5}$$

$$\underset{\sim}{q}_s = \underset{\sim}{q}_s\ (\theta,\phi_s,\underset{\sim}{D}_s,\nabla\theta). \tag{4.6}$$

It is of interest to observe that constitutive equations of this type have been employed previously to study fluidized beds (cf. Murray [13] and Anderson and Jackson [14]). However, it is also important to note that these constitutive equations hold equally well for low concentrations and serve to describe the flow of dilute suspensions. In this case, assigning a viscous stress $\underset{\sim}{\tau}_s$ to the particles is motivated by the fact that the mere presence of the particles in the fluid results in an increased viscosity of the mixture [25].

Finally, we consider constitutive equations for momentum and energy exchange. Studies of the motion of a single particle in a viscous fluid have shown that the particle is subjected to a variety of forces [1]: drag forces due to the viscous effects in the neighborhood of the particle, lift forces which can move the particle laterally in the flow and result from the non-uniform pressure distribution around the particle, higher-order viscous forces called Faxen forces which can result in particle rotation, and local inertial forces, often referred to as virtual mass effects, which are associated with the acceleration of the fluid past the particle. Here we will simplify our analysis and consider only the consequences of lift and drag forces; that is, we assume Faxen forces and virtual mass effects are negligible.† It is well-known from Stokes flow that the drag on a particle is proportional to the slip velocity $\underset{\sim}{v}_s - \underset{\sim}{v}_f$. Thus, the extra momentum exchange $\overset{+}{\underset{\sim}{n}}_a$ must depend on both diffusion velocities $\underset{\sim}{u}_s$ and $\underset{\sim}{u}_f$. Saffmann's analysis [27] of the lift forces on a particle suggest that $\overset{+}{\underset{\sim}{n}}_a$ also depends on the rate of deformation of

† For studies of these effects, see Drew [18] and Drew, Cheng, and Lahey [26].

the fluid $\underset{\sim}{D}_f$. In addition, it is not unreasonable to expect a dependence on $\underset{\sim}{D}_s$ at high concentrations where collisions of particles are important. Thus, we shall assume the extra momentum exchange is given as

$$\underset{\sim}{n}_f^+ = \underset{\sim}{n}_f^+(\theta, \phi_s, \underset{\sim}{u}_f, \underset{\sim}{u}_s, \underset{\sim}{D}_f, \underset{\sim}{D}_s) \qquad (4.7)$$
$$= - \underset{\sim}{n}_s^+ \quad .$$

For the corresponding energy exchange, we assume simply that

$$e_f^+ = e_s^+ = 0 \qquad (4.8)$$

This is motivated by the fact that the common temperature θ can be computed from the energy equation for the mixture and the energy exchange e_a^+ is no longer an integral part of the theory.

With these constitutive equations, it is not difficult to see from (2.2), (3.9), (3.10), and (2.17) that there are only ten equations to solve for the twelve unknowns: θ, ϕ_f, ϕ_s, $\underset{\sim}{v}_s$, $\underset{\sim}{v}_f$, p_f, p_s, π. This dilemma describes the closure problem for two-phase flows. In the case of fluid-particle flows, the typical solution to this problem has been to set all the pressures equal [eg., 7,8,21]:†

$$p_f = p_s = \pi = p \quad . \qquad (4.9)$$

This proposal of pressure equilibrium leads to two conclusions. First, it is apparent from (3.10) that diffusive forces are no longer of consequence whether they be due to particle contacts at high concentrations or due to Brownian motion in dilute systems. Second, this assumption can reresult in ill-posed initial-value problems; i.e., problems with unstable solutions [28]. Both of these results are aphysical and not very appealing. In view of this, we consider further an idea first suggested by Stuhmiller [28] in which the interface pressure was associated with the pressure distribution over a particle. For example, Lamb's solution for the pressure distribution over a single isolated particle [29] in an inviscid fluid may be interpreted as suggesting a relation between the interface pressure π, the fluid

† If the particles are liquid or gas bubbles, models including surface tension are sufficient for closure; cf. Drew and Lahey [8]. Another approach has been to use additional microstructural force balance equations arising in granular material theories; cf. Nunziato, Passman, and Walsh [9-12].

pressure p_f, and the relative motion of the particle with respect to the fluid:

$$\pi = p_f + \omega(\mathbf{u}_f - \mathbf{u}_s)\cdot(\mathbf{u}_f - \mathbf{u}_s).$$

In terms more descriptive than precise, the interface pressure π represents the integrated average of the pressure distribution over the particle. Expanding this notion to a system of particles moving through a viscous fluid, it is conceivable that, in addition to dynamic pressure effects, the pressure distribution on any given particle (represented by π) may also be influenced by local volume fraction changes due to particles moving in and out of close proximity of it (bulk viscous effects), and by lift and drag forces acting on neighboring particles. On the other hand, the solid pressure p_s will accommodate the forces acting on the particle surface including the pressure distribution (represented by π), interparticle contact forces, and forces resulting from the collisions with fluid molecules (Brownian forces). This physical picture provides a reasonable basis for closure in that it suggests constitutive equations for the <u>diffusive pressures</u> $(\pi - p_a)$:

$$\pi - p_f = P_f(\theta, \phi_f, \dot{\phi}_f, \mathbf{u}_f, \mathbf{u}_s, \mathbf{D}_f, \mathbf{D}_s) \quad , \tag{4.10}$$

$$\pi - p_s = P_s(\theta, \phi_s, \dot{\phi}_s, \mathbf{u}_f, \mathbf{u}_s, \mathbf{D}_f, \mathbf{D}_s) \quad . \tag{4.11}$$

With the response of the mixture completely prescribed, we can now record the restrictions imposed on the constitutive equations by the dissipation inequality (3.6). In particular, if the mixture free energy ψ and entropy η are defined by

$$\rho\psi = \rho_f\psi_f + \rho_s\psi_s \quad , \tag{4.12}$$

$$\rho\eta = \rho_f\eta_f + \rho_s\eta_s \quad , \tag{4.13}$$

then it follows from (3.6), (3.11), (4.1) and (4.4) that

$$\psi = \psi(\theta, \phi_s) \quad , \tag{4.14}$$

$$\eta = -\,\partial\psi/\partial\theta \ , \tag{4.15}$$

and

$$\begin{aligned} &\boldsymbol{\tau}_f\cdot\mathbf{D}_f + \boldsymbol{\tau}_s\cdot\mathbf{D}_s - \mathbf{n}_f^{+}\cdot(\mathbf{v}_f - \mathbf{v}_s) - (\bar{\mathbf{q}}\cdot\nabla\theta)/\theta \\ &\quad - (\pi - p_f)\dot{\phi}_f - (\pi - p_s + \beta_s)\dot{\phi}_s \geq 0 \end{aligned} \tag{4.16}$$

where

$$\bar{\mathbf{q}} = \Sigma\ \mathbf{q}_a + \rho_a\theta\ (\partial\psi_a/\partial\theta + \eta_a)\mathbf{u}_a \quad , \tag{4.17}$$

and

$$\beta_s = \rho_s(\partial\psi_s/\partial\phi_s) \ . \tag{4.18}$$

It is important to note that the inequality (4.16) shows four sources of dissipation in a fluid-particle mixture; viscous effects associated with both the fluid and solid phases, drag forces on the particles, heat conduction in each phase, and pressure non-equilibrium due to the hydrodynamic flow field around the particles. Obviously, when the mixture is at rest ($\underset{\sim}{v}_f = \underset{\sim}{v}_s = 0$) and the temperature is uniform ($\nabla\theta = 0$), (4.16) is trivially satisfied and it can be shown that

$$\underset{\sim}{\tau}_f = \underset{\sim}{\tau}_s = \underset{\sim}{0} \ , \tag{4.19}$$

$$\underset{\sim}{n}_f^+ = \underset{\sim}{n}_s^+ = \underset{\sim}{0} \ , \tag{4.20}$$

$$\underset{\sim}{q}_f = \underset{\sim}{q}_s = \underset{\sim}{0} \ , \tag{4.21}$$

$$\pi = p_f, \ p_s = p_f + \beta_s \ . \tag{4.22}$$

The result (4.22) is significant in that it shows that even when the phases are at rest, the pressures are not all equal. Rather the solid pressure differs from that in the continuous fluid phase by the amount β_s which we call the <u>equilibrium diffusive pressure</u>. Physically, it is clear from (4.22) that β_s accounts for interparticle contact forces in mixtures with high particulate concentrations. On the other hand, for dilute suspensions, β_s is related to the 'thermodynamic' force associated with the translational Brownian motion of particles whose velocity fluctuations result in zero mean motion. This will become more apparent during our discussion of Brownian motion in Section 6.

5. ISOTHERMAL FLOWS OF DILUTE SUSPENSIONS

Up to this point, our discussion of fluid-particle mixtures has been rather general and we have attempted to establish some guidelines for developing a theory for such systems. However, the forms of the constitutive equations discussed are far too general to solve any specific problems which might delineate the salient features of the theory. Therefore, it would be instructive to consider a simple model and we do that here. In particular, the remainder of this paper will focus on isothermal flows of dilute suspensions.

In this case, the constitutive equation for the free energy of the fluid is a constant and, assuming the fluid is linearly viscous, the extra stress $\underset{\sim}{\tau}_f$ is of the form

$$\underset{\sim}{\tau}_f = 2(1-\phi_s)\mu_f \underset{\sim}{D}_f \tag{5.1}$$

where μ_f is the constant fluid viscosity. We also assume that the mixture is sufficiently dilute and the particles sufficiently small that Brownian motion may be important, but that the collisions between particles are negligible. As we shall see in Section 6, the effects of Brownian motion can be accounted for by letting the free energy of the particles depend on the volume fraction ϕ_s. Thus, for the solid particles, we have

$$\psi_s = \psi_s(\phi_s), \tag{5.2}$$

$$\underset{\sim}{\tau}_s = 2\phi_s \mu_s \underset{\sim}{D}_s \ . \tag{5.3}$$

where μ_s is a constant viscosity. In writing (5.1) and (5.3), we have omitted a term involving the bulk viscosity commonly encountered in theories of fluidized beds [13,14]. This, however, is motivated by the fact that we choose to view such effects as contributions to the diffusive pressures $(\pi - p_a)$. Thus bulk viscous effects will be part of our model for pressure differences (cf. equation (5.10)). Note that since $\underset{\sim}{\tau}_s$ is proportional to the volume fraction ϕ_s, this contribution to the total shear stress in the mixture represents a first-order correction for dilute systems.

Also, in the absence of collisions, the extra momentum exchange $\underset{\sim}{n}_a^+$ only needs to account for lift and drag forces and we assume

$$\underset{\sim}{n}_f^+ = -\phi_s \underset{\sim}{S}(\underset{\sim}{v}_f - \underset{\sim}{v}_s) = -\underset{\sim}{n}_s^+ \tag{5.4}$$

where the <u>drag tensor</u> $\underset{\sim}{S}$ is a linear, isotropic tensor function of $\underset{\sim}{D}_f$:

$$\underset{\sim}{S} = \alpha_1 \underset{\sim}{1} + 2\alpha_2 \underset{\sim}{D}_f \ . \tag{5.5}$$

The coefficients α_i may be scalar-valued functions of ϕ_s and the invariants of $\underset{\sim}{D}_f$; tr $\underset{\sim}{D}_f$, tr $\underset{\sim}{D}_f^2$. The drag forces are characterized by the diagonal of $\underset{\sim}{S}$, diag $\underset{\sim}{S}$, while $(\underset{\sim}{S} - \text{diag}\ \underset{\sim}{S})$ represents the lift forces. Of course, as $\phi_s \to 0$, α_1 must approach the low concentration limit for the drag on individual particles deduced from Batchelor's analysis [2] for settling spheres, i.e.,

$$\alpha_1 \to 9\mu_f(1 + 5.55\ \phi_s)/2a^2 \tag{5.6}$$

where a is the particle radius. Likewise, the coefficient α_2 is assumed to reduce to Saffman's result for the lift on a single particle as $\phi_s \to o$;

$$\alpha_2 \to 3(3.23)\ (\gamma_f\mu_f/|D_f|)^{1/2}/4\pi a. \tag{5.7}$$

It remains only to specify the diffusive pressures $\pi - p_f$ and $\pi - p_s$. Motivated by our discussion in Section 4, we write

$$\pi - p_f = \xi_f \dot{\phi}_f + \omega_f(\underset{\sim}{v}_f - \underset{\sim}{v}_s)\cdot(\underset{\sim}{v}_f - \underset{\sim}{v}_s) \tag{5.8}$$

$$\pi - p_s = -\beta_s + \xi_s \dot{\phi}_s \tag{5.9}$$

where β_s is defined by (4.18). Physically, the terms whose coefficients are ξ_f, ξ_s account for pressure changes due to local changes in the volume fractions, and the term whose coefficient is ω_f accounts for the dynamic pressure on a particle due to local inertial effects. Clearly, we can combine (5.8) and (5.9), along with mass balance (3.9), to obtain the pressure difference relation:

$$p_s - p_f = \beta_s - (\xi_f + \xi_s)(1-\phi_s)\ \mathrm{tr}\ \underset{\sim}{D}_f + \xi_s \nabla\phi_s\cdot(\underset{\sim}{v}_f - \underset{\sim}{v}_s) + \omega_f(\underset{\sim}{v}_f - \underset{\sim}{v}_s)\cdot(\underset{\sim}{v}_f - \underset{\sim}{v}_s) \tag{5.10}$$

The form of this expression suggests that we call ξ_f, ξ_s, bulk viscosities.

Finally, using these constitutive equations, we can write a more useful form of the equations of motion (3.9) and (3.10). Assuming that the only body force is due to gravity ($\underset{\sim}{b}_a = -\underset{\sim}{g}$), (2.3) and (5.1) - (5.4) can be substituted into (3.9) and (3.10) and after some rearrangement, we have

$$\nabla\cdot[\phi_s \underset{\sim}{v}_s + (1-\phi_s)\underset{\sim}{v}_f] = 0 \tag{5.11}$$

$$\dot{\phi}_s + \phi_s\ \nabla\cdot\underset{\sim}{v}_s = 0 \tag{5.12}$$

$$\begin{aligned}\gamma_f(1-\phi_s)\ \dot{\underset{\sim}{v}}_f = &-(1-\phi_s)\nabla p_f - \gamma_f(1-\phi_s)\underset{\sim}{g} \\ &+ \mu_f\nabla\cdot((1-\phi_s)(\nabla\underset{\sim}{v}_f + \nabla\underset{\sim}{v}_f^T)) \\ &- \phi_s\ \underset{\sim}{S}(\underset{\sim}{v}_f - \underset{\sim}{v}_s) - (\pi - p_f)\nabla\phi_s\end{aligned} \tag{5.13}$$

$$\begin{aligned}\gamma_s\phi_s\ \dot{\underset{\sim}{v}}_s = &-\phi_s\ \nabla p_s - \gamma_s\phi_s\underset{\sim}{g} \\ &+ \mu_s\ \nabla\cdot(\phi_s(\nabla\underset{\sim}{v}_s + \nabla\underset{\sim}{v}_s^T)) \\ &- \phi_s\ \underset{\sim}{S}(\underset{\sim}{v}_s - \underset{\sim}{v}_f) + (\pi - p_s)\nabla\phi_s\end{aligned} \tag{5.14}$$

where π, p_s, and p_f are related by (5.8) - (5.10) and $\underset{\sim}{S}$ is given by (5.5). Furthermore, it is not difficult to show that the dissipation inequality (4.16) requires

$$\mu_f > 0,\ \mu_s > 0,\ \xi_f > 0,\ \xi_s > 0\quad , \tag{5.15}$$

and the drag tensor $\underset{\sim}{S}$ to be positive semi-definite.

6. APPLICATION TO TRANSLATIONAL BROWNIAN MOTION

The classical theory of translational Brownian motion is concerned with the random migration of isolated colloidal particles (less than 10 μm) in a suspending fluid. This random motion is a result of the collisions between the particles and the molecules of the fluid which are undergoing thermal fluctuations [25,30]. In general the inertial forces acting on the particles are assumed small in comparison to the viscous forces and the resulting phenomenon is diffusive in nature. Recently, Batchelor [19] extended the classical theory to include the effect of the particles interacting hydrodynamically. In particular he showed that for neutrally buoyant particles the particle flux, due to the fluctuating velocity field, is the same as that which would be produced by a 'thermodynamic' force acting on each particle. Furthermore, this 'thermodynamic' force is equal to the gradient of the chemical potential of the particles; a result familiar to physical chemists but not previously applied to Brownian motion. Then, assuming that the drag force on the particles is the same as that for a suspension of identical particles falling under gravity, Batchelor [19] calculated a diffusivity λ, which in the present context, has the form

$$\lambda = (2a^2 K(\phi_s)/9\mu_f)\ \phi_s \gamma_s\ (1-\phi_s)^{-1}(\partial c_s/\partial \phi_s) \qquad (6.1)$$

where c_s is the chemical potential of the particles (per unit mass) and $K(\phi_s)$ is called the mobility coefficient. Values of $K(\phi_s)$ can be estimated either theoretically or from experimental observations of particulate sedimentation. On the basis of theoretical results for dilute suspensions [2], Batchelor suggested that

$$K(\phi_s) = 1 - 6.55\ \phi_s + O(\phi_s{}^2) \quad . \qquad (6.2)$$

What is most striking about Batchelor's results are the observations that the effects of Brownian motion can be described in the context of thermodynamics and that the dependence of a thermodynamic potential on the solid volume fraction is capable of capturing the essence of its diffusive character. For a dilute suspension, the Helmholtz free energy ψ_s and the chemical potential c_s are related by

$$c_s = \psi_s + p_s/\gamma_s \tag{6.3}$$

and, consequently, in permitting ψ_s to depend on ϕ_s, the theory outlined in the previous section should embody these effects. Although we have alluded to this previously, we demonstrate it more clearly here.

Consider a dilute suspension ($\phi_s << 1$) and assume that the viscous forces in the fluid and all inertial forces are negligible. Furthermore, neglect all second-order effects such as lift forces, dynamic pressure forces, etc. Then, the equations of (5.11) - (5.14), along with (5.8) - (5.10) and (5.5) reduce to

$$\nabla \cdot \underset{\sim}{v}_f = 0 , \tag{6.4}$$

$$\partial \phi_s/\partial t + \nabla \cdot (\phi_s \underset{\sim}{v}_s) = 0 , \tag{6.5}$$

$$\nabla p_f + \gamma_f \underset{\sim}{g} = 0 \tag{6.6}$$

$$\phi_s \nabla p_s + \gamma_s \phi_s \underset{\sim}{g} + \alpha_1 \phi_s (\underset{\sim}{v}_s - \underset{\sim}{v}_f) + \beta_s \nabla \phi_s = 0 , \tag{6.7}$$

where

$$p_s = p_f + \beta_s . \tag{6.8}$$

The interesting equation here is (6.7). Combining (6.6) - (6.8), we obtain an expression for the particle flux relative to the fluid:

$$\phi_s(\underset{\sim}{v}_s - \underset{\sim}{v}_f) = - (1/\alpha_1) \nabla(\phi_s \beta_s) + \phi_s (\gamma_f - \gamma_s)\underset{\sim}{g}/\alpha_1 . \tag{6.9}$$

Then, by (6.3), (4.18), and (6.8) it follows that

$$\nabla(\phi_s \beta_s) = \phi_s \gamma_s \nabla c_s ; \tag{6.10}$$

this represents the 'thermodynamic' force on the particles discussed by Batchelor [19] and suggests calling $\phi_s \beta_s$ the Brownian stress [30]. By making use of (6.10) and the low concentration limit for α_1 given by (5.6), (6.9) can be rewritten as

$$\phi_s(\underset{\sim}{v}_s - \underset{\sim}{v}_f) = - \lambda \nabla \phi_s + \phi_s(\gamma_f - \gamma_s)\underset{\sim}{g}/\alpha_1 \tag{6.11}$$

where λ is the diffusivity:

$$\begin{aligned} \lambda &= (1/\alpha_1) \, \partial(\phi_s \beta_s)/\partial \phi_s , \\ &= 2a^2(1-5.55\phi_s) \, \phi_s \gamma_s \, (\partial c_s/\partial \phi_s)/9\mu_f . \end{aligned} \tag{6.12}$$

This latter expression agrees with Batchelor's results (6.1) and (6.2) to within terms of order ϕ_s^3.

Using results from the statistical-mechanical theory of dilute solutions to evaluate the chemical potential c_s [19], (6.12) can be simplified to

$$\lambda = \kappa\theta \, (1 + 1.45 \, \phi_s)/6\pi\mu_f a \tag{6.13}$$

where κ is Boltzman's constant. For micron size particles, λ will have values of the order 10^{-6} mm^2/sec. As one would expect in Brownian motion, the diffusivity increases with decreasing particle size. It is also of interest to note that the diffusivity λ given by (6.13) increases with concentration for dilute systems. This may at first seem surprising; however, statistical-mechanical calculations of particle interactions have shown that this is due to the long-range repulsion of particles. Furthermore, the results are corroborated by experiment.†

Finally, equations (6.11), (6.5), and (6.4) combine to yield the familiar convection-diffusion equation governing the particle motion:

$$\partial\phi_s/\partial t = \nabla\cdot[\lambda\nabla\phi_s] - [\underset{\sim}{v}_f - (\gamma_s-\gamma_f)\underset{\sim}{g}/\alpha_1]\cdot\nabla\phi_s \ . \qquad (6.14)$$

The last term in (6.14) accounts for the forced convection of the particles by the fluid motion as well as that due to buoyancy. This diffusion equation was derived previously by Drew [18] in the context of a two-phase flow model and was subsequently used to study a centrifuge problem. However, Drew's development of (6.14) made no attempt to give the diffusivity λ a thermodynamic interpretation as we have here.

7. POISEUILLE FLOW OF DILUTE SUSPENSIONS

In order to further illustrate our theory for dilute suspensions, we next consider the problem of steady Poiseuille flow. This problem was chosen for two reasons; first, it is a shear flow which is convenient to analyze and it offers a basis on which to compare previous analyses [18,31]; and second, it represents the flow field encountered in a capillary viscometer and hence the solution will permit an explicit calculation of the rheological properties of the suspension which might prove useful in the analysis of experimental data (Section 8).

We are interested in the flow of a dilute suspension with neutrally buoyant particles through a fixed, infinite circular pipe of radius R. At any point on the axis of

---

† See the review by Russel [30].

symmetry, we can locate a triad of unit vectors $\{\underset{\sim}{e}_r, \underset{\sim}{e}_\theta, \underset{\sim}{e}_z\}$ aligned with the radial direction -r, the azimuth direction -θ, and the axis of symmetry -z. The geometry is shown in Figure 2.

We assume that the flow is steady and driven by a constant pressure gradient, dp/dz, in the axial direction where p is the force per unit area of the cross section of the pipe. Thus, p is related to the pressure in each phase by

$$p = (1-\phi_s)p_f + \phi_s p_s , \tag{7.1}$$

and we note that

$$dp/dz < 0 . \tag{7.2}$$

In the absence of gravitational forces, the path lines for such flows are straight and parallel to the axis. Consequently, the velocity fields can be represented by

$$\underset{\sim}{v}_f = v_f(r)\underset{\sim}{e}_z , \tag{7.3}$$

$$\underset{\sim}{v}_s = v_s(r)\underset{\sim}{e}_z . \tag{7.4}$$

In addition, we impose the boundary conditions

$$v_f(R) = v_s(R) = 0 , \tag{7.5}$$

$$v_f'(0) = v_s'(0) = 0 , \tag{7.6}$$

where

$$v_f'(r) = \partial v_f(r)/\partial r . \tag{7.7}$$

The condition (7.6) is equivalent to assuming that the shear stresses $\underset{\sim}{\tau}_f$, $\underset{\sim}{\tau}_s$ are continuous at r = 0; while (7.5) represents a no slip condition at the wall. Finally, we assume that the solid volume fraction ϕ_s, which measures the concentration of particles, only varies in the radial direction:

$$\phi_s = \phi_s(r) . \tag{7.8}$$

With these assumptions, mass balance for both the fluid and the solid phase, (5.11) and (5.12), is trivially satisfied. Moreover, balance of momentum, (5.13) and (5.14), reduces to

$$0 = -(1-\phi_s)(\partial p_f/\partial r) - \phi_s\alpha_2(v_f - v_s)v_f' - (\pi - p_f)\phi_s' \tag{7.9}$$

$$0 = -(1-\phi_s)(\partial p_f/\partial z) + \mu_f(r(1-\phi_s)v_f')'/r - \phi_s\alpha_1(v_f - v_s) \tag{7.10}$$

for the fluid phase, and

$$0 = -\phi_s(\partial p_s/\partial r) + \phi_s\alpha_2(v_f - v_s)v_f' + (\pi - p_s)\phi_s' \tag{7.11}$$

$$0 = -\phi_s(\partial p_s/\partial z) + \mu_s(r\phi_s v_s')'/r + \phi_s\alpha_1(v_f - v_s) \tag{7.12}$$

for the solid phase. In writing these expressions, we have

used the drag tensor $\tilde{S}$ given in (5.5). Finally, the diffusive pressures, (5.8) and (5.9), become

$$\pi - p_f = \omega_f \, (v_f - v_s)^2 \,, \tag{7.13}$$

$$\pi - p_s = - \beta_s \,. \tag{7.14}$$

Eliminating the interface pressure π, (7.13) and (7.14) assert that

$$p_s - p_f = \beta_s + \omega_f \, (v_f - v_s)^2 \,. \tag{7.15}$$

This set of equations can be written in a more manageable form by noting some properties of the pressure fields. It should be evident from (7.3), (7.4), (7.8), (7.15) and (7.1) that

$$\partial p_s / \partial z = \partial p_f / \partial z = dp/dz \,. \tag{7.16}$$

Then, adding (7.10) and (7.12), with (7.16), yields the result

$$\mu_f (r(1-\phi_s) v_f')'/r + \mu_s (\phi_s v_s')'/r = (dp/dz) \,. \tag{7.17}$$

Another useful result is obtained by multiplying (7.10) by ϕ_s and (7.12) by $(1-\phi_s)$ and subtracting:

$$\begin{aligned} \phi_s \, [r(v_f - v_s)']'/r - \phi_s \alpha_1 (v_f - v_s)/\mu_s \\ = - \phi_s \, (1/\mu_s - 1/\mu_f) \, (dp/dz) \\ + \phi_s' \, (\phi_s v_f' - (1-\phi_s) v_s') \,. \end{aligned} \tag{7.18}$$

Equations (7.17) and (7.18) represent a coupled set of ordinary differential equations for the velocities v_f, v_s, and it remains only to specify the volume fraction ϕ_s. The appropriate additional equation can be obtained by subtracting (7.9) and (7.11) and utilizing the pressure relations (7.13) and (7.14):

$$\begin{aligned} [\phi_s \omega_f (v_f - v_s)^2 - (1-\phi_s) \, \alpha_1 \, \lambda] \, \phi_s' \\ = \phi_s (v_f - v_s) \, [(2\omega_f - \alpha_2) v_f' - 2\omega_f v_s'] \end{aligned} \tag{7.19}$$

where we recall the definition of the Brownian diffusivity λ given in (6.12). In writing (7.18) and (7.19), we have neglected terms of order ϕ_s^2.

Before continuing further, it is important to emphasize two points. First, it is clear from (7.17) - (7.19) that even though they are highly coupled and nonlinear, the equations for the velocity fields are dominated by viscous and drag forces; while the equation governing the particle distribution is dominated by Brownian motion, dynamic pressure forces, and lift forces. The second point concerns boundary conditions. Equations (7.17) and (7.18), along with (7.5) and (7.6), constitute well-posed two-point boundary-value problems which can be solved for a given ϕ_s. However, no boundary condition has been given for the ordinary differential equation (7.19). Of course, we could specify a value of ϕ_s at some point of the flow; however, this is not very practical. Instead, it makes more sense to specify the mean concentration of the dispersed phase, ϕ^*, since this quantity would be controllable in any given experiment. Thus, we shall require that

$$(2/R^2) \int_0^R \phi_s(r)\, r\, dr = \phi^*. \tag{7.20}$$

Due to the nonlinearity of (7.17) - (7.19), it is appropriate to consider some specific cases. One case of interest is when the particle radius approaches zero ($a \to 0$). In this situation, we would physically expect perfect entrainment of the particles and the particles to move with the same velocity as the fluid. To show that the present theory does indeed predict this result, we utilize the low concentration limit for the drag coefficient α_1, (5.6), in (7.18). Taking the limit as $a \to 0$, it is easily seen that $v_s \to v_f$. With $v_s = v_f$, (7.19) asserts that the distribution of particles across the pipe is uniform. Furthermore, with $v_s = v_f$, (7.17) can be integrated along with the boundary conditions (7.5) and (7.6) to obtain the classical parabolic velocity profile

$$v_f(r) = v_s(r) = (r^2 - R^2)\,(dp/dz)/4\mu \tag{7.21}$$

where μ is the mixture viscosity defined by

$$\mu = (1-\phi_s)\,\mu_f + \phi_s \mu_s \,. \tag{7.22}$$

Recalling that the Einstein formula for the effective viscosity of a suspension [25],

$$\mu^* = \mu_f\,(1 + 2.5\,\phi_s + \ldots.)\,, \tag{7.23}$$

is also based on no slip between the phases ($v_s = v_f$), it is clear from a comparison with (7.22) that we can now evaluate the viscosity associated with the solid particles; that is, in the low concentration limit,

$$\mu_s = 3.5\,\mu_f \,. \tag{7.24}$$

It is interesting to contrast our results with those of Drew [18] who assumed the phase pressures were equal and set the viscosity μ_s to zero. In particular, he found that the <u>only</u> possible plane parallel flow with uniform particle distribution is plane Couette motion with the particles moving with the fluid. The conclusion that Couette flow is the only such flow would not seem, however, to be consistent with experience. This underscores the appealing nature of our results and it emphasizes the importance of pressure differences and of assigning viscous effects to the particles.

To explore in more detail the character of Poiseuille flow of suspensions, we return to the expressions (7.17) - (7.19) and obtain an approximate solution based on a linearization of equation (7.18) which governs the slip velocity, $v_f - v_s$. If terms involving the squares of velocity gradients are assumed to be small, then it follows from (7.19) that the last term of (7.18) can be neglected. In this case, (7.18) reduces to a modified Bessel equation of the form

$$[r(v_f-v_s)']'/r - \alpha_1(v_f-v_s)/\mu_s = -(1/\mu_s-1/\mu_f)(dp/dz) \ . \tag{7.25}$$

By virtue of the boundary conditions (7.5) and (7.6), this equation has the solution

$$v_s(r) = v_f(r) - (\mu_s-\mu_f)(dp/dz)(F(r)-1)/\alpha_1\mu_f \tag{7.26}$$

where

$$F(r) = I_o(r(\alpha_1/\mu_s)^{1/2})/I_o(R(\alpha_1/\mu_s)^{1/2}) \ . \tag{7.27}$$

The properties of the Bessel function I_o imply that $F(r) \leq 1$ for all $r \leq R$. Thus, by (7.24) and (7.2), it is clear that $v_s < v_f$ for all r. That is, in general, <u>the particles lag the fluid</u>. Moreover, noting the low concentration limits for α_1 and μ_s, the slip velocity $(v_f - v_s)$ is, to leading order, proportional to a^2; a result consistent with single particle analyses [31,32].†

---

† There are known examples involving colloidal particles in which, on the average, a suspended particle moves faster than the fluid. In these cases, the particles are large in comparison to the pipe radius and thus wall effects are important. Brownian motion also has a major influence on the flow field. See, for example, Brenner [32] and Brenner and Gaydos [33].

It is now a straightforward procedure to compute the appropriate fluid velocity $v_f(r)$ and particle distribution $\phi_s(r)$ from (7.17) and (7.19). Integrating (7.17) once, subject to the boundary condition (7.6), and using (7.26), we find that

$$v_f'(r) = [r+2\phi_s(\mu_s-\mu_f)\mu_s F'(r)/\alpha_1\mu_f]\ (dp/dz)/2\mu \tag{7.28}$$

where μ is the mixture viscosity (7.22) and

$$F'(r) = (\alpha_1/\mu_s)^{1/2}\ I_1(r(\alpha_1/\mu_s)^{1/2})/I_o(R(\alpha_1/\mu_s)^{1/2})\ . \tag{7.29}$$

Retaining terms only to order ϕ_s, integration of (7.28) with (7.5) yields

$$\begin{aligned} v_f(r) &= (r^2-R^2)\ (dp/dz)/4\mu_f \\ &\quad - (\mu_s/\mu_f-1)\ (dp/dz)\int_r^R [2\mu_s F'(r)/\alpha_1-r]\phi_s(r)dr/2\mu_f \end{aligned} \tag{7.30}$$

for the velocity profile in the continuous fluid phase. Notice that the solution (7.30) shows a correction to the classical parabolic profile which depends on both the drag forces and the viscous forces assigned to the dispersed particulate phase. Of course, to evaluate $v_f(r)$, the particle distribution must be known. This distribution can be computed directly from (7.19). Using (7.26) and (7.28) and again neglecting terms of order ϕ_s^2, (7.19) becomes

$$\begin{aligned} \phi_s'(r) &= \phi_s(F(r)-1)\ (dp/dz)^2\ (\mu_s-\mu_f)\ [\alpha_2 r \\ &\quad - 4\omega_f(\mu_s-\mu_f)F'(r)/\alpha_1]/2\alpha_1\mu_f\lambda\ . \end{aligned} \tag{7.31}$$

Assuming the parameters ω_f, α_1, and λ are independent of ϕ_s and noting that the lift coefficient α_2 is a function of r; i.e., in the low concentration limit (5.7),

$$\alpha_2(r) = (9.69\mu_f/4\pi a)\ (2\gamma_f/r|dp/dz|)^{1/2}\ , \tag{7.32}$$

integration of (7.31) yields

$$\begin{aligned} \phi_s(r) = A\ \exp\Big\{ &- \Big[(dp/dz)^2(\mu_s-\mu_f)/2\alpha_1\mu_f\lambda\Big] \\ &\cdot\int (F(r)-1)(4\omega_f F'(r)/\alpha_1-\alpha_2 r)dr \Big\} \end{aligned} \tag{7.33}$$

where the constant of integration A is chosen to satisfy (7.20).

The expression (7.33) includes the effect of Brownian diffusion on the particle distribution and points out an interesting interplay between lift forces and dynamic pressure forces. Again using the properties of Bessel functions, inspection of (7.31) clearly shows that there exist <u>three</u> extrema in the particle distribution; one at the wall ($r = R$), one on the axis ($r = 0$), and one off-axis at $r = r^*$ where r^* is the solution of[†]

$$4\omega_f F'(r^*)\ (\mu_s - \mu_f)/\alpha_1 = \alpha_2(r^*)\ r^* \ . \tag{7.34}$$

Furthermore, it can be shown that the off-axis extremum is a maximum; a result suggestive of the Segre-Silberberg effect where particles concentrate in an annular ring at an equilibrium position approximately 0.6 of the tube radius [5]. This equilibrium is achieved by a balance between lift forces and dynamic pressure forces. Whenever the particles lag the fluid, the lift forces tend to cause the migration of particles to the axis, the region of low shear; while the dynamic pressure forces cause migration toward the walls, the region of highest shear. Since (7.34) is independent of diffusivity λ, the equilibrium position for particle concentration is independent of Brownian motion. Brownian motion does, however, affect the overall shape of the particle distribution. Inspection of (7.31) clearly shows that for large Brownian diffusion (small particles), ϕ_s' becomes small and thus the distribution of particles tends to broaden. In essence, the particles have the opportunity to sample more positions in the cross-section and the resulting concentration profile becomes more uniform. On the other hand, for small Brownian diffusion where the flow is inertia-controlled, ϕ_s' becomes large and the profile is sharply-peaked around r^* where the particles have collected. A qualitative representation of the velocity and concentration profiles are shown in Figure 3. These results are in good qualitative agreement with those obtained by Ho and

---

† Of course, this extremum will only exist within the confines of the pipe, $r^* < R$, for appropriate values of the constants ω_f, μ_f, γ_f, a, and the pressure gradient (dp/dz).

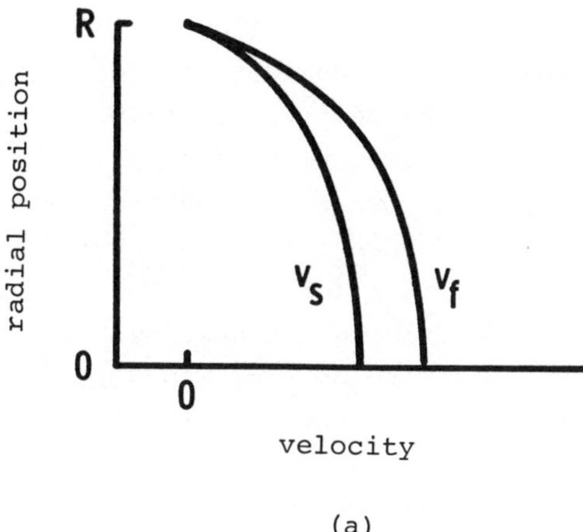

(a)

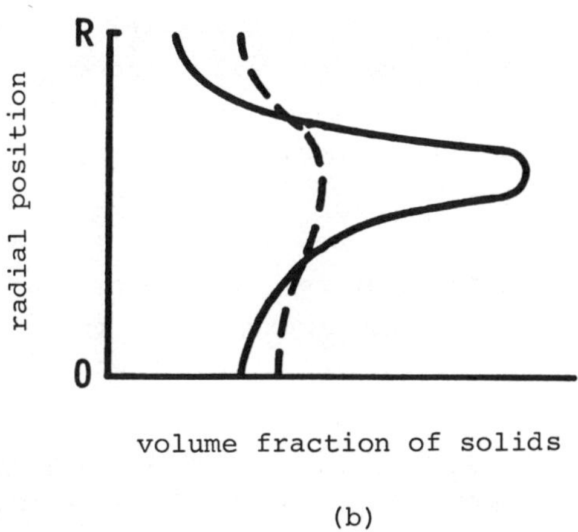

(b)

Figure 3. (a) Velocity profiles for the continuous fluid and dispersed particle phases in Poiseuille flow. The particles lag the fluid and there is no slip at the wall. (b) Particle distribution across the pipe; ——...inertia dominated flow, - - -...diffusion dominated flow.

Leal [31] in an analysis of the lateral migration of a single particle in Poiseuille flow.

Before leaving our consideration of the effect of Brownian motion on the particle distribution, we would like to mention one curious result. That is, the volume fraction solution in the neighborhood of the wall is significantly different for $\lambda = 0$ than for λ small but finite. To investigate the solution for $\lambda = 0$, it is necessary to go back to (7.19). Retaining the next highest order term in ϕ_s', it follows that with no Brownian diffusion, the slope of the profile goes to minus infinity at the wall in order to enforce the no slip condition (7.5). The rest of the profile remains essentially the same as for small diffusion (Figure 3). This result is difficult to interpret; however, it might suggest the existence of an exclusion zone near the wall which can only be occupied by particles too large to be influenced by Brownian forces. Whether or not this is the case remains an open question. Clearly the matter rests on higher-order expansions in the solid volume fraction and a more detailed study of wall effects.

Finally, it should be evident that if dynamic pressure forces are neglected ($\omega_f = 0$), the particle distribution $\phi_s(r)$ is significantly different. In particular, returning to (7.31), a straightforward analysis reveals ϕ_s is a maximum on the axis; that is, lift forces cause the particles to migrate to the center of the tube. This result is similar to that obtained previously by Drew [18].

8. RHEOLOGICAL PROPERTIES OF DILUTE SUSPENSIONS

As we have previously noted, there are many applications where it is of interest to know how dilute suspensions behave in response to applied forces and moving boundaries. In this situation, the main concern is the macroscopic behavior of the mixture and it sufficies to treat the suspension as a single fluid with an effective viscosity μ^*. Although experimental measurements of μ^* are not decisive, it does appear that the Einstein formula (7.23), assuming a uniform particle distribution ϕ^*, predicts the behavior of very dilute suspensions ($\phi^* < 0.05$). However, as has been noted by Happel and Brenner [1], Ho and Leal [31], Jeffrey and Acrivos [34], and Cox and Mason [35], measurements of effective viscosity can be clouded by wall effects (i.e., the type of viscometer and the ratio of the particle size to the characteristic length of the viscometer (a/R)), buoyancy effects, and apparent non-Newtonian behavior. A summary of effective viscosity data obtained for a variety of suspensions using capillary and cone-and-plate viscometers has been collected by Thomas [36] and compared with the predictions of the Einstein formula.

Here we will consider the macroscopic flow properties of dilute suspensions from a somewhat different point of view; that is, in the context of the two-phase mixture theory. In general, it is not possible to calculate explicitly an equivalent constitutive equation for the mixture and its effective viscosity. However, by recognizing that the flow in a capillary viscometer is steady Poiseuille flow, we can use the results of Section 7 to calculate explicitly the rheological properties of the suspension which would be measured by such a device. Furthermore, how these properties are influenced by the various forces acting on the particles will be apparent.

For a capillary tube viscometer, the effective viscosity μ^* of a suspension is determined from pressure gradient/volume flow rate data assuming the Newtonian relation [35]

$$\mu^* = \pi R^4 |dp/dz| / 8Q \quad . \tag{8.1}$$

The volume flow rate Q is defined in terms of the barycentric velocity of the mixture:

$$Q = \int_0^R 2\pi v \; r \; dr \; , \tag{8.2}$$

and, for neutrally buoyant particles, we can use (2.7) to arrive at

$$Q = \int_0^R 2\pi \left(v_f - \phi_s (v_f - v_s) \right) r \; dr. \tag{8.3}$$

Approximate expressions for the velocity of the continuous phase v_f and for the slip velocity $(v_f - v_s)$ in a Poiseuille flow are given in (7.30) and (7.26), respectively. Using these results, we can compute the corresponding flow rate Q and hence, the effective viscosity μ^* by (8.1). Carrying out the straightforward but tedious computations, we find that in the low concentration limit,

$$\begin{aligned}\frac{\mu_f}{\mu^*} - 1 = &- 5\phi^*/2 \\ &+ (5\phi^*/2)\Big(40a^2/9R^2 \\ &\quad - 40\sqrt{7}\,a^3\, I_1(3R/a\sqrt{7})/27R^3 I_o(3R/a\sqrt{7})\Big) \\ &+ 20\int_o^R \{4a^2F(r)\,(\phi_s(r)-\phi^*)/9 \\ &\quad + \int_r^R (\xi - 7a^2F'(\xi)/9)\,(\phi_s(\xi) - \phi^*)d\xi\}\ r\,dr/R^4 .\end{aligned} \tag{8.4}$$

where the particle distribution $\phi_s(r)$ is given by (7.33).

Each of the three terms on the right side of (8.4) can be interpreted physically. The first term is, of course, Einstein's correction to the viscosity and represents the mere presence of the particles in the suspension. The second term involves a correction proportional to the length scale a/R. This contribution to the viscosity is due to drag and viscous forces and results from the slip between phases. Clearly, this contribution includes wall effects. The last term, which depends on the difference between the particle distribution and its mean value, includes the effects of Brownian diffusion and lift and dynamic pressure forces. This term is definitely flow-rate dependent and hence we can conclude that the rheological behavior of the suspension is non-Newtonian.

If the particle distribution is nearly uniform, as would be the case if dynamic pressure and lift forces were negligible or if the Brownian diffusion were large, then the last term of (8.4) can be neglected and the effective viscosity becomes

$$\mu^* = \mu_f\Big(1 + (5\phi^*/2)\,(1 - 40a^2/9R^2)\Big) \tag{8.5}$$

to within terms of order $(a/R)^2$. This result is rather interesting in that it predicts that the effective viscocity decreases with increasing a/R; an observation which is well-known experimentally [35], but, as yet, not reproduced by single particle analyses.

Quite often, non-Newtonian rheological behavior also involves the existence of normal stress differences which are measurable and useful in characterizing the fluid medium of interest. To evaluate the normal stresses in the context of our theory of dilute suspensions, it is necessary to calculate the total stress in the mixture $\underset{\sim}{\sigma}$. Using the summing rule (2.18), along with (3.7), the constitutive equations for the extra stresses in each phase, (5.1) and (5.3), the equation for the pressure difference (5.10), we have

$$\begin{aligned} \underset{\sim}{\sigma} = &- p_f \underset{\sim}{1} + \phi_s \left[\beta_s - (\xi_f+\xi_s)\mathrm{tr}\, \underset{\sim}{D}_f + \xi_s\, \nabla\phi_s \cdot (\underset{\sim}{v}_f-\underset{\sim}{v}_s) \right. \\ &\left. + \omega_f\, (\underset{\sim}{v}_f-\underset{\sim}{v}_s)\cdot(\underset{\sim}{v}_f-\underset{\sim}{v}_s)\right] \underset{\sim}{1} + 2(1\phi_s)\mu_f\, \underset{\sim}{D}_f \\ &+ 2\, \phi_s\mu_s\, \underset{\sim}{D}_s - \gamma_s\phi_s(\underset{\sim}{v}_f-\underset{\sim}{v}_s)\otimes(\underset{\sim}{v}_f-\underset{\sim}{v}_s) \end{aligned} \tag{8.6}$$

to within terms of order ϕ_s^2. Note that if there is no slip between phases ($\underset{\sim}{v}_f=\underset{\sim}{v}_s$), then (8.6) does reduce to the constitutive equation for a Newtonian fluid with the viscosity given by the Einstein formula (7.23) and the pressure being the sum of the fluid pressure, the Brownian stress, and the bulk viscous stress. However, in general, such a reduction is not possible.

For a capillary viscometer, we can use the solution to the Poiseuille flow problem to evaluate the normal components of the stress $\underset{\sim}{\sigma}$. Of principle interest is the difference in the radial and axial components, $\sigma_{rr}-\sigma_{zz}$. Using (7.3) and (7.4), along with (8.6), the normal stress difference is

$$\sigma_{rr}-\sigma_{zz} = \gamma_s\phi_s(v_f-v_s)^2 \ . \tag{8.7}$$

Clearly, the normal radial stress exceeds the axial stress by an amount related to the dynamic pressure on the particle. Furthermore, this stress difference is maximum on the axis and vanishes at the boundary as a result of the no slip condition. This latter point is quite important for it asserts that a capillary viscometer will be unable to discern normal stress effects unless there is slip flow at the wall. Of course, this conclusion is only valid for dilute suspensions.

9. CONCLUSION

In this paper we have presented a theory for suspensions of particles in fluids based on the continuum theory of multiphase mixtures. The theory embodies three new features not included in previous theories of two-phase flows. First, we have utilized the ideas of Batchelor [19] to develop a thermodynamic setting in which to include translational Brownian motion. Second, we ascribed a viscosity to the particles even when the suspension is dilute; a viscosity which permits the recovery of Einstein's result when the particles move with the fluid. Third, we admitted the possibility of distinct pressures in each phase and modeled the diffusive pressures $\pi - p_a$ in terms of the Brownian pressure and the dynamic pressure acting on a particle due to local inertial effects. These aspects of the theory were then illustrated for dilute suspensions by considering two example problems: the motion of particles subject to Brownian and gravitational forces, and steady Poiseuille flow. In the first problem, we showed that the present theory leads to the familiar convection-diffusion equation governing the particle motion where the diffusivity is deriveable from a thermodynamic potential. The Poiseuille flow problem provided a useful vehicle in which to demonstrate the significance of the viscosity of the particles and the dynamic pressure effects. By including the viscosity of the particles, our analysis revealed that the particles will lag the fluid in most cases, a result which is physically appealing. Moreover, we showed that the maximum concentration of particles can occur off the axis of symmetry because of a balance between lift forces and dynamic pressure forces. This provides a possible explanation of the Segre-Silberberg effect in the context of mixture theory and the suggestion that the phenomenon is due to local inertial effects is consistent with the analysis of single particle motion.

The theory presented here appears to be quite rich in that it embodies many of the features necessary to describe the range of transport processes observed in dilute suspensions. Certainly, the theory offers the means of studying shear flows in more detail than in the past, especially those arising in viscometers. In the context of the present theory, it should be possible to further an understanding of the role of viscous forces, lift and drag forces, diffusive (Brownian) forces, and dynamic pressure forces in determining the rheological properties of the mixture, independent of viscometer geometry.

Finally, we believe that this work provides an additional step in the development of a unified approach to fluid-particle systems. The general structure of the present theory contains several of the theoretical models proposed previously for fluidized beds and thus should provide a useful basis for further studies of two-phase flows with arbitrary concentrations of particles.

REFERENCES

1. Happel, J. and H. Brenner, Low Reynolds Number Hydrodynamics, Prentice-Hall, New Jersey, 1965.
2. Batchelor, G. K., "Sedimentation in a Dilute Dispersion of Spheres," J. Fluid Mech. 52, 1972, 245-268.
3. Batchelor, G. K. and J. T. Green, "The Determination of the Bulk Stress in a Suspension of Spherical Particles to Order c^2," J. Fluid Mech. 56, 1972, 401-427.
4. Leal, L. G., "Particle Motions in a Viscous Fluid", Ann. Rev. Fluid Mech. 12, 1980, 435-476.
5. Segre, G. and A. Silberberg, "Behavior of Macroscopic Rigid Spheres in Poiseuille Flow, Parts 1 and 2," J. Fluid Mech. 14, 1962, 115-157.
6. Ishii, M., Thermo-Fluid Dynamic Theory of Two-Phase Flow, Eyrolles, Paris, 1975.
7. Drew, D. A. and L. A. Segel, "Averaged Equations for Two-Phase Flows," Studies in Appl. Math. L, 1971, 205-231.
8. Drew D. A. and R. T. Lahey, "Application of General Constitutive Principles to the Derivation of Multidimensional Two-Phase Flow Equations," Int. J. Multiphase Flow 5, 1979, 243-264.
9. Passman, S. L., "Mixtures of Granular Materials," Int. J. Engng. Sci. 15, 1977, 117-129.

10. Nunziato, J. W. and E. K. Walsh, "On Ideal Multiphase Mixtures with Chemical Reactions and and Diffusion," Arch. Rational Mech. Anal. 73, 1980, 285-311.
11. Nunziato, J. W. and S. L. Passman, "A Multiphase Mixture Theory for Fluid-Saturated Granular Materials," in: Mechanics of Structured Media, Elsevier, Amsterdam, 1981, 243-254.
12. Passman, S. L., J. W. Nunziato, and E. K. Walsh, "A theory for Multiphase Mixtures," in: Rational Thermodynamics (C. Truesdell, ed.), Springer, Berlin, 1983.
13. Murray, J. D., "On the Mathematics of Fluidization, Part 1," J. Fluid Mech. 21, 1965, 465-493.
14. Anderson, T. B. and R. Jackson, "A Fluid Mechanical Description of Fluidized Beds," I & E C Fundametals 6, 1967, 527-539.
15. Bedford, A. and C. D. Hill, "A Mixture Theory for Particulate Sedimentation with Diffusivity," AIChE J. 23, 1977, 403-404.
16. Hill, C.D., A. Bedford, and D. S. Drumheller, "An Application of Mixture Theory to Particulate Sedimentation," J. Appl. Mech., 102, 1980, 261-265.
17. Thacker, W. C. and J. W. Lavelle, "Two-Phase Flow Analysis of Hindered Settling," Phys. Fluids 20, 1977, 1577-1579.
18. Drew, D. A., "Two-Phase Flows: Constitutive Equations for Lift and Brownian Motion and Some Basic Flows," Arch. Rational Mech. Anal. 62, 1976, 149-163.
19. Batchelor, G. K., "Brownian Diffusion of Particles with Hydrodynamic Interaction," J. Fluid Mech. 74, 1976, 1-29.
20. Truesdell, C., Rational Thermodynamics, McGraw-Hill, New York, 1969.
21. Soo, S. L., Fluid Dynamics of Multiphase Systems, Blaisdell, Massachusetts, 1967.
22. Truesdell, C., and W. Noll, "The Non-Linear Field Theories of Mechanics," in: Handbuch der Physik (S. Flugge, ed.), Springer, Berlin, 1965.

23. McTigue, D. F., "A Nonlinear Continuum Model for Flowing Granular Materials," Ph.D Dissertation, Stanford University, 1979.

24. Savage, S. B. and D. J. Jeffrey, "The Stress Tensor in a Granular Flow at High Shear Rates," J. Fluid Mech. 110, 1981, 255-272.

25. Einstein, A., Investigations on the Theory of Brownian Movement, Dover, 1956.

26. Drew, D. A., L. Y. Cheng, and R. T. Lahey, "The Analysis of Virtual Mass Effects in Two-Phase Flow," Int. J. Multiphase Flow 5, 1979, 232-242.

27. Saffman, P. G., "The Lift on a Small Sphere in a Slow Shear Flow," J. Fluid Mech. 22, 1965, 383-400.

28. Stuhmiller, J. H., "The Influence of Interfacial Pressure Forces on the Character of Two-Phase Flow Model Equations", Int. J. Multiphase Flow 3, 1977, 551-560.

29. Lamb, H., Hydrodynamics, Dover, New York, 1945.

30. Russel, W. B., "Brownian Motion of Small Particles Suspended in Liquids," Ann. Rev. Fluid Mech. 13, 1981, 425-455.

31. Ho, B. P. and L. G. Leal, "Inertial Migration of Rigid Spheres in Two-Dimensional Unidirectional Flows," J. Fluid Mech. 65, 1974, 365-400.

32. Brenner, H., "Dynamics of Neutrally Buoyant Particles in Low Reynolds Number Flows," Proceedings of the International Symposium on Two-Phase Systems, 1972, 509-574.

33. Brenner, H., and L. T. Gaydos, "The Constrained Brownian Movement of Spherical Particles in Cylindrical Pores of Comparable Radius," J. Coll. Interface Sci. 58, 1977, 312-356.

34. Jeffrey, D. J. and A. Acrivos, "The Rheological Properties of Suspensions of Rigid Particles," AIChE J. 22, 1976, 417-432.

35. Cox, R. G. and S. G. Mason, "Suspended Particles in Fluid Flow Through Tubes," Ann. Rev. Fluid Mech. 3, 1971, 291-316.

36. Thomas, D. G., "Transport Characteristics of Suspensions: VIII. A Note on the Viscosity of Newtonian Suspensions of Uniform Spherical Particles," J. Coll. Sci. 20, 1965, 267-277.

Acknowledgement. The author would like to express his sincere thanks to D. F. McTigue for his thought-provoking ideas on this subject and his willingness to share them. Thanks are also due to D. B. Hayes and S. L. Passman for the many instructive conversations we have had on mixture theories, and to H. Brenner and W. Russel for their comments at the Symposium which helped to improve the final manuscript. This work was supported by the U. S. Department of Energy under contract number DE-AC04-76DP00789.

Jace W. Nunziato
Sandia National Laboratories
Fluid Mechanics and Heat
Transfer Division I
Albuquerque, NM 87185

Mixture Theory for Turbulent Diffusion of Heavy Particles

D. F. McTigue

The diffusion of negatively buoyant particulate material in turbulent flows has been extensively studied, most notably in the context of transport in the atmosphere and natural waters. Many of these studies simply assume at the outset that a Fickian gradient-diffusion model is appropriate for the turbulent mass flux. Motivation for this approach has been sought in the equations of motion in a number of different ways.

One line of reasoning assumes that the particle concentration is governed by a convection-diffusion equation (e.g., [11], pp. 263-264). Reynolds decomposition and averaging applied to the convective term yields a correlation that can be identified with a turbulent mass flux. Another approach starts with either the fluid phase or the dispersed particulate phase mass balance, applies the turbulence averaging, and again results in an equation of similar form (e.g., [10]). However, as Monin and Yaglom [11] note, these developments are appropriate only for a "passive" mixture; that is, one in which the fluid-particle interaction does not affect the dynamics of the flow. In fact, there are no considerations of dynamics in these models, and it should be obvious that the sedimentation problem, involving body forces, can not be treated within such schemes.

THEORY OF DISPERSED MULTIPHASE FLOW

ISBN 0-12-493120-0

A number of studies have recognized the need to include momentum considerations in order to properly represent certain two-phase flows. Large contributions have been made by, among others, Hinze [5,6], Soo and coworkers (e.g., [14]), and Drew [3]. However, none of these studies identifies both buoyancy effects and the source of the diffusive flux in the manner presented here.

This paper examines the classical sedimentation problem in view of the continuum theory of mixtures. Stated briefly, the central object is to identify the source of the diffusive flux that balances the gravitational settling flux of negatively buoyant particles in a steady, uniform mean flow. It is desirable to seek the actual driving mechanism for this phenomenon in the _interaction_ of the fluid and particulate phases, rather than to simply build "turbulent diffusion" into the governing equations in some _ad hoc_ fashion. Mixture theory offers a powerful tool for examination of this problem. It provides a rigorous axiomatic framework for the dynamics of two-phase flow, explicitly including, for example, exchanges of momentum between the phases. The theory very clearly demarcates the behavior of the fluid phase, the dispersed particulate phase, and the total mixture.

The theoretical treatment presented here duplicates that of an earlier paper [10], correcting several minor errors in that version. The principal result of that development was the identification of turbulent diffusion with a correlation of concentration and fluid velocity fluctuations, arising from averaging of the drag interaction term in the momentum balances. In addition, the present work develops more thoroughly the analogy between the turbulent mass flux which emerges from the mixture theory and the turbulent momentum flux (Reynolds stress) familiar from single-phase flows. An exact correspondence is shown between the two correlations for both linear (Boussinesq) and nonlinear (mixing length) models, in both inner and outer regions of a wall-bounded shear flow. Finally, more data are presented in support of this view of mass transport as a multilayer problem identical in structure to the conventional approach in momentum transport. For the Boussinesq model, the addition of a third layer is proposed and tested, suppressing the diffusive mass flux near the free surface in accord with experimental observations.

Theory

The foundations of the continuum theory of mixtures are presented by Truesdell [15]. Atkin and Craine [1,2] give a general review and cite numerous applications. The development presented here follows closely that of Drew and Segel [4] and Drew [3].

Both the fluid and the dispersed particulate phases are assumed to behave as continuous media. Variations in the concentration of discrete particles are thus manifested as (smooth) variations in the density of each phase. For incompressible fluid and solid, the bulk densities of the two phases are $\rho^f(1-c)$ and $\rho^s c$, where ρ^f = fluid density = constant, ρ^s = solid density = constant, and c = concentration or volume fraction of solids. The mass balances then become

$$\frac{\partial}{\partial t}(1-c) + \nabla\cdot\left[(1-c)\underset{\sim}{u}^f\right] = 0 \qquad (1)$$

$$\frac{\partial c}{\partial t} + \nabla\cdot(c\underset{\sim}{u}^s) = 0 \qquad (2)$$

where $\underset{\sim}{u}^f$ = fluid velocity and $\underset{\sim}{u}^s$ = dispersed particulate phase velocity.

The momentum balances are

$$\rho^f(1-c)\left(\frac{\partial \underset{\sim}{u}^f}{\partial t} + \underset{\sim}{u}^f\cdot\nabla\underset{\sim}{u}^f\right) = \nabla\cdot\left[(1-c)\underset{\approx}{T}^f\right] + \rho^f(1-c)\underset{\sim}{g} - \underset{\sim}{f} \qquad (3)$$

$$\rho^s c\left(\frac{\partial \underset{\sim}{u}^s}{\partial t} + \underset{\sim}{u}^s\cdot\nabla\underset{\sim}{u}^s\right) = \nabla\cdot(c\underset{\approx}{T}^s) + \rho^s c\underset{\sim}{g} + \underset{\sim}{f} \qquad (4)$$

where $\underset{\approx}{T}^f$ = fluid phase stress, $\underset{\approx}{T}^s$ = dispersed particulate phase stress, $\underset{\sim}{g}$ = gravitational vector, and $\underset{\sim}{f}$ = force of the fluid on the dispersed phase per unit volume. Note that the stresses, $\underset{\approx}{T}^f$ and $\underset{\approx}{T}^s$ are interpreted as force per unit area of the given phase.

Closure for equations (1) to (4) is obtained by posing constitutive equations for the stresses, $\underset{\approx}{T}^f$ and $\underset{\approx}{T}^s$, and the interaction force, $\underset{\sim}{f}$. The usual choice of properly invariant independent variables is

$$c,\ \nabla c,\ \underset{\sim}{u}^f - \underset{\sim}{u}^s,\ \underset{\approx}{D}^f,\ \underset{\approx}{D}^s,\ \underset{\approx}{W} \tag{5}$$

where $\underset{\approx}{D}^f = 1/2(\nabla\underset{\sim}{u}^f + \underset{\sim}{u}^f\nabla)$, $\underset{\approx}{D}^s = 1/2(\nabla\underset{\sim}{u}^s + \underset{\sim}{u}^s\nabla)$, and $\underset{\approx}{W} = 1/2[(\nabla\underset{\sim}{u}^f - \underset{\sim}{u}^f\nabla) - (\nabla\underset{\sim}{u}^s - \underset{\sim}{u}^s\nabla)]$. In the present special case of incompressible constituents, the dependence on c and ∇c accounts for dependence on the densities and density gradients of both phase.

Because concern is limited here to flows in which inertia dominates the momentum transfer, it is assumed at the outset that the rate dependent terms, $\underset{\approx}{D}^f$, $\underset{\approx}{D}^s$, and $\underset{\approx}{W}$ can be neglected. That is, the characteristic material time scales are assumed to be long compared to the characteristic inertial time, L/U, where L and U are length and velocity scales. The assumption is certainly valid at low concentrations ($c \ll 1$) and high mean flow Reynolds numbers, in which case the flow behavior must approach that of a single phase ideal fluid.

Under the assumptions of linearity and isotropy, the remaining independent variables c, ∇c, and $\underset{\sim}{u}^f - \underset{\sim}{u}^s$, lead to

$$\underset{\approx}{T}^f = -\, p^f\, \underset{\approx}{1} \tag{6}$$

$$\underset{\approx}{T}^s = -\, p^s\, \underset{\approx}{1} \tag{7}$$

$$\underset{\sim}{f} = \alpha(\underset{\sim}{u}^f - \underset{\sim}{u}^s) + \beta\nabla c\ , \tag{8}$$

where α and β are, in general, functions of c. From (6) and (7), both phases are modeled as ideal fluids. The two terms in the interaction force (8) have straightforward physical interpretations. The first is a drag force due to relative motion of the two phases. The linear case treated here is valid for small particle Reynolds numbers, and, for non-interacting ($c \ll 1$) spherical particles, α is easily shown to be

$$\alpha = \frac{9\mu}{2a^2}\, c\ , \tag{9}$$

where a = particle radius and μ = fluid viscosity. The second term in (8) is identified with the net force due to fluid pressure in the presence of concentration gradients (e.g., [4]), from which

$$\beta = p^f \tag{10}$$

Note that the concentration gradient term in (8) has <u>not</u> been identified with diffusion, although a term representing Brownian diffusion could be included here. Such a process, however, is easily shown to be of negligible importance in the present case.

Substituting from (6) - (10) into (3) and (4), the momentum balances become

$$\rho^f(1-c)\left(\frac{\partial \underset{\sim}{u}^f}{\partial t} + \underset{\sim}{u}^f\cdot\nabla\underset{\sim}{u}^f\right) = -(1-c)\nabla p^f + \rho^f(1-c)\underset{\sim}{g} - \frac{9\mu}{2a^2}\left(\underset{\sim}{u}^f - \underset{\sim}{u}^s\right) \tag{11}$$

$$\rho^s c\left(\frac{\partial \underset{\sim}{u}^s}{\partial t} + \underset{\sim}{u}^s\cdot\nabla\underset{\sim}{u}^s\right) = -c\nabla p^f + \nabla\left[c(p^f - p^s)\right] + \rho^s c\underset{\sim}{g} + \frac{9\mu}{2a^2}\left(\underset{\sim}{u}^f - \underset{\sim}{u}^s\right) \tag{12}$$

The pressure difference, appearing in (12), is commonly taken to represent the interfacial jump in normal stress between the phases [8]. Any such jump here is assumed to be negligible, so that

$$p^f = p^s = p \tag{13}$$

Adding (11) and (12) to obtain a total momentum balance for the mixture, and using (13):

$$\rho\left(\frac{\partial \underset{\sim}{u}}{\partial t} + \underset{\sim}{u}\cdot\nabla\underset{\sim}{u}\right) = -\nabla p + \nabla\cdot\left[-\rho^f(1-c)\underset{\sim}{v}^f\underset{\sim}{v}^f - \rho^s c\underset{\sim}{v}^s\underset{\sim}{v}^s\right] + \rho\underset{\sim}{g} \tag{14}$$

where $\rho = (1-c)\rho^f + c\rho^s$. The barycentric velocity, $\underset{\sim}{u}$, is defined by $\rho\underset{\sim}{u} = \rho^f(1-c)\underset{\sim}{u}^f + \rho^s c\underset{\sim}{u}^s$, and the diffusion velocities are given by $\underset{\sim}{v}^f = \underset{\sim}{u} - \underset{\sim}{u}^f$ and $\underset{\sim}{v}^s = \underset{\sim}{u} - \underset{\sim}{u}^s$. Thus, when the diffusion velocities are small, p may be identified approximately with the mean normal stress of the mixture, $1/3 \text{ tr } \underset{\approx}{T}^m$, where $\underset{\approx}{T}^m = (1-c)(\underset{\approx}{T}^f - \rho^f\underset{\sim}{v}^f\underset{\sim}{v}^f) + c(\underset{\approx}{T}^s - \rho^s\underset{\sim}{v}^s\underset{\sim}{v}^s)$.

<u>Decomposition and averaging</u>

Drew [3] adopts a Reynolds decomposition and averaging scheme used for compressible fluids (Favre averaging), in which $\underset{\sim}{u}^f = \bar{\underset{\sim}{u}}^f + \underset{\sim}{u}^{f'}$, $\underset{\sim}{u}^s = \bar{\underset{\sim}{u}}^s + \underset{\sim}{u}^{s'}$, $c = \bar{c} + c'$, and $p = \bar{p} + p'$. The average velocities are defined in terms of mean momenta, such that

$$\rho^f(1-\bar{c})\bar{\underset{\sim}{u}}^f = \rho^f\overline{(1-c)\underset{\sim}{u}^f}\ , \tag{15}$$

$$\rho^s\bar{c}\ \bar{\underset{\sim}{u}}^s = \rho^s\ \overline{c\underset{\sim}{u}^s} \tag{16}$$

Note that the average of the velocity fluctuations does not vanish, but $\overline{\underset{\sim}{u}^{f'}} = \overline{c\underset{\sim}{u}^{f'}}$ and $\overline{c\underset{\sim}{u}^{s'}} = 0$. Applying these definitions to equations (1), (2), (11), and (12) yields

$$\frac{\partial(1-\bar{c})}{\partial t} + \nabla\cdot[(1-\bar{c})\bar{\underset{\sim}{u}}^f] = 0 \tag{17}$$

$$\frac{\partial\bar{c}}{\partial t} + \nabla\cdot(\bar{c}\bar{\underset{\sim}{u}}^s) = 0 \tag{18}$$

$$\rho^f(1-\bar{c})\left(\frac{\partial\bar{\underset{\sim}{u}}^f}{\partial t} + \bar{\underset{\sim}{u}}^f\cdot\nabla\bar{\underset{\sim}{u}}^f\right) = -(1-\bar{c})\nabla\bar{p} + \overline{c'\nabla p'} - \nabla\cdot\left[\rho^f\overline{(1-c)\underset{\sim}{u}^{f'}\underset{\sim}{u}^{f'}}\right] + \rho^f(1-\bar{c})\underset{\sim}{g} - \frac{9\mu}{2a^2}\bar{c}(\bar{\underset{\sim}{u}}^f-\bar{\underset{\sim}{u}}^s) - \frac{9\mu}{2a^2}\overline{c\underset{\sim}{u}^{f'}} \tag{19}$$

$$\rho^s\bar{c}\left(\frac{\partial\bar{\underset{\sim}{u}}^s}{\partial t} + \bar{\underset{\sim}{u}}^s\cdot\nabla\bar{\underset{\sim}{u}}^s\right) = -\bar{c}\nabla\bar{p} - \overline{c'\nabla p'} - \nabla\cdot(\rho^s\overline{c\underset{\sim}{u}^{s'}\underset{\sim}{u}^{s'}}) + \rho^s\bar{c}\underset{\sim}{g} + \frac{9\mu}{2a^2}\bar{c}(\bar{\underset{\sim}{u}}^f-\bar{\underset{\sim}{u}}^s) + \frac{9\mu}{2a^2}\overline{c\underset{\sim}{u}^{f'}} \tag{20}$$

It is particularly important to note the new correlation $\overline{c\underset{\sim}{u}^{f'}}$, which arises from the drag interaction term in the exact equations. The correlation $\overline{c'\nabla p'}$ also appears in the form of an exchange of momentum between the two phases.

Plane gravity flow

The classical sedimentation problem, in which the force balancing the buoyant weight of the particles is sought, can now be formulated. For steady plane rectilinear flow (Figure 1), $\bar{\underset{\sim}{u}}^f = \{\bar{u}^f(y),0,0\}$, $\bar{\underset{\sim}{u}}^s = \{\bar{u}^s(y),0,0\}$, $\underset{\sim}{u}^{f'} = \{u^{f'}, v^{f'}, w^{f'}\}$, $\underset{\sim}{u}^{s'} = \{u^{s'}, v^{s'}, w^{s'}\}$, $\partial/\partial x = \partial/\partial z = 0$, $\bar{c} = \bar{c}(y)$, and $\underset{\sim}{g} = \{g\sin\beta, -g\cos\beta, 0\}$, where β = bed slope. This flow satisfies equations (17) and (18) identically.

For $c \ll 1$, the x-momentum of the mixture is, as expected, dominated by the fluid, and the streamwise component of (19) is approximated by

$$0 = \rho^f g\sin\beta - \rho^f\frac{\partial}{\partial y}(\overline{u^{f'}v^{f'}}) \tag{21}$$

This is, of course, the balance between the gravitational driving force and the gradient of the fluid Reynolds stress. Models for the Reynolds stress term in (21) have been exhaustively studied, and require little comment here. However, it is instructive to summarize the simplest approach commonly taken to this closure problem, because the parallels with the mass flux models to follow render that less familiar problem more transparent. So-called "mean-field" or "zero equation" methods are the most widely used closures for the momentum transport equations [13] and include Boussinesq (eddy viscosity) and mixing length models. In either case, neglecting the viscous sublayer, it can be argued that due to the dominance of different length scales near the bed and far away, the flow field must be divided into inner and outer layers. These models are summarized in Table 1, and the velocity profiles resulting from their substitution into (21) are given in Table 2. All the results for momentum transport are well known.

The y-components of the fluid and particulate phase momentum balances are

$$0 = -(1-\bar{c})\frac{\partial \bar{p}}{\partial y} + \overline{c'\frac{\partial p'}{\partial y}} - \frac{\partial}{\partial y}\left[\rho^f \overline{(1-c)v^{f\prime}v^{f\prime}}\right] - \rho^f(1-\bar{c})g\cos\beta - \frac{9\mu}{2a^2}\overline{cv^{f\prime}} \tag{22}$$

$$0 = -\bar{c}\frac{\partial \bar{p}}{\partial y} - \overline{c'\frac{\partial p'}{\partial y}} - \frac{\partial}{\partial y}\left(\rho^s \overline{cv^{s\prime}v^{s\prime}}\right) - \rho^s\bar{c}g\cos\beta + \frac{9\mu}{2a^2}\overline{cv^{f\prime}} \tag{23}$$

Eliminating the mean pressure gradient between (22) and (23) and approximating for small concentrations ($1-\bar{c} \cong 1$) and small slope ($\cos\beta \cong 1$) leads to:

$$0 = \overline{p'\frac{\partial c'}{\partial y}} - \frac{\partial}{\partial y}\left(\overline{p'c'}\right) - \rho^s\frac{\partial}{\partial y}\left(\overline{cv^{s\prime}v^{s\prime}}\right) + \bar{c}\frac{\partial}{\partial y}\left(\rho^f\overline{v^{f\prime}v^{f\prime}}\right) - \bar{c}(\rho^s-\rho^f)g + \frac{9\mu}{2a^2}\overline{cv^{f\prime}} \tag{24}$$

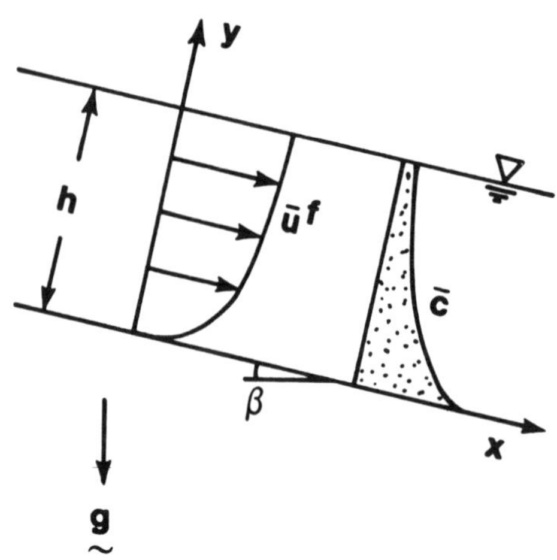

Figure 1. Definition sketch for plane rectilinear gravity flow.

turbulent flux	correlation	Boussinesq model	mixing length model
momentum	$-\rho^f \overline{\underset{\sim}{u}^{f\prime} \underset{\sim}{u}^{f\prime}}$	$\rho^f u_* \ell \; 2\underset{\approx}{\overline{D}}^f$	$\rho^f \ell^2 (4II^{1/2}) \; 2\underset{\approx}{\overline{D}}^f$
mass	$-\rho^s \overline{c \underset{\sim}{u}^{f\prime}}$	$\rho^s u_* \ell \; \nabla \bar{c}$	$\rho^s \ell^2 (4II^{1/2}) \; \nabla \bar{c}$

$$\underset{\approx}{\overline{D}}^f \equiv \frac{1}{2}(\nabla \overline{\underset{\sim}{u}}^f + \overline{\underset{\sim}{u}}^f \nabla) \; , \qquad II \equiv \frac{1}{2} \, \mathrm{tr}(\underset{\approx}{\overline{D}}^f)^2$$

Table 1. Zero equation closure models for turbulent momentum and mass flux correlations.

Table 2a. Predicted velocity and concentration based on Boussinesq model.

	region, scale	profile	"constant stress" approximation
momentum	inner $\ell_i = \kappa_i y$	$\overline{u}^f = \frac{u_*}{\kappa_i}(\ln \frac{y}{h} - \frac{y}{h}) + A_1$	$\overline{u}^f = \frac{u_*}{\kappa_i} \ln \frac{y}{h} + A_1'$
	outer $\ell_o = \kappa_o h$	$\overline{u}^f = \frac{u_*}{\kappa_o} \frac{y}{h}(1 - \frac{1}{2}\frac{y}{h}) + A_2$	
mass	inner $\ell_i = \kappa_i^s y$	$\overline{c} = A_3 (\frac{y}{h})^{-U_\infty/\kappa_i^s u_*}$	
	middle $\ell_o = \kappa_o^s h$	$\overline{c} = A_4 \exp\left(- \frac{U_\infty}{\kappa_o^s u_*} \frac{y}{h}\right)$	
	surface $\ell_s = \kappa_s^s (h-y)$	$\overline{c} = A_5 (1 - \frac{y}{h})^{U_\infty/\kappa_s^s u_*}$	

A_1, A_1', A_3 fixed by near-bed condition

A_2, A_4, A_5 fixed by matching with underlying layer

Table 2b. Predicted velocity and concentration profiles based on mixing length model.

	region, scale	profile	"constant stress" approximation
momentum	inner $\ell_i = \kappa_i y$	$\bar{u}^f = \frac{u_*}{\kappa_i}\left[2(1-\frac{y}{h})^{1/2} - \tanh^{-1}(1-\frac{y}{h})^{1/2}\right] + A_6$	$\bar{u}^f = \frac{u_*}{\kappa_i}\ln\frac{y}{h} + A_6'$
	outer $\ell_o = \kappa_o h$	$\bar{u}^f = A_7 - \frac{2u_*}{3\kappa_o}(1-\frac{y}{h})^{3/2}$	
mass	inner $\ell_i = \kappa_i^s y$	$\bar{c} = A_8\left[\frac{1-(1-\frac{y}{h})^{1/2}}{1+(1-\frac{y}{h})^{1/2}}\right]^{-U_\infty \kappa_i/(\kappa_i^s)^2 u_*}$	$\bar{c} = A_8' (\frac{y}{h})^{-U_\infty \kappa_i/(\kappa_i^s)^2 u_*}$
	outer $\ell_o = \kappa_o^s h$	$\bar{c} = A_9 \exp\left[\frac{2U_\infty \kappa_o}{(\kappa_o^s)^2 u_*}(1-\frac{y}{h})^{1/2}\right]$	

A_6, A_6', A_8, A_8' fixed by near-bed condition

A_7, A_9 fixed by matching with underlying layer

Equation (24) may be viewed as a balance for the excess vertical momentum of the dispersed phase over that of the fluid phase. The first two terms result from the conventional rearrangement of the correlation $-\overline{c'\nabla p'}$ into a source/sink term, $\overline{p'\nabla c'}$, and a term representing transport of c by pressure fluctuations, $-\nabla\cdot(\overline{p'c'\underset{\approx}{1}})$. The third and fourth terms represent transport of c by the Reynolds stresses. The last two terms are sources/sinks due to the buoyant weight of the particles and the "Reynolds drag," respectively.

The principal object here is to identify the dominant term or terms that balance the buoyant weight. From (21), the fluid velocity fluctuations clearly scale like $\mathcal{U} = (\ell/h)^{1/2} u_*$, where $u_* = (\tau_o/\rho^f)^{1/2}$, $\tau_o = \rho^f gh \sin \beta$, and ℓ is an appropriate length scale. It is assumed that the dispersed phase velocity fluctuations also scale with $\mathcal{U}$, c scales with some value c_o, and $\partial/\partial y$ with ℓ^{-1}. Arranging the dimensionless equivalent of (24) so that the buoyant weight term is of order unity,

$$0 = -\bar{c} + \frac{\mathcal{U}}{U_\infty}\overline{cv^{f\prime}} + (Fr')^2 \frac{h}{\ell}\left[\overline{p'\frac{\partial c'}{\partial y}} - \frac{\partial}{\partial y}(\overline{p'c'}) - \frac{\rho^s}{\rho^f}\frac{\partial}{\partial y}(\overline{cv^{s\prime}v^{s\prime}}) + \bar{c}\frac{\partial}{\partial y}(\overline{v^{f\prime}v^{f\prime}})\right] \tag{25}$$

where $U_\infty = 2a^2(\rho^s-\rho^f)g/9\mu$ = settling velocity of a single particle in an infinite fluid, $Fr' = \mathcal{U}/(g'h)^{1/2}$ = turbulent densiometric Froude number, and $g' = (\rho^s-\rho^f)g/\rho^f$.

Consider a typical wide stream carrying very fine quartz sand ($a = 5.0 \times 10^{-5}$m, $U_\infty \cong 1.0 \times 10^{-2}$ m/s) over a flat bottom. For slope $\beta \cong 10^{-4}$ and depth $h = 1.0$ m, $u_* = 3.1 \times 10^{-2}$ m/s. In the outer part of the flow, "far" from the bed, the dominant length scale is h. Thus, $\mathcal{U} = u_*$ and $\mathcal{U}/U_\infty \cong 3.1$, while $(Fr')^2 \cong 6.3 \times 10^{-5}$. Clearly, then, the "Reynolds drag" is the dominant balancing term for the buoyant weight in the outer region. Near the bed, the characteristic length is $\ell = y$, the distance from the bed. It is possible, then, that the transport terms may become important in this region. However the ratio

$(Fr')^2\,(h/y)/(u_*/U_\infty)$ is, in the present example, approximately $(2 \times 10^{-5})(y/h)^{-1/2}$. Thus, the drag term dominates everywhere; the remaining terms become significant only at a distance from the bed smaller than the height of typical roughness elements in a stream. For a smooth bed, the drag term would dominate in the postulated flow everywhere outside the buffer layer ($y > 40\nu/u_*$), and the inviscid theory posed here would in any event break down at smaller y. Returning to dimensional variables, then, the important terms in (24) are

$$0 = -\,\bar{c}\,(\rho^s - \rho^f)g + \frac{9\mu}{2a^2}\,\overline{cv^{f'}} \tag{26}$$

Dividing (26) by $9\mu/2a^2$ results in a flux balance:

$$0 = -\,U_\infty\bar{c} + \overline{cv^{f'}} \tag{27}$$

It becomes quite clear at this point that the correlation $\overline{cv^{f'}}$ is associated with a turbulent volume flux. The result is simple and intuitively appealing: it is the drag associated with vertical fluid velocity fluctuations that balances the buoyant weight of the particles. Correlations of the form $\overline{cv^{f'}}$ are routinely modeled as turbulent diffusion (e.g., [6]), but typically arise from averaging convective transport terms of the form $\nabla\cdot(c\underset{\sim}{u}^f)$. In the present case, the correlation arises from averaging of the fluid-particle interaction forces, and, indeed, it seems obvious in retrospect that this <u>must</u> be the origin of the balance to the buoyant weight.

Modeling the correlation $\overline{c\underset{\sim}{u}^{f'}}$

There exist several well established and extensively explored approaches to modeling turbulent correlations of the sort arising here. They are most familiar in the form of zero equation models for turbulent momentum fluxes (Reynolds stresses), as mentioned previously, and that parallel is emphasized in the following. Although the extension to mass transport considerations is also common knowledge, it appears that this exercise has not been carried out systematically in the context of the sediment transport problem.

The Boussinesq model for the turbulent mass flux is given by

$$\overline{c\underset{\sim}{u}^{f'}} = -\varepsilon_B \nabla \bar{c} \qquad (28)$$

where ε_B is a diffusion coefficient, and isotropy has been assumed. From (28) it is apparent that the diffusivity must scale like $u_*\ell$, so that

$$\varepsilon_B = u_*\ell \qquad (29)$$

The mixing length model is given by

$$\overline{c\underset{\sim}{u}^{f'}} = -\varepsilon_P \; 4II^{1/2}\nabla\bar{c} \qquad (30)$$

where II is the invariant 1/2 $tr(\underset{\approx}{\bar{D}}^f)^2$, and $\underset{\approx}{\bar{D}}^f = 1/2\,(\nabla\underset{\sim}{\bar{u}}^f + \underset{\sim}{\bar{u}}^f\nabla)$. Inspection of (30) indicates that ε_P scales like ℓ^2:

$$\varepsilon_P = \ell^2 \qquad (31)$$

These assumptions are summarized in Table 1 along with the analogous expressions for the fluid Reynolds stress.

It is also a routine procedure in momentum calculations to note that, for either the Boussinesq or mixing length models, different length scales dominate near and far from a wall. At the boundary, the correlations must vanish because the normal velocity fluctuations must vanish. Thus, it is commonly assumed that the appropriate length scale in the "inner" region near the bed is

$$\ell_i = \kappa_i y \qquad (32)$$

where κ_i is an empirical constant. In the "outer" part of the flow, where the turbulence is more nearly homogeneous, the dominant scale is the boundary layer thickness, in the present case equal to the flow depth, h:

$$\ell_o = \kappa_o h \qquad (33)$$

It has been noted in data from both laboratory experiments and natural streams that the diffusive mass flux appears to be damped near the free surface. An example is shown in the following section (Figure 3), where (33) clearly fails in the Boussinesq model for $y/h > 0.8$. Following a suggestion made in previous work [10], this is remedied by posing a third layer near the free surface in which the dominant length scale is h-y, and

$$\ell_s = \kappa_s(h-y) \qquad (34)$$

This difficulty does not arise for the mixing length model because the velocity gradient, and consequently II (Eq. 30), vanishes at the surface. In the following, κ values associated with application of (32) - (34) to mass transport considerations will be indicated by a superscript $\underline{s}$.

Equations (27) - (34) together lead to ordinary differential equations for the mean concentration in each of three layers in the Boussinesq scheme and in each of two regions in the mixing length approach. These are all easily solved, and the results are compiled in Table 2. The boundary condition for the outer layers is fixed by matching with each underlying region. That for the innermost layer must be fixed by some independent argument, perhaps analogous to "roughness" postulates used in calculating velocity profiles.

Comparison to experiment

In an earlier paper [10], the prediction of inner and outer layers for the Boussinesq model was tested against a single experimental run by Montes and Ippen [12]. This is extended here to include the third (surface) layer posed in the preceding. In addition, predictions based on the mixing length model are tested against the same data.

Six experimental velocity and concentration profiles by Montes and Ippen [12] are considered here. These individual runs all fall within a narrow range of experimental conditions. The suspension is fine quartz sand in water ($a = 6.5 \times 10^{-4}$ m, $U_\infty = 1.33 \times 10^{-2}$ m/s), the flow depths vary from 5.7×10^{-2} m to 7.7×10^{-2} m, and the shear velocities from 6.1×10^{-2} m/s to 7.0×10^{-2} m/s. The predicted velocity and concentration profiles were fit to the data for both the Boussinesq and mixing length models for each layer, and the constant, κ, scaling the characteristic length for each was calculated. Linear regression was used to determine the best fit of each profile within the appropriate region. The inner region was arbitrarily defined by $y/h < 0.2$. The surface layer invoked for the Boussinesq mass flux model was similarly defined by $y/h > 0.8$.

Results are summarized in Table 3. The mean value of κ for the momentum flux in the inner layer is, for both models, $\kappa_i = 0.34$. For the outer layer, mean values are $\kappa_o = 0.060$ and $\kappa_o = 0.088$ for the eddy viscosity and mixing length models, respectively. These values compare favorably with the typical values for particle-free flows given by Reynolds and Cebeci [13] of $\kappa_i = 0.4$, $\kappa_o = 0.075$.

	velocity profile				concentration profile				
	Boussinesq		mixing length		Boussinesq			mixing length	
	inner	outer	inner	outer	inner	middle	surface	inner	outer
run no.	κ_i	κ_o	κ_i	κ_o	κ_i^s	κ_o^s	κ_s^s	κ_i^s	κ_o^s
37	.38	.061	.38	.088	.34	.118	.93	.36	.133
38	.32	.064	.32	.092	.27	.107	.34	.29	.122
39	.35	.070	.35	.103	.26	.094	.35	.30	.122
40	.32	.057	.32	.085	.32	.088	.30	.32	.104
41	.32	.055	.32	.080	.28	.086	.56	.30	.106
42	.33	.055	.33	.082	.28	.091	.38	.30	.106
mean	.34	.060	.34	.088	.29	.097	.48	.31	.116

Table 3. Values of κ (see Table 2) obtained for experiments by Montes and Ippen [12].

It has long been observed, of course, that the presence of particles tends to reduce κ (e.g., [7]). Results from fitting the concentration profiles indicate mean κ^s values for the Boussinesq model of $\kappa_i^s = 0.29$, $\kappa_o^s = 0.097$, $\kappa_s^s = 0.48$, and the mixing length model yields comparable values of $\kappa_i^s = 0.31$ and $\kappa_o^s = 0.116$. Note that the consistency of the results shown in Table 3 is not intended to suggest that the values obtained are "universal". The range of experimental conditions is too limited to establish such a claim.

Plots of the concentration data for a single run (Montes and Ippen [12], run no. 42) are shown in Figures 2-5, in various graphical forms for comparison to the predictions discussed in the foregoing. Figure 2 shows the inner solution for both the Boussinesq and the mixing length models. Figure 3 shows the fit of the Boussinesq model for the middle layer, $0.2 < y/h < 0.8$. The outer solution for the mixing length model is shown in Figure 4, and the surface layer based on the Boussinesq scheme is illustrated in Figure 5. In all of these examples, it is clear that the data depart markedly from the predicted profile outside the layer for which it was derived.

Finally, it is of interest to calculate turbulent Schmidt numbers for these flows. The Schmidt number is analogous to the Prandtl number, and represents the ratio of momentum flux to mass flux, defined by (e.g., [9]):

$$Sc' = \frac{\overline{u^{f\prime}v^{f\prime}} \Big/ \left(\partial\bar{u}^f/\partial y\right)}{\overline{cv^{f\prime}} \Big/ \left(\partial\bar{c}/\partial y\right)} \tag{35}$$

For the Boussinesq model, the Schmidt numbers in the inner, middle, and surface layers are, then, κ_i/κ_i^s, κ_o/κ_o^s, and κ_o/κ_s^s, respectively. For the mixing length model, Sc' is given by $(\kappa_i/\kappa_i^s)^2$ and $(\kappa_o/\kappa_o^s)^2$ for the inner and outer layers, respectively. Calculated values are collected in Table 4. Sc' is of order unity throughout most of the flow, as anticipated in the original scaling arguments. However, $Sc' > 1$ near the boundary and decreases to values substantially less than unity in the outer region, reaching

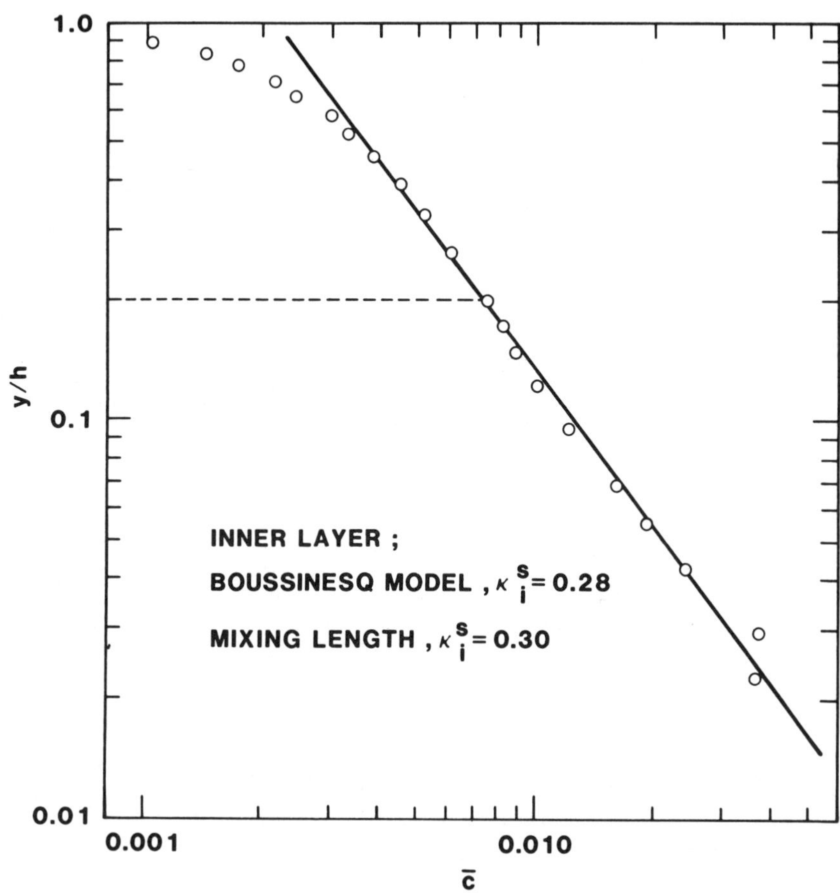

Figure 2. Comparison of inner solution for Boussinesq and mixing length models to experimental data of Montes and Ippen [12], run no. 42.

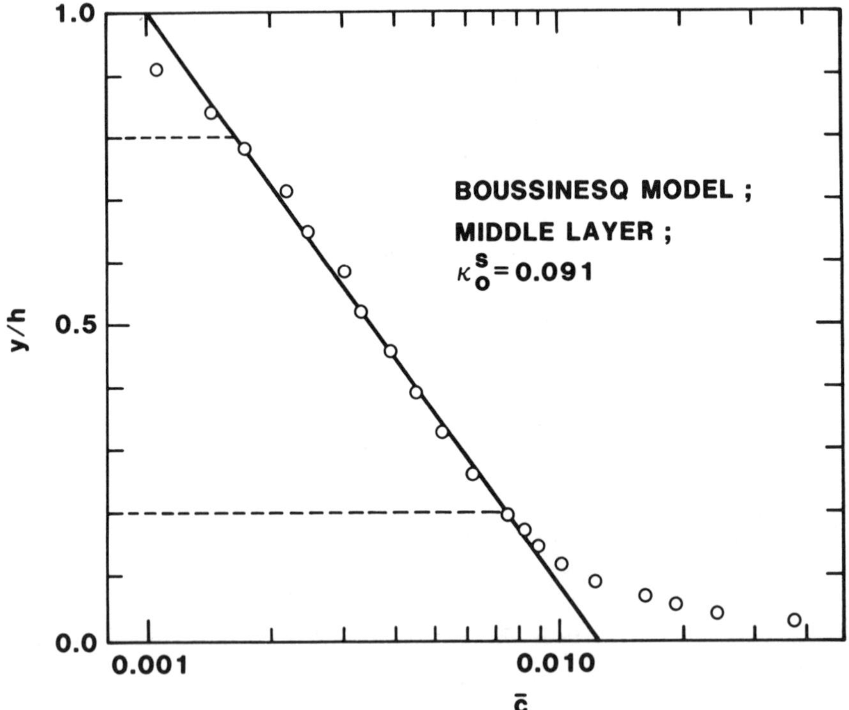

Figure 3. Comparison of middle layer solution for Boussinesq model to experimental data of Montes and Ippen [12], run no. 42.

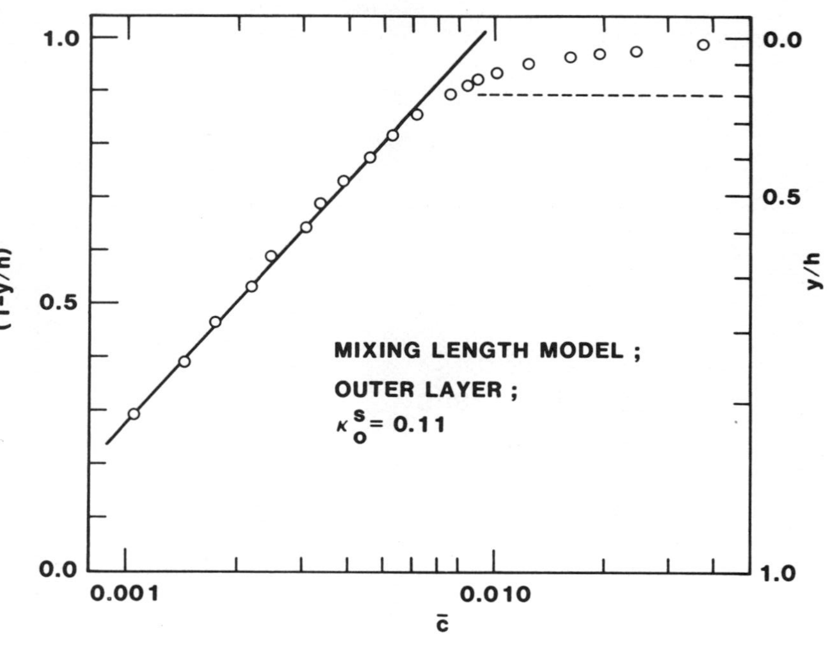

Figure 4. Comparison of outer solution for mixing length model to experimental data of Montes and Ippen [12], run no. 42.

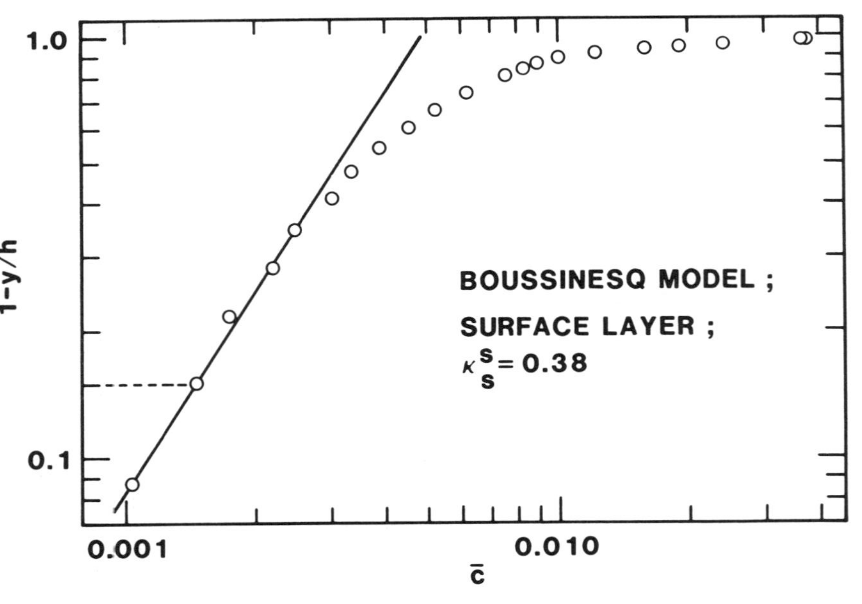

Figure 5. Comparison of surface layer solution for Boussinesq model to experimental data of Montes and Ippen [12], run no. 42.

	Boussinesq			mixing length	
	inner	middle	surface	inner	outer
run no.	κ_i/κ_i^s	κ_o/κ_o^s	κ_s/κ_s^s	$(\kappa_i/\kappa_i^s)^2$	$(\kappa_o/\kappa_o^s)^2$
37	1.12	.52	.07	1.12	.44
38	1.19	.60	.19	1.20	.57
39	1.35	.74	.20	1.36	.71
40	1.00	.65	.19	1.01	.67
41	1.14	.64	.10	1.12	.57
42	1.18	.60	.14	1.19	.60
mean	1.16	.62	.15	1.17	.59

Table 4. Turbulent Schmidt numbers, Sc', calculated for experiments by Montes and Ippen [12].

$Sc' = 0.15$ in the Boussinesq surface layer. The implication is that the mass flux is more strongly retarded near the boundary. Launder [9] reviews numerous heat transfer studies in which the turbulent Prandtl number also decreases away from the wall, and ascribes the phenomenon to the expectation that $\overline{u^{f'}v^{f'}}$ drops more rapidly away from the wall than does $\overline{v^{f'2}}$. It is interesting to note, however, that few of the studies reviewed indicate turbulent Prandtl numbers greater than unity, with the exception of experiments in stably stratified flows. Indeed, this may explain the large values of Sc' calculated for the near-bed region in the present study. The steep gradient in $\bar{c}$ in the inner region represents a strong stable stratification, and vertical velocity fluctuations can be expected to be strongly damped.

<u>Summary</u>

Mixture theory for a dilute suspension of small negatively buoyant particles provides a physically sound and intuitively satisfying interpretation of turbulent diffusion. The phenomenon can be identified with a correlation $\overline{c\underset{\sim}{u}^{f'}}$ arising from decomposition and averaging applied to the drag interaction term in the momentum balances. The classical sedimentation problem, in which a balance of gravitational settling and diffusive fluxes is posed, is embodied in this theory. The balance has most often been put forth from purely kinematic arguments, whereas the mixture theory identifies this supposition in the <u>dynamics</u> of fluid-particle interaction.

Earlier treatments of the sediment transport problem appear to neglect developments that are standard procedure in considerations of momentum transport in turbulent boundary layers. In particular, it is expected that the flow must be divided into layers in which different length scales predominate. The analogy between momentum and mass transport has been fully developed here for both Boussinesq and mixing length models for multilayer flows. The procedure is not new, but apparently has not been carried through systematically in the context of sediment transport. The resulting theoretical concentration profiles compare very well with data from Montes and Ippen [12], and consistent results are obtained within a set of experiments covering a narrow range of conditions. Both the Boussinesq and mixing length models yield similar results. Calculated turbulent Schmidt numbers for these flows average 1.16 in the near-wall region, and decrease to less than unity in the outer region. Thus, the mass flux is more strongly damped than the momentum flux near the boundary, but that relationship is reversed far away. This may reflect the stable stratification due to steep concentration gradients near the wall.

REFERENCES

1. Atkin, R. J., and Craine, R. E., 1976a, "Continuum Theories of Mixtures: Basic Theory and Historical Development," Quarterly Journal of Mechanics and Applied Mathematics, Vol. 29, pp. 209-244.
2. Atkin, R. J., and Craine, R. E., 1976b, "Continuum Theories of Mixtures: Applications," Journal of the Institute of Mathematics and its Applications, Vol. 17, pp. 153-207.
3. Drew, D. A., 1975, "Turbulent Sediment Transport Over a Flat Bottom Using Momentum Balance," Journal of Applied Mechanics, Vol. 42, pp. 38-44.
4. Drew, D. A., and Segel, L. A., 1971, "Averaged Equations for Two-Phase Flows," Studies in Applied Mathematics, Vol. 50, pp. 205-231.
5. Hinze, J. O., 1961, "Momentum and Mechanical-Energy Balance Equations for a Flowing Homogeneous Suspension with Slip Between the Two Phases," Applied Scientific Research, Section A, Vol. 11, pp. 33-46.
6. Hinze, J. O., 1972, "Turbulent Fluid and Particle Interaction," Progress in Heat and Mass Transfer, Vol. 6, pp. 433-452.
7. Ippen, A. T., 1971, "A New Look at Sedimentation in Turbulent Streams," Journal of the Boston Society of Civil Engineers, Vol. 58, pp. 131-163.
8. Ishii, M., 1975, Thermo-Fluid Dynamic Theory of Two-Phase Flow, Eyrolles, Paris.

9. Launder, B. E., 1978, "Heat and Mass Transport" in Turbulence, P. Bradshaw, ed., Springer-Verlag, Berlin, pp. 231-287.

10. McTigue, D. F., 1981, "Mixture Theory for Suspended Sediment Transport," Journal of the Hydraulics Division, ASCE, Vol. 107, No. HY6, pp. 659-673; discussions and reply, to appear, 1982.

11. Monin, A. S., and Yaglom, A. M., 1979, Statistical Fluid Mechanics: Mechanics of Turbulence, Vol. 1, MIT Press, Cambridge.

12. Montes, J. S., and Ippen, A. T., 1973, "Interaction of Two-Dimensional Turbulent Flow with Suspended Particles," Ralph M. Parsons Laboratory for Water Resources and Hydrodynamics, Report No. 164, Massachusetts Institute of Technology, Cambridge.

13. Reynolds, W. C., and Cebeci, T., 1978, "Calculation of Turbulent Flows," in Turbulence, P. Bradshaw, ed., Springer-Verlag, Berlin, pp. 193-229.

14. Soo, S. L., and Tung, S. K., 1971, "Pipe Flow of Suspensions in Turbulent Fluid; Electrostatic and Gravity Effects," Applied Scientific Research, Vol. 24, pp. 83-97.

15. Truesdell, C., 1969, Rational Thermodynamics, McGraw-Hill, New York.

This work was performed at Sandia National Laboratories, supported by the U.S. Department of Energy under contract DE-AC04-76DP00789.

David F. McTigue
Fluid Mechanics and Heat
Transfer Division I
Sandia National Laboratories
Albuquerque, New Mexico 87185

Waves in Gas–Liquid Flows

L. van Wijngaarden

§1. Introduction

At present there does not exist a general theory for dispersed two phase flow. That is to say a theory which predicts a reasonably broad range of phenomena in such flows. Apparently our physical insight in what happens in two phase flow is not developed far enough to enable us to contruct satisfactory models. At best we can hope that certain model equations, tested experimentally in a particular situation, apply as well to other, similar, situations. There is definitely a lack of experimental data obtained in systematic and controlled experiments. Such experiments are impeded by the fact that both in this country and elsewhere research contracts seem to be granted primarily to experiments testing out computer codes with a large number of parameters. In such experiments there are too many variables and the results have learned us virtually nothing. In this lecture I would like to discuss two kind of waves for which sufficient experimental data are available to confirm their existence and to enable verification of theoretical reasoning. In the first case, pressure waves, a satisfactory theory is available as well. In the second case, kinematic waves or void fraction waves, a satisfactory theory has as yet not been given. In this case the lecture will emphasize on the theoretical problems which do exist here.

ISBN 0-12-493120-0

§2. Pressure waves in two-phase flow.

2.1. Theory for homogeneous medium.

The simplest model for a gas/liquid flow is a (fictitious) homogeneous medium with effective properties obtained from considerations regarding the properties of the involved inhomogeneities.

For example, a suspension of small air bubbles in water may be thought of as a homogeneous fluid with density ρ, given by

$$\rho = (1-\alpha)\rho_{\ell} + \alpha\rho_{g} \tag{2.1}$$

α being the concentration of air by volume, ρ_{ℓ} and ρ_{g} the density of liquid and gas respectively. Such a concept makes sense when the length scale of the phenomena which we wish to consider is large with respect to the characteristic length parameters of the dispersed flow. First the typical dimension of the dispersed particles, the effective radius, R, say. Then the average distance between particles for which $n^{-1/3}$ is representative where n is the number density. With a macroscopic length scale λ we must require therefore

$$\lambda \gg R \quad , \quad \lambda \gg n^{-1/3} \; . \tag{2.2}$$

Just how much λ must be larger than $n^{-1/3}$ in other words how many bubbles must be covered by the macroscopic length scale, can only be found by experiment. For spherical bubbles the void fraction α is related to the mentioned lengths by

$$\alpha = \frac{4}{3}\pi n R^{3} \; . \tag{2.3}$$

It follows from (2.1) that gas/liquid flows have, unless α is close to unity, approximately the density of the liquid. The compressibility is mainly due to the gas content because liquids are much harder to compress. If we disregard for the time being a possible difference between the velocity of the gas and that of the liquid (indicated by many workers in the field as slip) the mass of gas in a unit mass of liquid is a constant. This is expressed mathematically as

$$\frac{\alpha\rho_g}{\alpha\rho_g+(1-\alpha)\rho_\ell} = \text{constant} \ . \tag{2.4}$$

Expressing α in terms of ρ and ρ_g with this relation and introducing in (2.1) gives with

$$\frac{dp}{d\rho_g} = c_g^2 \ ; \ \frac{dp}{d\rho_\ell} = c_\ell^2 \tag{2.5}$$

for the compressibility

$$\frac{d\rho}{dp} = \frac{\alpha^2}{c_g^2} + \frac{(1-\alpha)^2}{c_\ell^2} + \alpha(1-\alpha)\left\{\frac{c_\ell^2\rho_\ell^2+c_g^2\rho_g^2}{c_\ell^2 c_g^2 \ \rho_\ell\rho_g}\right\} \sim \frac{\alpha^2}{c_g^2} + \frac{(1-\alpha)^2}{c_\ell^2} + \frac{\alpha(1-\alpha)\rho_\ell}{\rho_g c_g^2} \ . \tag{2.6}$$

The quantity on the left hand side of this relation is the reciprocal sound speed, squared. From the terms on the right hand side it follows that the sound speed c, is determined mainly by the third term when α is not too close to zero or unity. Neglecting the first and the second term gives

$$c^2 = \frac{\gamma p}{\rho_\ell\alpha(1-\alpha)} \quad \text{for isentropic waves}$$
$$= \frac{p}{\rho_\ell\alpha(1-\alpha)} \quad \text{for isothermal waves} \ . \tag{2.7}$$

In figure 1 the sound speed c is given as a function of α at given pressure under isothermal behaviour. Most striking is its low value, as compared with the sound speed in both liquid and gas, at values of α off the sides $\alpha=1$ and $\alpha=0$. Relation (2.7) has been amply confirmed in experiments (see for a survey Van Wijngaarden 1972). The low speed of sound of a bubbly suspension has great practical significance. Even for a value of α as low as 1% the sound speed in a mixture of air and water is only 100m/s. Near the surface large amounts of bubbles are dispersed in the oceans which severely affect the propagation of acoustic waves. As another example we mention that the emission of sound by a turbulent liquid flow is enhanced by a factor as large as $(c_\ell/c)^4$ when gas is dispersed in the liquid (Crighton and Ffowcs Williams 1969). Twenty years ago acoustic properties of bubbly flows were used by the navy to direct sound generated by the propeller. To this end the propeller was shrouded and bubbles were ejected from the shroud to form a cylindrical screen. In this way the sound was radiated directly from the stern into the wake of the ship. I don't know whether this practice still exists. Waves of finite amplitude can be analysed just like in single phase flow with the classical theory of characteristics, described e.g. in Whitham (1974). Shock waves result from linear steepening of compression waves in much the same way as in a single phase fluid. Only the mechanism which opposes the steepening is quite different, in most circumstances, as we shall see.

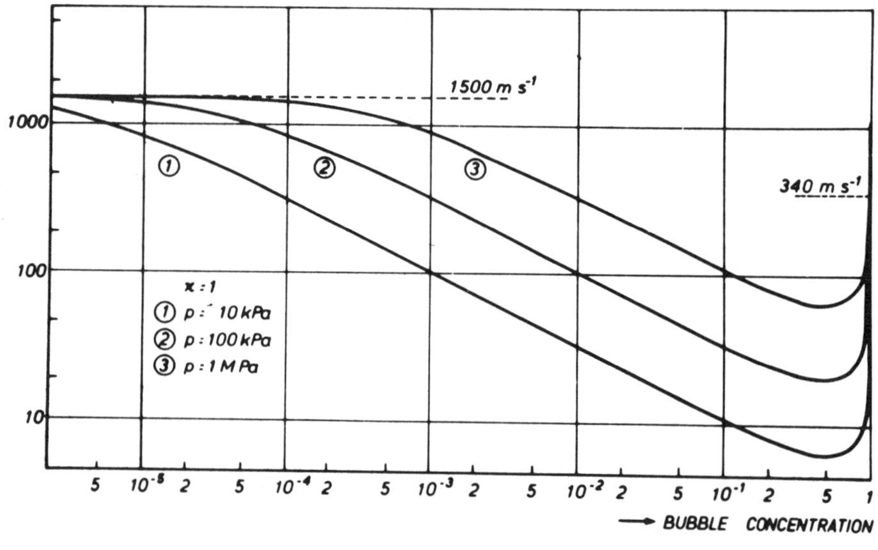

Figure 1

2.2. Dispersion caused by volume oscillations.

Waves in multiphase flow are always dispersed by the presence of the inhomogeneities. The underlying mechanism is known for dusty gases (Marble 1970), vapour with small liquid droplets (Marble and Wooten 1970) and bubbly flows (Van Wijngaarden 1968). In the latter case it is as follows: Bubbles are capable of executing both volume oscillations and shape oscillations. The latter are relatively unimportant for sound waves because they are heavily damped. The former are important. With surface tension denoted with σ, the response of one single bubble with pressure p_g to a time varying pressure p is mathematically described by the Rayleigh-Plesset equation

$$\frac{p_g - p}{\rho_\ell} = \left(R\ddot{R} + \frac{3}{2}\dot{R}^2\right) + \frac{2\sigma}{R} + \frac{4\nu}{R}\dot{R} \,. \tag{2.8}$$

In this equation the dot means differentiation with respect to time. For air-water interfaces $\sigma \sim 10^{-1}$ N/m and surface tension is important therefore only for bubbles as small as 1 μm at atmospheric conditions. Linearization of (2.8) leads to an angular frequency for free oscillations

$$\omega_B = \frac{1}{R}\left(\frac{3\gamma p}{\rho_\ell}\right)^{1/2} \tag{2.9}$$

for isentropic bubbles. For bubbles of radius $R \sim 10^{-3}$m the associated frequency is of the order of 1 KHz. Air bubbles in water emit therefore sound in the audible range. This was first described quantitatively by Minnaert (1933). The frequency given in (2.9) is therefore often called Minnaert frequency. When sound waves pass through a bubbly liquid volume oscillations are induced the amplitude of which sharply increases when the Minnaert frequency of the bubbles is approached. This causes dispersion of the wave. Mathematically this can be incorporated in the theory by identifying the pressure p occurring in (2.8) as the local liquid pressure. In contrast with the theory in the previous section there is now a pressure difference between the gas and the liquid associated with the radially accelerated liquid.

For one-dimensional motion we have further as equations of motion

$$\frac{d\rho}{dt} + \rho\frac{\partial u}{\partial x} = 0 \,, \tag{2.10}$$

which is the continuity equation,

$$\rho\frac{du}{dt} + \frac{\partial p}{\partial x} = 0, \tag{2.11}$$

the equation of momentum,

$$\frac{dn}{dt} + n\frac{\partial u}{\partial x} = 0 \,, \tag{2.12}$$

which expresses the conservation of numbers of bubbles.

Finally we need the relation (2.3) between α, n and R and an energy equation for the bubbles. A simple one is

$$p_g R^3 = \text{constant} \tag{2.13}$$

which holds for isothermal changes, or

$$p_g R^{3\gamma} = \text{constant} \tag{2.14}$$

for isentropic behaviour, γ being the ratio of specific heats. From linearization of these equations and seeking solutions in the form of travelling waves we obtain for waves with angular frequency ω and wave number k

$$\left(\frac{k}{\omega}\right)^2 = \frac{1}{c_\ell^2} + \frac{1-\omega^2/\omega_B^2 + i\delta\omega/\omega_B}{c^2\{(1-\omega^2/\omega_B^2)^2 + \delta^2\omega^2/\omega_B^2\}} . \tag{2.15}$$

In this equation c is the sound velocity as given in (2.7) whereas δ is the logarithmic decrement associated with the attenuation. This is here due to the viscous term in (2.8). In fact we should have formulated an energy equation more realistic than (2.14). Notably the heat conduction from the gas to the liquid should have been taken into account. This leads to a net loss of energy which can be accounted for by a thermal coefficient, D_h being the thermal diffusivity of the gas phase,

$$\delta_{th} = \frac{3(\gamma-1)}{2(\omega/2D_h)^{1/2} R_0} \tag{2.16}$$

which is under most conditions larger than the value $4\nu/\omega_B R_o^2$ following from the viscous term in (2.8). The dispersion equation (2.15) has been verified in many experiments, notably Silberman (1957).

It should be noted that one bubble size only is considered. A theory valid for a distribution of bubble sizes is formally possible but hardly tractable in practice. Still the one size theory can be used to obtain data for bubble distributions, e.g. in the ocean. This is caused by the fact that the scattering cross section and the extinction cross section is largest at resonance. When sound waves are transmitted through a bubbly suspension the bubble distribution can be obtained from measurement of the atttenuation under the assumption that attenuation comes mainly from resonant bubbles. Such measurements have in this country been carried out for ocean bubble spectra by Medwin (see Medwin 1980).

2.3. Waves of finite amplitude.

There is an interesting analogy with gravity waves on a fluid of finite depth. Both such waves and the pressure waves that we have discussed in the foregoing sections display dispersion and steepening of compression waves. For gravity waves of moderate amplitude the Boussinesq equations incorporate both these phenomena. They reduce for waves travelling in one direction only to the famous Korteweg-de Vries equation. This equation holds (Van Wijngaarden 1968) also for waves of moderate amplitude in bubbly flows. When the pressure perturbation is indicated with $\tilde{p}=p-p_o$, we have, the subscript o referring to the undisturbed state

$$\frac{\partial \tilde{p}}{\partial t} + c_o \frac{\partial \tilde{p}}{\partial x} + c_o \tilde{p} \frac{\partial \tilde{p}}{\partial x} + \frac{1}{2} \frac{c_o^3}{\omega_B^2} \frac{\partial^3 \tilde{p}}{\partial x^3} = o . \qquad (2.17)$$

This equation predicts the existence of solitons in bubbly flows. These have indeed been measured by several workers, Kuznetsov (1978), Roelofsen (1981).

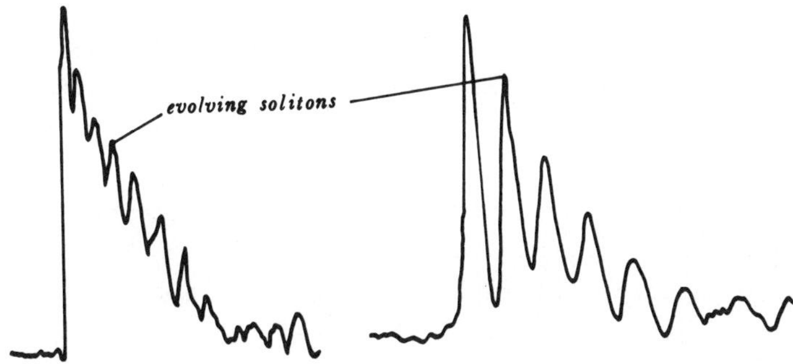

Figure 2. Evolution of solitons from a triangular pressure disturbance in a bubbly liquid. (From Roelofsen 1981)

In figure 2 an example from the latter reference is shown. It is clear that attenuation is significant. A better agreement with reality is therefore obtained when an attenuation term is added to (2.17), resulting in the Korteweg-de Vries-Burgers equation

$$\frac{\partial\tilde{p}}{\partial t} + c_o \frac{\partial\tilde{p}}{\partial x} + c_o\tilde{p}\,\frac{\partial\tilde{p}}{\partial x} + \frac{1}{2}\,\frac{c_o^3}{\omega_B^2}\,\frac{\partial^3\tilde{p}}{\partial x^3} - \frac{1}{2}\,\frac{\delta c_o^2}{\omega_B}\,\frac{\partial^2\tilde{p}}{\partial x^2} = o\ . \qquad (2.18)$$

This equation, apart from allowing for the attenuation of solitons, has steady solutions of shock wave type. While shock waves in single phase gases are determined by a balance between nonlinear steepening and diffusion, here it is rather the dispersion which balances the nonlinearity. These effects are represented by the third and fourth term in the above equation. Their balance gives for the thickness d of the shock wave the estimate

$$d \sim \frac{R_o}{\alpha^{1/2}}\left(\frac{p_o}{\Delta p}\right)^{1/2}\ , \qquad (2.19)$$

Δp being the pressure jump over the shock wave. This wave is similar to the undular bore in hydraulics, steep at the front and wavy at the back. Its existence in bubbly flows has been amply verified (Noordzij 1973, Van Wijngaarden 1970, Kuznetsov et al. 1978). An example, taken from Noordzij (1973) is shown in figure 3. In connection with (2.2) it was observed that experiments should decide how large the length scale λ should be with respect to $n^{-1/3}$. In the experiments on shock waves it turned out that the theory presented here holds good even when the shock covers only a few bubbles. We have laid emphasis on bubbly flows. A theory supported by experiments is available also for waves of small amplitude in separated flow, (Matsui 1975, Morioka & Matsui 1975).

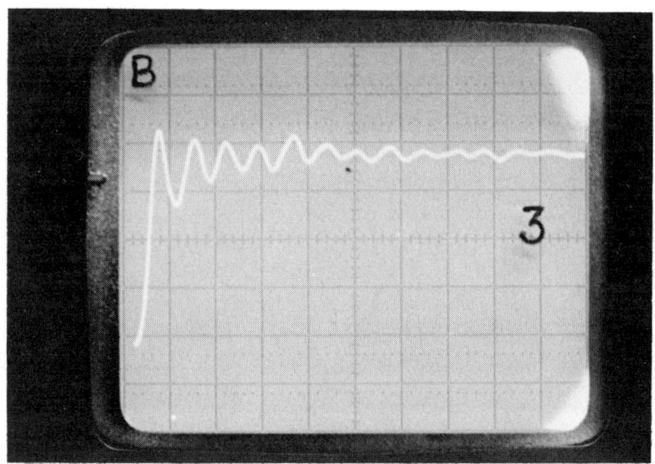

Figure 3. Pressure profile in shock wave in a bubbly liquid. The shock wave is dispersion controlled. (From Noordzij 1973)

3. Separated flows with relative velocity

The notion of a fictitious homogeneous medium becomes less easy to maintain when both phases behave differently. In the previous section we were able to account for a pressure difference by taking the pressure p in the liquid to be the external pressure as seen from the point of view of a single bubble.

Similar concepts are not so obvious when the gas phase moves at a velocity which differs from that of the surrounding liquid. It seems more plausible then to formulate equations of mass and momentum conservation for the separate phases. This has the advantage that velocities and pressures appear in a natural way in the equations of the pertinent species. The disadvantage is that there are interactions between them. These are hard to describe since their mechanisms have not been studied far enough to permit a mathematical description. Only the interaction force exerted by the isotropic part p of the stress tensor is easily incorporated. Denoting all other interactions together with I we have, indicating the velocity of the gas phase with v, as momentum equations

$$\frac{\partial}{\partial t}\rho_\ell(1-\alpha)u + \frac{\partial}{\partial x}\rho_\ell(1-\alpha)u^2 - (1-\alpha)\frac{\partial p}{\partial x} = I_\ell \tag{3.1}$$

$$\frac{\partial}{\partial t}\rho_g\alpha v + \frac{\partial}{\partial x}\rho_g\alpha v^2 + \alpha\frac{\partial p}{\partial x} = I_g \ . \tag{3.2}$$

For simplicities' sake we have assumed here that significant pressure differences like considered in the foregoing section do not occur. Nor do we consider processes as condensation or evaporisation. We suppose that thermodynamic changes can be summarized by

$$p = f(\rho_g) \ . \tag{3.3}$$

When we add to these the two relations expressing conservation of mass:

$$\frac{\partial}{\partial t}\alpha\rho_g + \frac{\partial}{\partial x}\alpha\rho_g v = o \ , \tag{3.4}$$

$$\frac{\partial}{\partial t}(1-\alpha)\rho_\ell + \frac{\partial}{\partial x}(1-\alpha)\rho_\ell u = o \ , \tag{3.5}$$

we have five equations for the unknown quantities α, p, ρ_g, u and v. However, we have still to determine the terms indicated with I. These may depend on the particular topology of the flow and as said above their determination in a particular flow is a major problem in two-phase flow. The easiest thing to do is to ignore them completely and to put the right hand sides of (3.1) and (3.2) equal to zero. We are left then with what seems at first sight a reasonable set of equations for a first try. Unfortunately, a serious problem turns up now. Suppose we want to do, and this happens all the time for instance in calculations concerning nuclear reactor safety, a Cauchy type of problem, that is to say a problem in which some set of initial values is given and the development in space and time is asked for. It turns out that such a problem is ill posed. This follows from considering the characteristics of the equations which remain from (3.1) - (3.5) if we take $I = o$ and specify (3.3) by

$$p/\rho_g = c_g^2 \ . \tag{3.6}$$

The characteristics are found by writing $\partial/\partial t = -\xi\partial/\partial x$. For ξ we find the following equation

$$\{(\xi-u)^2 - \frac{p}{\rho_\ell}\frac{1-\alpha}{\alpha}\}\{(\xi-v)^2 - c_g^2\} = \frac{pc_g^2}{\rho_\ell}\frac{1-\alpha}{\alpha} \ . \tag{3.7}$$

This equation possesses two real roots (which appear to be associated with sound waves) and in addition two complex roots, which are the cause of trouble. Their effect is to make the solution depend on the initial conditions in a non-continuous way. This can also be seen in the following way. Imagine small perturbations of the form $\exp i(kx-\omega t)$ superposed on a solution of (3.1) - (3.5) (with $I=o$ and (3.3) specified by (3.6)). Then the equation which one obtains for the quantity ω/k is the same as (3.7) for ξ. This means that for real ω the complex roots give conjugate complex values for k, leading to exponential growth of the disturbance. The practical consequences of this may be not as bad as it seems because the growth rate may be small. Nevertheless one would like to avoid having complex characteristics. They are related to Helmholtz instability which becomes apparent even more when like Stuhmiller (1977) did also ρ_g is taken as a constant. In that case all characteristics become complex. One way to get rid of them is to investigate the nature of the terms indicated with I in (3.1) and (3.2) in the hope that they contain such terms as necessary to render all characteristics real. The study of possible effects to be incorporated in the right hand side of (3.1) and (3.2) is central in two phase flow research today.

Many proposals and suggestions have been made in the recent past, none of them altogether satisfactory. The terms indicated with I contain forces exerted by one phase on the other but also effects from fluctuations in the variables of the considered phase and due to the other. If we for the moment leave out the fluctuations we could formulate equations for the whole flow. Then the interaction forces drop out and as momentum equation for a bubbly flow we could use

$$\rho_\ell(1-\alpha)\,\frac{du}{dt} + \rho_\ell(1-\alpha)u\,\frac{\partial u}{\partial x} = -\,\frac{\partial p}{\partial x}\,. \tag{3.8}$$

If we follow this approach we need to complete our set of equations with a relation between the local gas velocity v and the liquid velocity u. In the theory of suspensions at low Reynolds number such a relation is provided by

Faxen's law which expresses that in a liquid velocity field $\underset{\sim}{u}(\underset{\sim}{x})$ a particle obtains under influence of a distribution of surface forces $\underset{\sim}{F}(\underset{\sim}{x})$ a velocity

$$\underset{\sim}{U} = \frac{1}{6\pi\mu R} \int \underset{\sim}{F}(\underset{\sim}{x})dA + \frac{1}{4\pi R^2} \int \underset{\sim}{u}(\underset{\sim}{x})dA \ . \tag{3.9}$$

Can we find a similar relation in a suspension which is governed by inertial effects? To answer this question we consider the flow around a spherical gas bubble. In the absence of surface active agents the flow around the bubble can to good accuracy be described by potential flow. The calculation of the viscous dissipation in the flow is relatively easy (one should be reminded here that although the divergence of the deviatoric stress is zero the stress itself and therefore the dissipation is not) and putting this equal to the relative velocity times a drag force, gives for the latter

$$D = 12\pi\mu R(v-u) \ . \tag{3.10}$$

The potential flow provides also the concept of virtual mass m. If the potential for the relative motion is represented by ϕ we have

$$m(\underset{\sim}{v}-\underset{\sim}{u}) = \int \phi d\underset{\sim}{A} \tag{3.11}$$

The virtual mass can be interpreted as the mass of liquid which has to be accelerated along with the body when its velocity is changed in time and it is immersed in a perfect liquid.

For a sphere $m = \frac{1}{2}\rho_\ell\tau$, where τ is the volume. Combining the idea of virtual mass with the drag force described in (3.10) gives for a massless sphere

$$\frac{d}{dt}\frac{1}{2}\rho\tau(\underset{\sim}{v}-\underset{\sim}{u}) = \rho\tau\frac{d\underset{\sim}{u}}{dt} - 12\pi\mu R(\underset{\sim}{v}-\underset{\sim}{u}) \ . \tag{3.12}$$

In fact, a time dependent term like the well-known Basset term in the relation for the drag of a rigid body should be included. Recent work by Pham Dan Tam (1981) has not resulted in a simple representation of the time dependent drag. For that reason a quasi-steady approach is supposed to be permitted here. When for $t \to -\infty$ we have $\underset{\sim}{v}=\underset{\sim}{u}$, we may integrate (3.12) to obtain

$$\underset{\sim}{v}-\underset{\sim}{u} = 2 \int_{-\infty}^{t} \frac{d\underset{\sim}{u}}{dt'} \exp\left(\frac{t'-t}{T}\right) dt' \quad , \tag{3.13}$$

where the time

$$T = \frac{R^2}{18\nu} \tag{3.14}$$

is introduced. This time, of order $10^{-1} - 10^{-2}$s in a suspension with $\nu \sim 10^{-6}\ m^2/s$ and $R \sim 10^{-3}m$, is a relaxation time telling us how long it takes the liquid to adjust the velocity of a single bubble to the velocity of the liquid. For our purpose, a relation between $\underset{\sim}{u}$ and $\underset{\sim}{v}$ for bubble flow, (3.12) is not good enough, because we need a relation in a spatially nonuniform flow. It turns out not to be possible to obtain a simple relation like (3.12) for such a flow. For the important case in which the nonuniformity is brought about by the motion of neighbouring bodies, at an average distance D, Voinov, Voinov & Petrov (1973) showed that, leaving out viscous forces,

$$\frac{D}{Dt}(m\underset{\sim}{v}) - \frac{d}{dt}(m\underset{\sim}{u}) = \rho t \frac{d\underset{\sim}{u}}{dt} + O\left(\frac{R}{D}\right)^5 \ . \tag{3.15}$$

Here D/Dt is the material derivative with respect to the gas-phase and d/dt with respect to the liquid. Relation (3.15) has been discussed by several writers on two phase flow, for instance Drew et al (1979). Adding the viscous term we arrive at

$$\frac{D}{Dt}(m\underset{\sim}{v}) - \frac{d}{dt}(m\underset{\sim}{u}) = \rho\tau \frac{d\underset{\sim}{u}}{dt} - 12\pi\mu R(\underset{\sim}{v}-\underset{\sim}{u}) \tag{3.16}$$

as an approximate relation to be used in bubbly flow equations. It must be emphasized that its use can only be justified by the lack of something better.

§4. Fluctuations

When (3.8) and (3.16) are used instead of two momentum equations for the separate phases and these are complemented with equations for mass conservation, we again find complex characteristics. The real ones are associated with sound waves, the nature of the complex ones is yet not clear. Recent experiments (Mercadier 1981, Bouré and Mercadier 1982, Brennen and Bernier 1982) have revealed the existence of a new type of wave. These can best be described as propagating perturbations in the void fraction α. In these waves pressure variations are small and they travel at a speed in the neighborhood of the gas velocity v. These waves seem to be due to interactions between bubbles. Therefore, and also within the more general scope of establishing satisfactory equations for bubbly flows it is of importance to study the effects of fluctuations resulting from interactions between the bubbles. Indeed, apparently the void fraction waves just described are associated with real characteristics. The study of fluctuations is relatively new in two phase flow although formal descriptions (Ishii 1975, Nigmatulin 1978, Buyevich & Shchelchkova 1978) have been given. In the majority of these works actual interactions have not been described. A longer tradition exists in the theory of suspensions at low Reynolds number, see e.g. Batchelor (1977). In the present lecture I would like to illustrate the type of problems that are met here by looking at the stress distribution in a bubbly liquid, ignoring for convenience volume oscillations and viscous stresses. Quantities like pressure, velocity etc. fluctuate as a result of varying velocities and relative positions of the constituent inhomogeneities. We are only interested in their mean or average values. These may be volume averages, ensemble averages or surface averages. There is no time here to discuss the relative merits of these various averaging methods and we refer to the works cited above. Suppose we want to talk about the average pressure, the bulk pressure $\langle p \rangle_{bulk}$ in a bubbly suspension, where $\langle \rangle$ is understood as an ensemble average.

The ensemble consists of all the realizations of a suspension of a large number of bubbles in a volume V over which the macroscopic variables are constant. Under statistical homogeneity the ensemble average is equivalent with a volume average. Carrying this out for the pressure gives (see Van Wijngaarden 1982)

$$\langle p \rangle_{bulk} = \frac{1}{V} \int p dV = \frac{1}{V} \int_{V_\ell} p dV + \Sigma \frac{1}{V} \int_{V_B} p dV . \tag{4.1}$$

In this expression V_ℓ denotes the part of V which is occupied by liquid. Similarly V_B is the volume of one bubble and the summation is over all the bubbles. Making use of

$$(1-\alpha) = \frac{V_\ell}{V} ,$$

and introducing $\langle p \rangle_\ell$ as the average over the liquid phase only, we write the right hand side of (4.1) as

$$(1-\alpha) \langle p \rangle_\ell + \Sigma \frac{1}{V} \int_{V_B} p dV = \langle p \rangle_\ell + \Sigma \frac{1}{V} \int_{V_B} \{p - \langle p \rangle_\ell\} dV .$$

Next we introduce the quantity $\langle S \rangle$ which is the particle contribution to the bulk pressure, by

$$S = \int_{V_B} (p - \langle p \rangle_\ell) dV . \tag{4.2}$$

Herewith the relation for the bulk pressure $\langle p \rangle_{bulk}$ becomes

$$\langle p \rangle_{bulk} = \langle p \rangle_\ell + n \langle S \rangle , \tag{4.3}$$

where n is the number density, as before. The interactions between bubbles appear in the determination of S. This contains the difference between $\langle p \rangle_\ell$ and the actual pressure p which is the result of the motion produced by all other bubbles. In principle we would like to know the velocity potential due to the motion of all these bubbles. Since this is impossible we work along the lines of successive approximations. In the lowest approximation a bubble finds itself isolated in an infinite liquid. In the next a bubble is in interaction with just one other bubble. This is analogeous to binary encounters in the kinetic theory of gases. The small parameter is the concentration α, and the first approximation gives equations accurate in α, the next pro-duces equations accurate in α^2.

In the lowest approximation $\langle S \rangle$ is equal to S, which is obtained in the following way. When the volume velocity for the suspension is $\underset{\sim}{U}_o$ we consider a sphere with radius R moving in an infinite liquid with this velocity, while the sphere has velocity $\underset{\sim}{U}_g$. The latter is the average velocity of the gas phase. For this motion the potential is

$$\phi = \underset{\sim}{U}_o \cdot \underset{\sim}{r} + \frac{(\underset{\sim}{U}_o - \underset{\sim}{U}_g) \cdot R^3}{2r^3} \underset{\sim}{r} . \tag{4.4}$$

If now the pressure at infinite distance is equal to the bulk pressure $\langle p\rangle_{bulk}$ we can calculate p in every point of the liquid by using (4.4) in Bernoulli's law. However (4.2) seems to make it necessary to know the pressure inside the bubble as well. This is however not so because the volume integral may be converted into

$$\frac{1}{3}\int (\langle p\rangle_\ell - p)\, \underset{\sim}{r}\cdot d\underset{\sim}{A} - \frac{1}{3}\int (\underset{\sim}{r}\cdot\nabla p)dV .$$

For a massless sphere the volume integral is zero and the remaining surface integral can be evaluated with help of Bernoulli's theorem and (4.4). The result is

$$\langle S\rangle = S = -\frac{\pi R^3}{3}\{(\underset{\sim}{U}_o - \underset{\sim}{U}_g)\}^2 \tag{4.5}$$

Inserted into (4.3) this gives for $\langle p\rangle_{bulk}$,

$$\langle p\rangle_{bulk} = \langle p\rangle_\ell - \frac{1}{4}\alpha\rho_\ell\{(\underset{\sim}{U}_o - \underset{\sim}{U}_g)\}^2 . \tag{4.6}$$

In a similar way fluctuations contribute to the average momentum flux. These contributions are analogeous to Reynolds stresses in turbulent flow. In the lowest approximation one obtains (Van Wijngaarden 1980)

$$-\langle u'_i u'_j\rangle = L_{ij}\alpha\{(\underset{\sim}{U}_o - \underset{\sim}{U}_g)\}^2 \tag{4.7}$$

where

$$L_{11} = L_{22} = 3/20;\ L_{33} = 1/5;\ L_{12} = L_{13} = L_{23} = o .$$

Here the fluctuating velocity components are indicated with a prime. The direction indicated with 3 coincides with the direction of $\underset{\sim}{U}_o$. Hence accurate in α the momentum equation in one dimensional (mean) flow is

$$\rho_\ell(1-\alpha)\left\{\frac{\partial U_\ell}{\partial t} + U_\ell\frac{\partial U_\ell}{\partial x}\right\} = -\frac{\partial}{\partial x}\langle p\rangle_\ell + \frac{1}{20}\frac{\partial}{\partial x}\{\alpha\,(U_g - U_\ell)^2\} . \tag{4.8}$$

Here U_ℓ is the average liquid velocity. In further approximations interactions between particles are taken into account. It is to be expected that these will contribute to the understanding of the void fraction waves mentioned earlier.

REFERENCES

Batchelor, G. K. 1970. The stress system in a suspension of force-free particles, J. Fluid Mech. 41, 545.

Batchelor, G. K. 1977. Developments in micro hydrodynamics. In 'Theoretical & Appl. Mech. (Proc. 14th IUTAM Congress, Delft 1976) Ed. W. T. Koiter, North Holland.

Bernier, R. J. N. 1981. Unsteady two-phase flow instrumentation and measurement. Rept. no. E 200.4, Div. Eng. & Appl. Sc., Cal. Inst. of Technology (Pasadena).

Boure, J. A. & Mercadier, Y. 1982. Existence and properties of flow structure waves in two-phase bubbly flows. Appl. Sc. Res. 38, 297.

Brennen, C. & Bernier, R. 1982. Private communication.

Buyevich, Yu. A. & Shchelchkova, I. N. 1978. Flow of dense suspensions. Progr. Aerosp. Sc. 18, 121.

Crighton, D. G. & Ffowcs Williams, J. E. 1969. Sound generation by turbulent two-phase flow. J. Fluid Mech. 36, 585.

Drew, D., Cheng, L. & Lahey, R. T. Jr. 1979. The analysis of virtual mass effects in two-phase flow. Int. J. Multiphase Flow, 5, 233.

Ishii, M. 1975. Thermo-fluid dynamic theory of two-phase flow. Eyrolles.

Kuznetsow, V. V., Nakoryakov, V. E., Pokusaev, B. G. & Schreiber, I. R. 1978. Propagation of perturbations in a gas-liquid mixture. J. Fluid Mech. 85, 85.

Marble, F. & Wooten, D. C. 1970. Sound attenuation in a condensing vapor. Phys. Fluid 13, 11.

Marble, F. 1970. Dynamics of dusty gases. Ann. Rev. Fluid Mech. 2, 397.

Minnaert, M. 1933. On musical air bubbles and the sound of running water. Phil. Mag. 16, 235.

Matsui, G. 1975. Pressure wave propagation through a separated gas-liquid system in a duct. Winter Annual Meeting ASME, Houston, Texas.

Medwin, H. 1980. Acoustical buble spectometry at sea. In "Caviation and inhomogeneities in underwater acoustics". Ed. W. Lauterborn. Springer-Verlag.

Mercadier, Y. 1981. Contribution a l'etude des propagations de perturbations de taux de vide dans les ecoulements diphasiques eau air a bulles. These Universite Grenoble.

Morioka, S. & Matsui, G. 1975. Pressure waves propagation through a separated gas/liquid layer in a duct. J. Fluid Mech. 70, 4.

Nigmatulin, R. I. 1979. Spatial averaging in the mechanics of heterogeneous and dispersed systems. Int. J. Multiphase Flow 5, 353.

Noordzij, L. 1973. Shock waves in mixtures of liquid and air. Thesis Technological University Twente.

Roelofsen, P. 1981. Solitons in liquid/bubble mixtures. Master's Thesis, Technological University Twente.

Silberman, E. 1957. Sound velocity and attenuation in bubble mixtures measured in standing wave tubes. J. Acoust. Soc. Am. 29, 925.

Stuhmiller, J. H. 1977. The influence of interfacial pressure forces on the character of two-phase flow model equations. Int. J. Multiphase Flow 3, 551.

Tam, P. D. 1981. De la trainee instationnaire sur une petite bulle. Thesis University of Grenoble, France.

Voinov, V. V., Voinov, O. V. & Petrov, A. G. 1973. Hydrodynamic interaction between bodies in a perfect incompressible fluid and their motion in nonuniform streams. Prikl. Math. Mech. 37, 680.

Whitham, G. B. 1974. Linear and nonlinear waves. Wiley & Sons, New York.

Wijngaarden, L. van 1968. On the equations of motion for mixtures of liquid and gas bubbles. J. Fluid Mech. 33, 465.

Wijngaarden, L. van 1970. Propagation of shock waves in bubble liquid mixtures. Progr. Heat & Mass transfer 6, 637.

Wijngaarden, L. van 1972. One-dimensional flow of liquids containing small gas bubbles. Ann. Rev. Fluid Mech. 4, 369.

Wijngaarden, L. van 1980. On the mathematical modelling of two phase flows. Proc. 4th Int. Meeting on water column separation. Cagliari, Sept. 1979, Ed. M. Reali.

Wijngaarden, L. van 1982. Bubble interactions in liquid/gas flows. Appl. Sc. Res. 38, 331.

L. van Wijngaarden
Technical University Twente
Enschede, The Netherlands

Frazil Ice

G. D. Ashton

BACKGROUND AND INTRODUCTION

The purpose herein is to examine the behavior of frazil ice both in the sense of a review of our current knowledge and in the sense of the viewpoint of it as a multiphase flow. The general order of presentation is a chronology beginning with the first nucleation of a particle and extending through the evolution, transport, deposition, and (sometimes) erosion stages. Other good reviews of the subject are available, most notably those by Williams (1959), Michel (1971) and Osterkamp (1978).

What is frazil? The term "frazil" is a French word and is used largely by the community of scientists and engineers in reference to ice that occurs in the form of small discoid crystals dispersed in the flowing water of rivers. It also forms in agitated water including lake water and saline water agitated by wind. For a good review of frazil in the ocean the reader is referred to Martin (1981). In other fields the term is absent and the simple term "ice crystals" is used for the identical phenomena. In its most common form, frazil consists of small crystals produced in flowing water that has been supercooled below 0°C. The number of crystals and their size increase with time, cluster together, then flocculate and accumulate near the water surface, are transported downstream, further accumulate into larger masses called pans or floes, and eventually are

THEORY OF DISPERSED MULTIPHASE FLOW

ISBN 0-12-493120-0

deposited (upwards) to contribute to the ice cover of the water body. Subsequently, the liquid water in the interstices may freeze completely to form a solid ice cover. Even that part of the solid cover originally derived from the accumulation of the particles is often termed frazil. Some writers reserve the term frazil only for the individual crystals suspended in the flow but most use the term in the larger generic context.

Frazil is not rare. It occurs every winter season in rivers at the beginning of the ice period and may continue to form and persist through the entire winter period. Because of certain of its characteristics and because of the immense volumes that can be produced, frazil is the cause of many problems. The most commonly identified problems are the clogging of water intakes, the blockage of river flows, the filling of reservoirs by it, and the impedance of navigation. The problem in some cases has been so severe as to cause abandonment of hydroelectric plants (Michel, 1965; Carstens, 1966).

GENERAL DESCRIPTION

Before examining the various details of the behavior of frazil it is useful to give a brief descriptive overview of its behavior from first formation to ultimate deposition. When a turbulent flow open to the atmosphere has been cooled to 0°C, further loss of heat results in supercooling of the water mass. If the velocity is slow enough the first ice formation will be in the form of thin sheets of ice that remain on the surface of the flow. Above some critical velocity, however, the thin sheet cannot form and instead the ice formation will be in the form of many crystals that are dispersed in the turbulent flow. These initial crystals are most generally in the form of small, thin discs on the order of 0.1 to 1 mm in diameter. These crystals are easily transported by the turbulent flow and actively grow as long as the matrix liquid water is supercooled. As they grow in size the water warms back towards 0°C and eventually may become 0°C or above after which no further growth occurs. Meanwhile, the individual crystals form clusters and flocs, the flocs rise to the surface and form larger masses having initially a weak structure. These larger masses often have the surface appearance of round pans and tend to develop somewhat raised edges due to collision with other similar masses.

Depending on the velocity and geometry of the stream, these frazil masses may interact with each other and the shorelines to bridge across the flow and become arrested to form an intact ice cover. Above certain velocities that are rather ill-defined but on the order of 0.6 m s^{-1} the cover cannot form in this way and the frazil continues to be transported downstream. The frazil may be drawn beneath an intact downstream cover and be carried along by the flow until eventually it will deposit. If the deposit is thick it is commonly termed a hanging dam. Entire reaches may become clogged with frazil that was produced upstream leaving only a small fraction of the cross section available for flow. The deposits of frazil tend to compress somewhat and porosities of the deposited frazil are typically 0.4 to 0.7. Eventually the top layers of the deposited frazil may become part of the solid cover that freezes from above. If the entire river becomes covered with ice the heat loss directly from the liquid water to the atmosphere will be interrupted and supercooling prevented. Viscous dissipation in the flow and heat from the bottom sediments may then gradually melt some of this frazil over time but, for the most part, it will remain in place until the warm inflows of a thaw period.

FRAZIL ICE AS A TWO-PHASE FLOW

Since the viewpoint herein centers on frazil behavior as a multiphase flow it is useful to summarize the particular characteristics that distinguish it from other multiphase flows. Frazil ice is a solid-liquid multiphase flow and therefore a two-phase flow. The solid is of the same material as the liquid and further the mixture is near the homologous melting point so that exchanges of mass take place between the two phases. These exchanges are driven by heat transfer so that the analysis of frazil flows requires consideration of the coupling between heat and mass transfer. This also means that the solid fraction cannot be treated as a conservative substance in all cases. The density of the solid fraction is smaller but very close to that of the liquid fraction and the solid particles are disc-like rather than of compact form so that the size has a large effect on the transport. The solid particles, being lighter than the liquid, tend to settle upwards. Often the upper surface of the flow is a free surface rather than a solid boundary so that the receiving boundary itself is in movement and a fairly erratic movement at that. With all these characteristics in mind we now turn to consideration of the details of the various frazil behaviors.

SUPERCOOLING

The necessary precursor to frazil formation is cooling of the water below the melting point. (The term "melting point" is somewhat preferred by the author to the term "freezing point" since it is extremely difficult if not impossible to superheat ice above the melting point but very easy to supercool water below the same temperature. However, when the term "freezing point" is used here it refers to the temperature at which equilibrium exists between the solid and liquid states.) A typical water temperature-time curve is shown in Figure 1 for a parcel of water moving with the flow. When above 0°C the decrease in temperature with time is the response due to heat loss to the atmosphere, and in this region the flow is single-phase (liquid). With further cooling the temperature decreases below 0°C and at this point the flow becomes metastable, i.e., if ice nucleates the temperature will seek the equilibrium temperature of 0°C by increase in the proportion of ice present. The initial nucleation occurs at the time when the line deviates from its previous constant slope. As the crystals grow and multiply the curve "bottoms out" at a maximum supercooling. Eventually the temperature returns to a level just below 0°C, termed the residual supercooling, and remains there as long as there is continual heat loss from the water body. The shape of the supercooling curves are generally the same. Carstens (1966) performed experiments in an oval racetrack type flume such that even though the temperatures were measured in the Eulerian sense the measurements effectively represented the experience of the water in the Lagrangean sense. With increased heat losses at the surface the maximum supercooling increased, the magnitude of the residual supercooling increased, and the time duration of the rise from maximum supercooling to residual supercooling decreased.

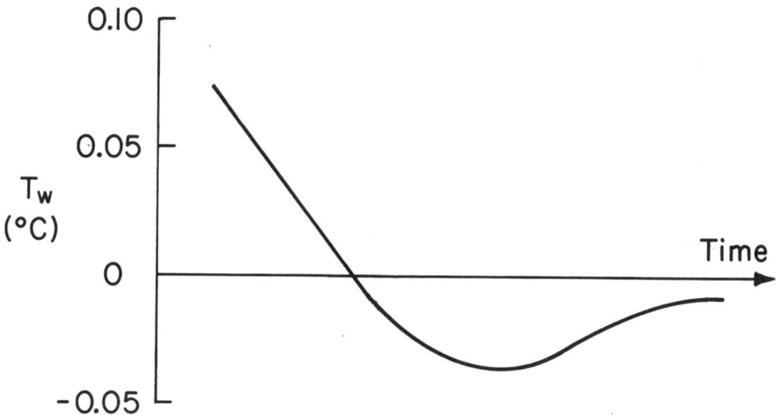

Figure 1 - Water temperature during frazil formation.

It is important to realize that the temperature-time curve shown in Figure 1 is for a parcel of water moving with the flow, rather than the temperature of the flow passing by a point. A point measurement may result in a curve such as shown and, in this case, simply reflects the situation in which the entire flow is being subjected to a similar heat loss history over the entire reach of river. Conversely, if the temperature of the water is measured downstream from a location of extreme cooling and the water temperature passes through 0°C in the course of transiting the cooling reach, then it is possible to measure a continual more-or-less constant supercooling depending on the location of the sensor.

Supercooling occurs at great depths and almost always when measured in rivers over the entire depth. One reason is simply that the time scale of vertical mixing is considerably shorter than the response time represented by the curve in Figure 1.

INITIAL NUCLEATION

Typical values of the maximum supercooling commonly observed in open channel flows are of the order of 0.01 to 0.1°C. This is a very small supercooling compared to the necessary supercooling required for nucleation of an initial ice particle in water (several °C below 0°C) as pointed out by Piotrovich (1956). This apparent paradox has led to the proposal of a number of mechanisms for production of the first crystals (see Osterkamp, 1978). Once an ice crystal is present, however, several processes can act to increase the number of crystals in the flow and include crystal multiplication processes (Chalmers and Williamson, 1965) and collision breeding of new crystal nuclei. For the case of rivers it is safe to assume that there will be initial nuclei available to start the process. The shape of the temperature curve is then determined by a balance between the overall heat loss and the growth of the individual crystals. To accomplish such a calculation requires some means of determining the number of crystals present, the rate of production of new crystals, and the crystal growth rates.

Both the growth rate of individual crystals and the rate of multiplication appear, from analogous studies of desalination crystallization processes, to depend in the magnitude of the supercooling ΔT and the turbulence characteristics of the flow. Analysis of continuous crystallizers in desalination processes (see Margolis et al., 1971, for example), suggest that the nucleation rate β depends on a separable way on ΔT and the specific power input ε such that

$$\beta = k \, \Delta T^{a} \, \varepsilon^{b} \qquad (1)$$

Such process crystallizers use specific power inputs orders

of magnitude greater than those associated with flowing rivers. For example, for a river flowing at a depth D = 1 m, with a velocity of 0.5 m s^{-1}, and a friction factor f the value of ε is

$$\varepsilon = \frac{f \rho u^3}{4D} \qquad (2)$$

and is on the order of 2 x 10^{-3} W kg^{-1}, while desalination crystallizers operate with power inputs on the order of 1 W kg^{-1}.

FRAZIL CRYSTALS

The individual frazil crystals, when in the stage of initial formation and before clustering and flocculating are dominantly disc-shaped and range in size from about 0.1 mm to 1 mm diameter and thicknesses much less than 0.1 of the diameter. The disc shape is the result initially of faster intrinsic kinetic growth rates perpendicular to the c-axis of the crystal than in the direction of the c-axis. (Thus, the c-axis is invariably normal to the plane of the disc.) Once the disc shape is established, the disc geometry is maintained by the higher rates of heat transfer in the regions of sharp curvature. Again studies motivated by desalination crystallization processes (Smith and Sarofim, 1979) provide experimental evidence on this point. It is found that an appropriate scaling parameter for the growth rates is the particle Reynolds number defined by

$$Re_d \equiv \left(\frac{\varepsilon\, d^4}{\nu}\right)^{1/3} \qquad (3)$$

Further, if the length parameter d is defined as the diameter of the sphere having the equivalent surface area as a disc, that is,

$$d_E = \left(\frac{A_s}{\pi}\right)^{1/2} \qquad (4)$$

then the transfer correlations for spheres may be used with good accuracy, not only for the disc total but also for the local coefficients appropriate for the edge of the disc (where d_E = edge diameter) and the face of the disk (where

d_f = face diameter). The result of the above findings is that it should be possible to construct a reasonably good model of the interactions between nucleation, crystal growth, and water temperature evolution during the formation stage of frazil. The test of such a model would be to reproduce the variations of the temperature-time curves that arise during frazil formation such as that obtained by Carstens (1966) shown in Figure 2.

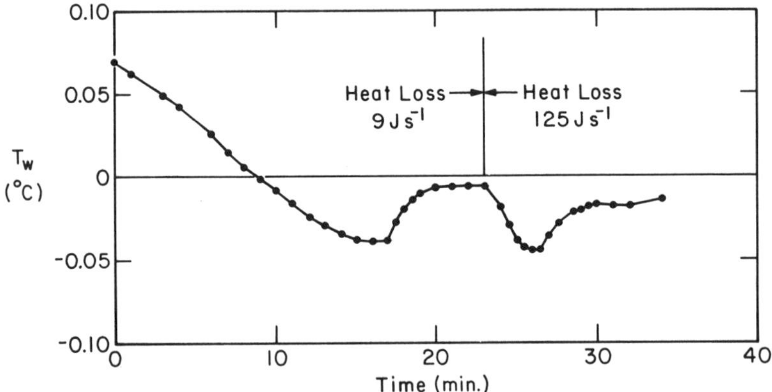

Figure 2 - Effect of change in rate of cooling on supercooling (after Carstens, 1966).

PARTICLE MECHANICS

The fall velocity of a sphere in the Stokes range ($wd/\nu < 0.1$) is

$$w = \frac{gd^2}{18\nu}\left(\frac{\rho_s - \rho_\ell}{\rho_\ell}\right) \tag{5}$$

where g is gravity, d is the sphere diameter, ν is the kinematic viscosity, and ρ_s and ρ_ℓ are the densities of the solid and liquid, respectively. A disc such as a frazil disc has a lower fall velocity than a sphere of equivalent volume and weight by a resistance factor K. A sphere with equivalent volume of a disc of diameter d_d and thickness t has an equivalent diameter d_E of

$$d_E = \left(\frac{3}{2}\right)^{1/3} \left(\frac{t}{d_d}\right)^{1/3} d_d \tag{6}$$

so that the fall velocity of a disc is given by

$$w_d = \frac{g d_E^2}{18 K \nu} \left(\frac{\rho_s - \rho_\ell}{\rho_\ell}\right) \tag{7}$$

The resistance factor K is also a function of the thickness to diameter ratio (see McNown et al., 1951) with typical values given below.

$\frac{t}{D}$	0.125	0.25	0.5	1.0
K	1.7	1.38	1.14	1.0

Substituting $g = 9.807\ m\ s^{-2}$, $\nu = 1.79 \times 10^{-6}\ m^2\ s^{-1}$, $\rho_s = 916\ kg\ m^{-3}$, $\rho_\ell = 1000\ kg\ m^{-3}$, and taking K = 1.38 corresponding to $t/d_d = 0.25$ results in

$$w_d = 9600\ d_d^2$$

where d_d is in meters and w_d is in $m\ s^{-1}$. For d_d = 0.1 mm, w_d is approximately 0.1 mm per second. For d_d = 1 mm, w_d = 1 cm per sec.

These fall velocities are very much less than those of the usual quartz-derived sediments found in rivers. The major reason is, of course, the small difference in specific gravity between ice and water as compared to quartz and water.

CLUSTERING AND FLOCCULATION

As the particles evolve they tend to form clusters that depend upon particles encountering each other. Assuming neutrally buoyant particles of radius d/2 and homogeneous isotropic turbulence, Saffman and Turner (1956) calculated collision frequencies N for n particles per unit volume as

$$N = 1.294 \left(\frac{d}{2}\right)^3 \left(\frac{\varepsilon}{\nu}\right)^{1/2} n^2 \tag{8}$$

From equation (8) we expect the likelihood of clustering to be proportional to d^3 and n^2 and that is consistent with the qualitative observation that clusters are composed of the larger frazil particles. The clusters further form flocs and these tend to rise to the surface. While the flocs seem to have a small amount of structural integrity and are able to resist the shearing forces of the surrounding turbulent flow, they do not have enough structural integrity to maintain their structure upon impact with a solid object placed in the flow. Upon reaching the surface the flocs tend to accumulate in roughly hemi-spherically shaped masses that tend to develop edges that rise slightly above the water surface. The edges are attributed to multiple collisions with other similar "pans." Particularly if these pans come to rest, but even if they don't, the surface water within the pans often freezes as a sheet of ice and provides, over time, an integrity greater than that of the simple accumulation of frazil crystals.

ACTIVE AND PASSIVE FRAZIL

When the frazil crystals are in a matrix of supercooled water they are actively growing and hence tend to cling, by freezing, to any object which they contact. When in this state the frazil is termed _active_ and distinguished from the _passive_ frazil that is characterized by the matrix water being at or above 0°C. The distinction between active and passive states of frazil is an extremely important one, since active frazil attaches to nearly all, if not all, submerged objects that do not have above 0°C surfaces. These points of attachment include the bottoms of streams where the attached ice is then termed anchor ice, objects protruding into the flow such as the bars of trashracks, and other objects such as anchor or mooring lines.

The active frazil that attaches to the bottom of streams and builds up, sometimes to considerable thickness, is termed "anchor ice." In the earlier part of this century it was speculated that anchor ice was the result of long wave radiation loss from the bottom materials but this is not the cause since water does not transmit long wave radiation. It now seems clear that the initial anchor ice is merely frazil that has contacted the bottom and frozen to it, and that it thickens by additional frazil attachment.

The masses of anchor ice may eventually float to the surface and often carry sediment material from the bottom with them. In many cases the sediment is merely fine-grained sediment but rocks of significant size have been found incorporated into the subsequent floating ice cover or, what is worse, incorporated into the turbine units of hydroelectric plants. The latter problem is a vexing one. On the one hand, plant operators should like to pull their trashracks to avoid blockage by frazil (see below) while on the other hand, they may need them present to avoid ingestion of rocks carried by frazil.

There have been more papers written on the problem of frazil ice accumulation on trashracks of water intakes than on all other aspects of frazil ice. Unfortunately, most merely repeat the same litany of cold, clear nights leading to frazil production in the river followed by flow blockage due to frazil accumulation on the bars of the trashracks. The remedies that have at times been successful include heating of the trashrack bars to above 0°C, avoidance through temporary shutdown of plant, and diversion of the floating frazil. Efforts to find an "icephobic" material either for constructing or coating the structural elements have not been very successful in terms of preventing frazil attachment but may be of some use to make easier its removal. The most common remedy of frazil blockage of trashracks is to heat the elements of the trashrack. The necessary heating required can be calculated by the usual methods of estimating the heat transfer from an object submerged in a turbulent flow with the major decision being determination of the desired surface temperature of the object; this need be only slightly above 0°C. Logan (1974) performed such calculations arbitrarily choosing surface temperatures of 0.11°C to 0.22°C. For typical trashrack geometries this led to a required power density on the order of 500 W m^{-3}. A review of actual field installations found power densities used to be on the order of 2000 to 7000 W m^{-2}.

Another means of changing the active frazil to passive frazil is to cover the flow (with an upstream ice cover), thus preventing or greatly reducing the heat loss to the atmosphere while simultaneously allowing viscous heat generation in the flow (or, to a lesser degree the heat flux from the bottom due to seasonal gain and loss of energy in the bottom sediments). The heat gain of a parcel of water moving with the flow is given by

$$\rho c \frac{dT_w}{dx} = -\rho g S \tag{9}$$

where T_w is the flow temperature, x is the streamwise distance coordinate, g is gravity, c_p is the specific heat capacity, and S is the slope. The slope may, in turn, be related to the friction factor f of a stream of depth D by

$$S = \frac{u^2}{(\frac{8g}{f}) \frac{D}{2}} \tag{10}$$

Substituting results in

$$\frac{dT_w}{dx} = \frac{fU^2}{4 c_p D}$$

for f = 0.07, c_p = 4215 J kg^{-1} $°C^{-1}$, U = 0.6 s^{-1} and D = 1 m, the rate of temperature rise is $dT_w/dx = 1.5 \times 10^{-6}$ °C m^{-1}. Thus a supercooling of 0.01°C is overcome by viscous dissipation in a distance of 6700 m. Clearly such lengths are not always practical.

Finally, even passive frazil can block intakes simply by clogging of the openings. Logan (1974) suggests that openings of the order of five to eight centimeters in trashracks are sufficiently large to prevent blockage by passive frazil accumulations.

ENTRAINMENT OF FRAZIL AT A SURFACE BARRIER

One of the fundamental concepts associated with ice in rivers is the threshold condition for passage of ice at a surface barrier such as a floating ice boom. The idealized situation is shown in Figure 3 with the barrier thickness t_c, an upstream velocity U, a depth H with ice of density ρ_i and porosity ε when deposited in the cover. The water density is denoted by ρ_w.

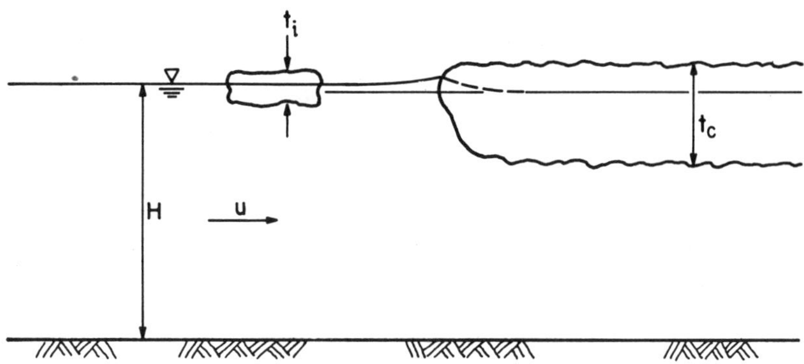

Figure 3 - Definition sketch for entrainment analysis.

In some analyses an arriving floe of thickness t_i is considered. If the floe has structural integrity it will either stop and form a cover by simple juxtaposition, or it will be entrained and form a cover in a more or less jumbled fashion, or it will be swept beneath the cover and transported downstream to a location more favorable for deposition. The solid floe case (or the frazil floe that has sufficient structural integrity to act as a unit) has been analyzed in a variety of ways (see Ashton, 1974; Michel, 1971; Pariset and Hausser, 1961; Uzuner and Kennedy, 1972). All of the analyses invoke some sort of incipient instability condition and generally it is the so-called "no-spill" condition that contends instability occurs when the top edge of the floe becomes submerged. All the data for individual floes may be reasonably represented by the result of Ashton's (1974) simple underturning moment analysis that finds incipient instability to correspond to the conditions of

$$\frac{u_c}{[gt_i(1 - \frac{\rho_i}{\rho_w})]^{1/2}} = \frac{2(1 - \frac{t_i}{H})}{[5 - 3(1 - \frac{t_i}{H})^2]^{1/2}} \qquad (11)$$

Equation (11) is shown plotted in two forms in Figures 4 and 5 with Figure 5 having the advantage that the vertical scale may be indexed as a Froude number with depth as the length parameter and the special case of $\rho_i/\rho_w = 0.916$ used for the ordinate on the right hand side.

The case of a loosely structured frazil mass is less clear. One simple analysis is to treat the accumulating cover as a stratified flow in the manner of a gravity current. The result is straightforward (Yih, 1980),

$$U_c = \left[2t_c\left(\frac{\rho_1 - \rho_2}{\rho_1}\right)\right]^{1/2} \quad (12)$$

where ρ_1 is the density of the water and ρ_2 is the (bulk) density of the deposited frazil making up the cover. If the frazil is deposited with a porosity ε then

$$\rho_2 = \rho_w \varepsilon + \rho_1(1-\varepsilon) \quad (13)$$

and equation (12) becomes

$$U_c = \left[2t_c g\left(1 - \frac{\rho_i}{\rho_w}\right)(1-\varepsilon)\right]^{1/2} \quad (14)$$

Equation (14) seems to accord well with field and laboratory experience with both frazil and ice pieces more usually termed blocks and the usual estimates of ε are in the range 0.4 to 0.6. Equation (14) is a local condition set by the local velocity and depth at the upstream edge of a cover. Thicker accumulations are possible by other mechanisms such as internal shoving and continued deposition of frazil at low velocity reaches.

DEPOSITION AND EROSION OF FRAZIL

This leads us to the difficult question of determining the conditions for deposition of frazil and of erosion of already-deposited frazil masses. It is tempting to use the classic diagram of Shields (Vanoni, 1975) that appears to be valid for a wide range of specific gravities. For a frazil particle of diameter d = 1 mm, and a specific gravity of 0.916 the quantity

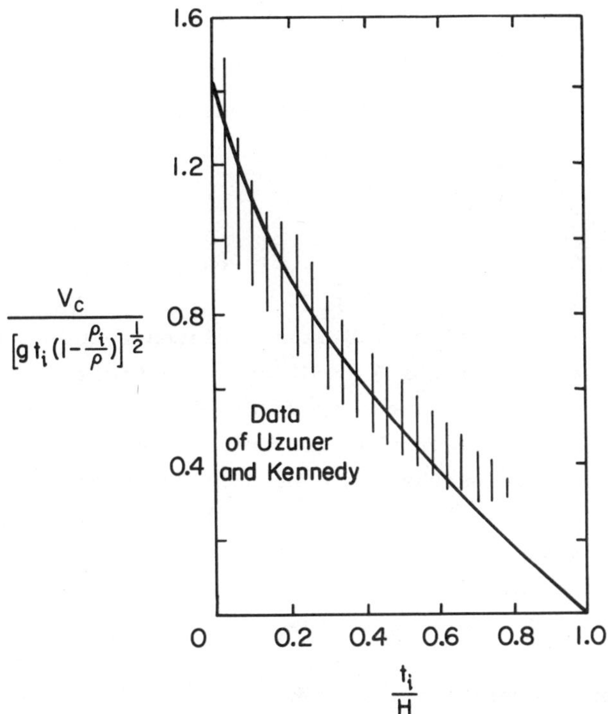

Figure 4 - Floe entrainment with thickness as length scale

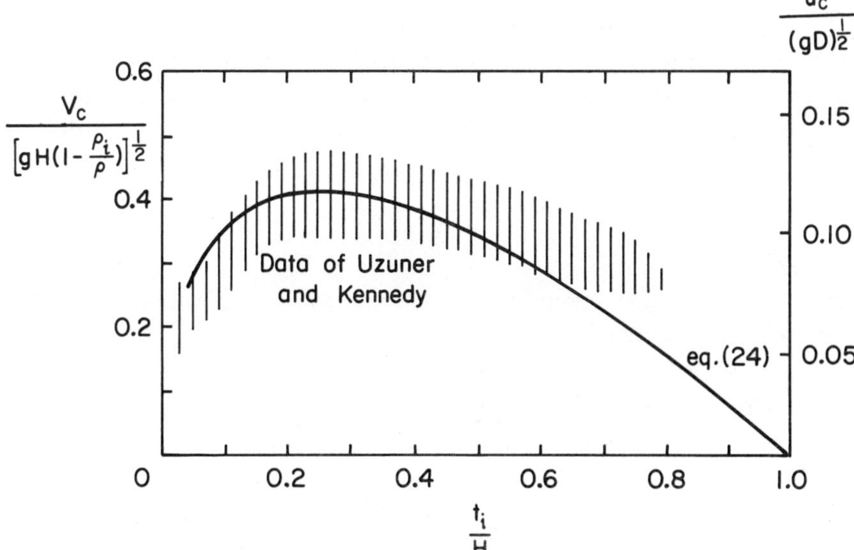

Figure 5 - Floe entrainment with depth as length scale

$$\frac{d_s}{\nu}\left[0.1\left(1-\frac{\rho_i}{\rho_w}\right)gd_s\right]^{1/2}$$

has a value of 5.07 and yields a dimensionless shear stress

$$\tau_* = \frac{\tau_o}{g(\rho_w - \rho_i)ds}$$

of $\tau_* = 0.035$. This corresponds, in turn, to $\tau_o = 0.029$ N m^{-2}. Using the relationship $U_* = \sqrt{\tau_o/\rho_w}$ and $U/U_* \approx 10$ the threshold velocity U is calculated to be on the order of 5 cm s^{-1}. For a 5 mm diameter particle the threshold velocity is on the order of 12 cm s^{-1}. These velocities are far lower than the critical erosion or deposition velocities that have been found by experience to be on the order of 1 m s^{-1}. Undoubtedly the higher critical velocities of erosion or deposition that are suggested by experience are characteristic of the flocs of frazil rather than the individual disc-like frazil crystals. It is also a commonly held opinion that the critical deposition velocity is less than the critical erosion velocity of a previously deposited frazil mass and presumably this is due to cohesion of the deposited frazil. The value of the critical erosion and/or deposition velocities is of some importance since that, together with the other hydraulic variables, determines the area of a frazil clogged channel that remains unobstructed by the frazil deposits. Similarly knowledge of the threshold velocities would allow simulation of the river-delta-like deposits of frazil that are often found where frazil enters reservoirs.

INSTRUMENTATION

Actual measurement of frazil behavior is difficult. The research instrumentation that is most needed is a device to monitor the volumetric concentration of frazil in a flow. Kristinsson (1970) reported a device consisting of a spiral, rod-shaped double element conductivity meter that detected the concentration of surface accumulations of frazil upstream from the Burfell hydroelectric project in

Iceland. The principal of operation was the change in conductivity as the ice content of the water increased. It is applicable to that case but not sensitive enough for the small concentrations that exist prior to clustering and flocculation. Osterkamp et al. (1975) observed changes in electrical conductivity of flowing water at the time of frazil production and, by associating the increase in conductivity to rejection of impurities during ice growth, estimated frazil concentrations on the order of 1.1 to 2.3%. Laser doppler velocimeters have been used in laboratory situations (Schmidt, 1974) but appear to be impractical for field or large scale laboratory studies. Calorimeter type measurements are difficult if intended to be used for continual sampling, again because of the low ice concentrations.

SUMMARY

Frazil ice in rivers constitutes a two-phase flow of solid particles in the liquid phase of the same material. The initial behavior is dominated by nucleation characteristics and thermal flows between the ice particles and the liquid at the microscale and the turbulent structure of the flow. As the crystals evolve, clustering and flocculation occur and the frazil portion of the flow tends to collect near the surface of the flow. Eventually the frazil is deposited and may be incorporated into the solid ice cover by ordinary thickening of the top surface ice sheet. While qualitative understanding of frazil behavior is adequate to provide "rules of thumb" the phenomena has not been considered rigorously within the context of two-phase flow theory. Such treatment should be useful in furthering our understanding of frazil ice and lead to better means of dealing with its associated problems.

REFERENCES

Ashton, G.D. (1974) Froude criterion for ice block stability. J. Glaciology, 13:307-13.

Carstens, T. (1966) Experiments with supercooling and ice formation in flow water. Geofys. Publ. 26:1-18.

Chalmers, B. and R.B. Williamson (1965) Crystal multiplication without nucleation. Science 148:1717-18.

Kristinsson, B. (1970) Ice monitoring equipment, Proceedings Symposium on Ice and Its Action on Hydraulic Structures, Intl. Assoc. Hydraul. Res., Reykjavik, Iceland, paper No. 1.1, 14 pp.

Logan, T.H. (1974) The prevention of frazil ice clogging of water intakes by application of heat. U.S. Bur. Reclam. Rept. REC-ERC-74-15, 21 pp.

Margolis, G., T.K. Sherwood, P.L.T. Brian and A.F. Sarofim (1971) Well stirred ice crystallizer, Ind. Eng. Chem. Fundam., 10:439-452.

Martin, S. (1981) Frazil ice in rivers and oceans, Ann. Rev. Fluid Mech. 13:379-97.

McNown, J.S., J. Malaika, and R. Pramanik (1951) Particle shape and settling velocity, Transactions, 4th Meeting of he International Association for Hydraluic Research, Bombay, India, pp. 511-522.

Michel, B. (1965) The metamorphosis of frazil ice in rivers, Trans. Engrg. Inst. Canada, 8: no. A-5, 9 pp.

Michel, B. (1971) Winter regime of rivers and lakes, Monograph III-B1a, Cold Reg. Res. Eng. Lab., Hanover, N.H. 131 pp.

Osterkamp, T.E., R.E. Gillfilian and C.S. Benson (1975) Observations of stage, discharge, pH, and electrical conductivity during periods of ice formation in a small subarctic stream, Water Resources Res., 11:268-272.

Osterkamp, T.E. (1978) Frazil ice formation: a review, J. Hydraul. Div. ASCE, 104:1239-1255.

Pariset, E. and R. Hausser (1961) Formation and evolution of ice covers on rivers, Trans. Engrg. Inst. Canada, 5:41-49.

Piotrovich, V.V. (1956) Formation of depth ice. Priroda (:94-95 (in Russian); Def. Res. Board, Canada. Translation T235R, 1956.

Saffman, P.G. and J.S. Turner (1956) On the collision of drops in turbulent clouds, J. Fluid Mech. 1:16.

Schmidt, C.C. (1974) Development and evaluation of a laser doppler velocimeter system for measuring frazil ice concentration, M.S. Thesis, Univ. Iowa.

Smith, K.A. and A.F. Sarofim (1979) Fundamental studies of desalination by freezing, Dept. Chem. Engrg., Mass. Inst. Tech., Final report to Office of Water Research and Technology, U.S. Dept. Interior, Wash. D.C., 69 pp.

Uzuner, M.S. and J.F. Kennedy (1972) Stability of floating ice blocks. J. Hydraul. Div. ASCE, 98:2117-33.

Vanoni, V.A. (1975) Sedimentation Engineering, ASCE Manual and Report No. 54, 745 pp.

Williams, G.P. (1959) Frazil Ice - A Review of its properties with a selected bibliography, Eng. J. Canada 42:55-60.

Yih, C-S (1980) Stratified Flows, Academic Press, New York, 418 pp.

ACKNOWLEDGMENT

This work was performed under U.S. Army Corps of Engineers Civil Works research project CWIS 31750 "Prediction of Ice Formation."

George D, Ashton
U.S. Army Cold Regions Research
and Engineering Laboratory
72 Lyme Road
Hanover, NH 03755

Some Mathematical and Physical Aspects of Continuum Models for the Motion of Granular Materials

R. Jackson

INTRODUCTORY IDEAS

Quantitative treatment of the mechanics of a granular material can be traced back to Coulomb [1], whose name is associated with a model of the material as an elastic-plastic continuum which yields by shearing on planes where the shear stress T first reaches a value related to the normal stress N by

$$T = N \tan \phi + c \tag{1}$$

where ϕ and c are parameters characteristic of the material.* If c takes a non-vanishing positive value the material is said to be cohesive. We shall be concerned mainly with non-cohesive materials (typically coarse sand) for which the only material parameter is ϕ, called the angle of internal friction.

In the non-cohesive case the Coulomb yield condition (1) reduces to a proportionality between T and N and can be shown to be equivalent to the following proportionality between the major and minor principal stresses

---

*Throughout this work the stress tensor is defined in the compressive sense, such that $\underline{n} \cdot \underline{T}\, dS$ is the traction exerted across a surface element dS with normal $\underline{n}$, acting on the material into which the vector $\underline{n}$ points. This is opposite to the conventional sense used in fluid mechanics.

ISBN 0-12-493120-0

$$\frac{\sigma_{maj}}{\sigma_{min}} = \frac{1 + \sin\phi}{1 - \sin\phi} \tag{2}$$

This is shown in Fig. 1 where the shaded region is that part of the plane of principal stresses (σ_1, σ_2) in which the material behaves elastically, while plastic yield occurs at the boundary, defined by eqn. (2).

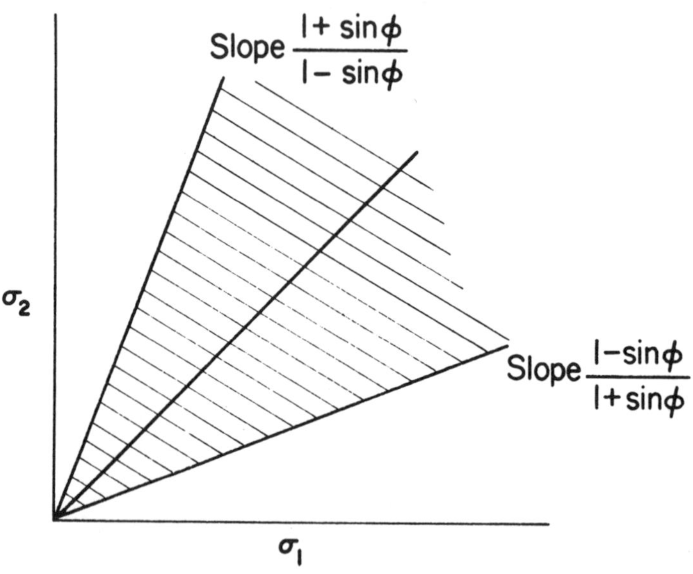

Figure 1. Coulomb yield condition

A convenient geometric representation of the state of stress at any point is provided by the Mohr construction, shown in Fig. 2. A circle is constructed about a center C on the horizontal co-ordinate axis, such that the abscissas of its intersections with this axis, σ_1 and σ_2, are equal to the principal stresses. Then the tangential and normal components of the traction on a plane whose normal makes an angle θ with the major principal stress axis are given by the distances PQ and OQ, respectively.

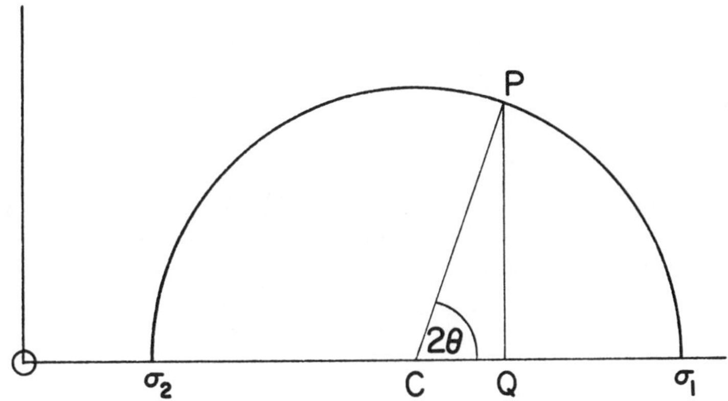

Figure 2. Mohr circle construction

In terms of this construction the Coulomb yield condition is met when the Mohr circle representing the state of stress touches a straight line of slope tan ϕ, as shown in Fig. 3a, and the ratio T/N then reaches the critical value tanϕ on slip planes disposed symmetrically about the principal axes of stress, with the orientations shown in Fig. 3b.

For a material which is about to yield, or is yielding slowly enough that inertial effects can be neglected, the components of force balance yield a pair of partial differential equations in the three elements of the stress tensor. The equations are closed by the algebraic relation between the elements of the stress tensor provided by the Coulomb yield condition. This awkward system can be simplified by a choice of variables introduced by Sokolovski [2] to describe the state of stress. The arithmetic mean σ of the principal stresses and the orientation of the principal stress axes are taken as independent stress variables, the orientation being specified by the angle γ measured clockwise from the x-axis to the direction of the major principal stress axis. Then it is easy to show that a stress tensor with elements

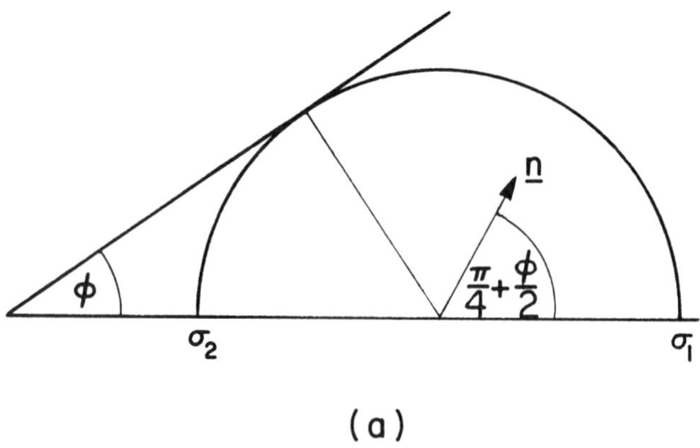

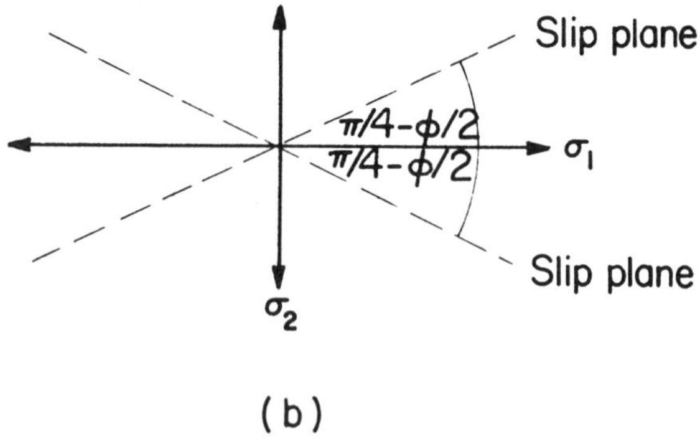

Figure 3. (a) Yield condition in relation to Mohr circle
(b) Orientation of slip planes

$$\sigma_{xx} = \sigma(1 + \sin\phi\cos 2\gamma), \quad \sigma_{yy} = \sigma(1 - \sin\phi\cos 2\gamma)$$
$$\sigma_{xy} = -\sigma\sin\phi\sin 2\gamma \tag{3}$$

satisfies the Coulomb yield condition.

In terms of the Sokolovski stress variables the x and y components of force balance take the form of a pair of first order hyperbolic partial differential equations:

$$(1 + \sin\phi\cos 2\gamma)\frac{\partial\sigma}{\partial x} - 2\sigma\sin\phi\sin 2\gamma\frac{\partial\gamma}{\partial x}$$
$$-\sin\phi\sin 2\gamma\frac{\partial\sigma}{\partial y} - 2\sigma\sin\phi\cos 2\gamma\frac{\partial\gamma}{\partial y} = \rho g_x \quad (4)$$

and

$$-\sin\phi\sin 2\gamma\frac{\partial\sigma}{\partial x} - 2\sigma\sin\phi\cos 2\gamma\frac{\partial\gamma}{\partial x}$$
$$+ (1 - \sin\phi\cos 2\gamma)\frac{\partial\sigma}{\partial y} + 2\sigma\sin\phi\sin 2\gamma\frac{\partial\gamma}{\partial y} = \rho g_y \quad (5)$$

These have been applied to a number of problems of incipient yield in soil mechanics, and an excellent account of this work is provided by Sokolovski [2].

The extension of these ideas to three dimensions is, in principle, straightforward. The pair of lines which defines the yield condition in the (σ_1, σ_2)-plane generalizes to a conical surface with the (1,1,1) direction as its axis in the space of the three principal stresses $(\sigma_1, \sigma_2, \sigma_3)$. The vertex of the cone lies on this axis, in the negative octant for cohesive materials and at the origin for non-cohesive materials. The section normal to (1,1,1) must be invariant under interchanges of the coordinate axes, since these correspond to a mere re-labeling of the principal stresses. The simplest sections with this property are circular (von Mises) or hexagonal (Tresca), as sketched in Fig. 4. For simplicity we shall confine attention to two-dimensional systems.

The yield condition, in itself, has nothing to say about the nature of the motion which is initiated at yield. Consequently alone it is not enough to provide the basis for a theory of granular media in continued motion, nor does it even permit problems of incipient yield to be solved if kinematic boundary conditions are involved. Clearly it must be supplemented by some constitutive relation involving the kinematics of motion at yield, and Coulomb's original concept contained ideas of this sort in its postulate of yield by shear on slip planes. This type of picture was formalized by Mandel [3], and later by de Jong [4], who regarded the motion following yield as compounded additively from plane shears on the two slip planes.

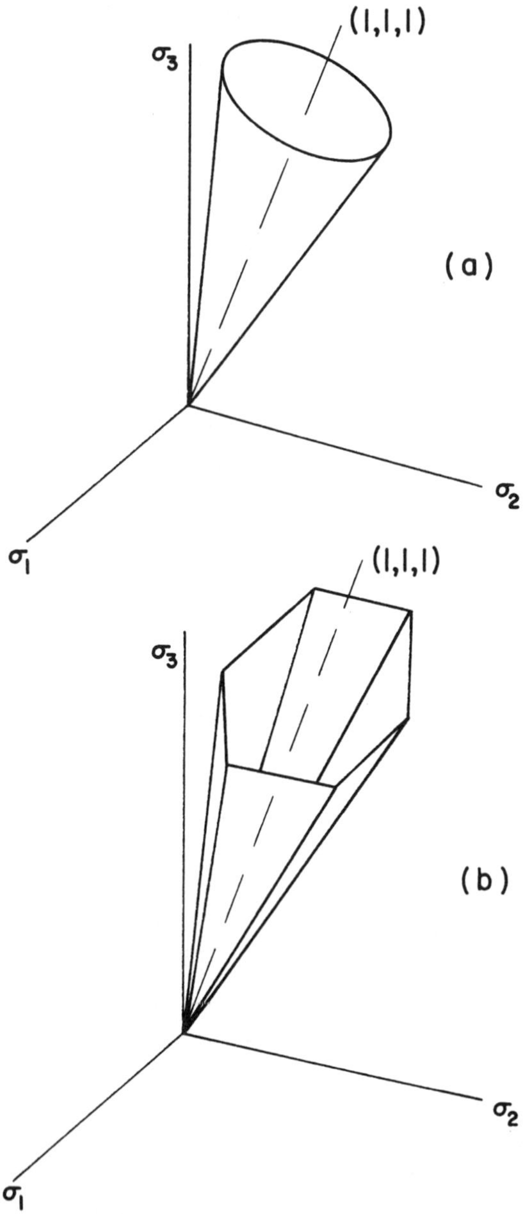

Figure 4. Three-dimensional yield surfaces
(a) von Mises type (b) Tresca type

A quite different feature of the kinematics of yield was revealed by two elegant experiments reported by Reynolds [5]. In the first of these a well-compacted sand was sealed in a tightly fitting, flexible bag, with the interstitial space completely filled with water. The bag was then found to be quite rigid, despite the fact that it could be deformed easily when the water was replaced by air.

In the second experiment the bag was sealed to a glass tube held vertically, and the interstices were again filled with water until a head of water stood in the tube above the bag. On squeezing the bag between wooden boards it was found that the water level in the tube fell initially, contrary to one's intuitive expectation. From these observations Reynolds concluded that deformation of a granular material is necessarily accompanied by a dilation, or reduction in bulk density. In the case of the sealed bag, since the total volume of water and solid material is constant, dilation cannot occur, so deformation is not possible.

Dilation does not appear as a natural consequence of Coulomb's kinematic ideas or their subsequent developments mentioned above, but is intrinsic in an alternative approach to the kinematics of yield, based on plasticity theory, proposed by Drucker and Prager [6]. In this picture the rate of deformation tensor and the stress tensor at yield are related through the plastic potential flow rule. This imposes the following conditions:

(a) The principal axes of rate of deformation are aligned with those of stress, with the major principal rate of deformation axis parallel to the minor principal stress axis.

(b) The ratio of the principal rates of deformation is equal to the ratio of the components of the inward normal to the yield surface in the space of the principal stresses.

Condition (b) is illustrated in Fig. 5 for the case of a Coulomb yield surface. The ratio of the major to the minor principal rate of deformation is then $-(1+\sin\phi)/(1-\sin\phi)$ and the trace of the rate of deformation tensor is always positive. Thus, deformation is always accompanied by dilation, and this arises naturally from the theory.

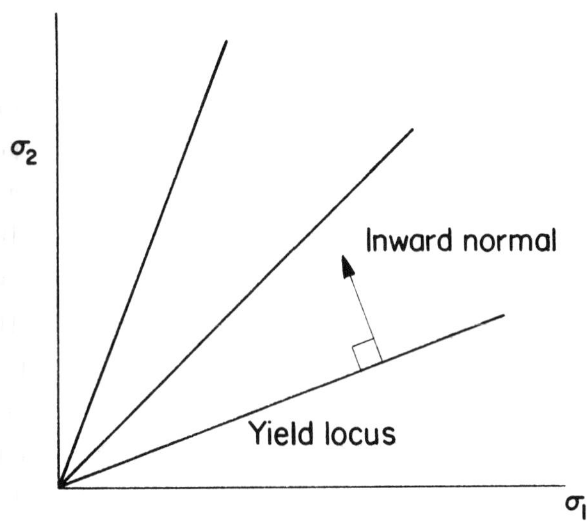

Figure 5. Plastic potential flow rule for a Coulomb powder

Despite this attractive feature the plastic potential flow rule generates other problems, which we will examine later when discussing constitutive relations in more detail.

Both types of constitutive relation sketched above have one feature in common; the stress tensor is unchanged if all elements of the rate of deformation tensor are multiplied by a common factor. This is the distinguishing feature of dry friction and is quite different from what is observed in the case of a Newtonian fluid, where the stress components increase in proportion to the rate of deformation components. More formally, the constitutive relation for a granular material at low rates of shear is of degree zero in the elements of the rate of deformation tensor, and this feature invalidates many intuitive ideas which one might be tempted to carry over from fluid mechanics.

CONSTITUTIVE RELATIONS

Constitutive relations based on Coulomb's idea and formalized by Mandel [3] and de Jong [4] have been developed further by Spencer [7], Mandl and Luque [8], de Jong [9, 10], and Mehrabadi and Cowin [11], [12]. The alternative Drucker-

Prager type of model was extended by Shield [13], and by Jenike and Shield [14], and was considerably elaborated by the Cambridge school of soil mechanics (Roscoe, Schofield and Wroth [15], Schofield and Wroth [16], Roscoe [17]), who showed how to circumvent some difficulties which had early been recognized [14].

A prime distinguishing feature of the two models is that the latter requires the principal axes of stress and rate of deformation to be aligned at all times, while the former predicts substantial misalignment in certain circumstances. Though there have been attempts to detect and quantify such misalignment (see, for example, Drescher [18]), the situation is still rather confused. While it seems certain that misalignment does occur, the available evidence is not sufficient to decide whether it is of the sort predicted by the constitutive theories based on Coulomb's idea. Thus, it does not provide an adequate basis for selecting this type of theory in preference to one based on the plastic potential, with co-axiality of the stress and rate of deformation tensors. In what follows we will formulate a constitutive model based on the "critical state" ideas of the Cambridge school, and therefore belonging to the second class of models. It is not intended, thereby, to imply that these are physically superior, but it does serve to give some insight into the difficulties, both mathematical and physical, which must be faced in formulating and solving problems of granular material flow.

First we describe qualitatively the results which might be obtained in a test where a layer of the material is sheared between infinite parallel plates under a fixed normal load. This can be regarded as an idealization of practical tests with a shear-cell type of apparatus, and the results will be viewed as salient features of the behavior of granular materials, which any constitutive theory should predict properly.

The plates are supposed to be rough, so that the material does not slip in contact with them, the lower plate is fixed, and the upper plate carries a normal load N per unit area. Initially it is assumed that the granular material between them has uniform bulk density ρ_o, and the shear stress T on the upper plate is gradually increased so that the sample between the plates deforms. The shear stress just needed to sustain a continuing shear of the material is then measured for increasing lateral displacement s of the upper plate. This arrangement is sketched in Fig. 6.

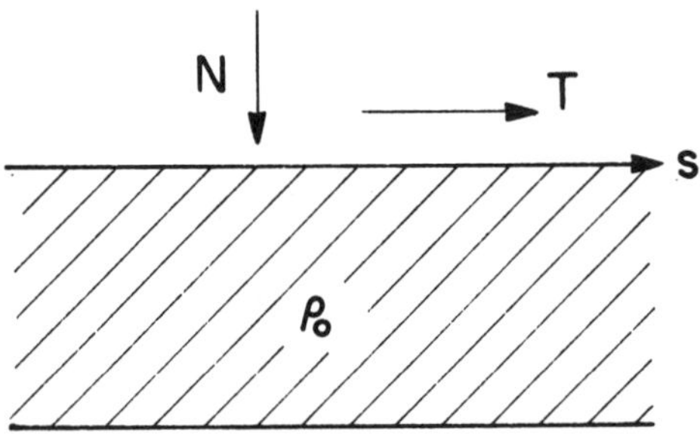

Figure 6. Plane shear test

The behavior observed depends on the relative values of the normal stress N and the initial bulk density ρ_o. When ρ_o is large or N is small the situation is as shown in Fig. 7. For small values of the displacement, T increases more-or-less linearly with s, and the curve is retraced if T is reduced. This is the domain of reversible elastic behavior, and it is terminated by the point denoted by F. The corresponding value of the shear stress, T_F, depends on both ρ_o and N. Beyond F, T decreases as shear continues, and the T v. s curve is no longer retraceable. At large values of s, T approaches an asymptotic value T_∞, which is independent of ρ_o, and is therefore a function of the normal stress only. This behavior is sketched in Fig. 7a, while Fig. 7b shows the displacement

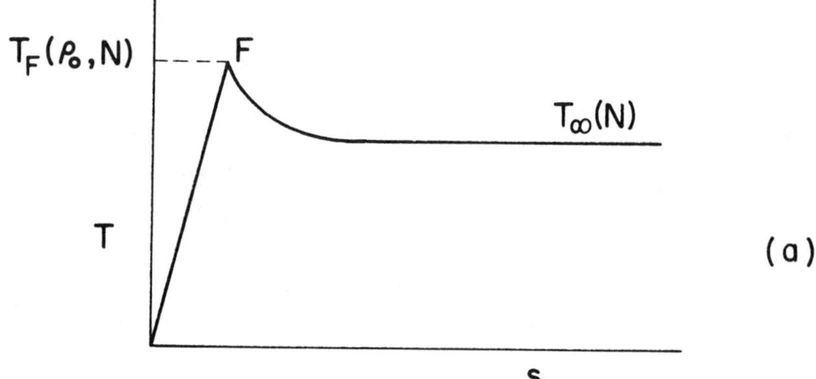

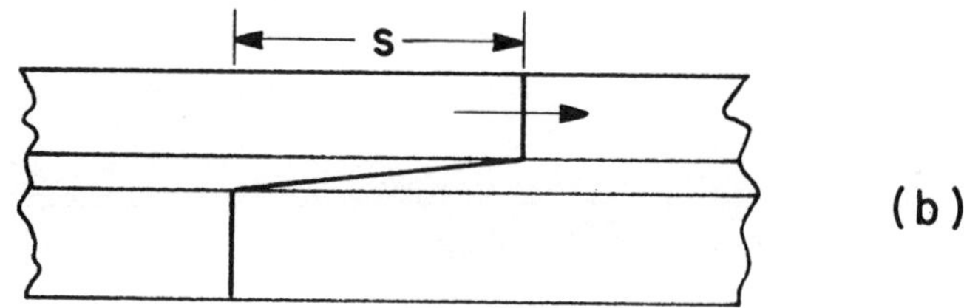

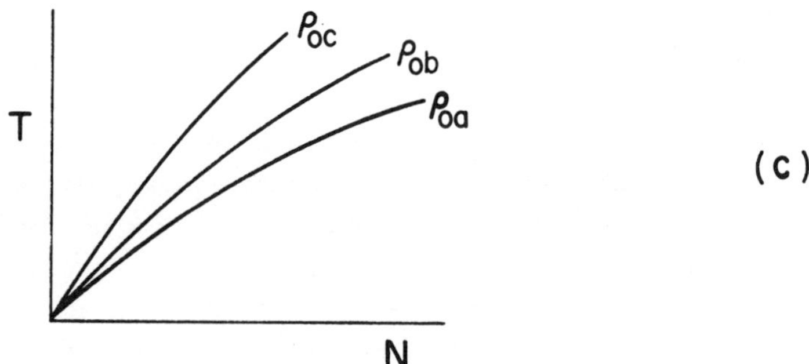

Figure 7. Failure in plane shear

of the material as a function of position between the plates, for a fixed and large value of s. Typically, the shear is confined to quite a thin layer, on each side of which the material moves en bloc.

The phenomenon described is referred to as failure of the material, and a curve relating T_F to N, for a fixed value of ρ_o, is called a failure locus. Fig 7c shows sketches of a set of failure loci, for different values of ρ_o, for a typical non-cohesive material.

Clearly the decrease in the value of T needed to maintain shear after passing the failure point is associated with weakening of the material during plastic deformation. This also accounts for the localization of the shear in a narrow layer. If measurements of local bulk density were made they would show a decrease in ρ within the shear layer but, because this is thin, the overall dilation, observed as a normal displacement of the upper plate, is quite small.

When ρ_o is small or N is large, the situation is quite different and is illustrated by Fig. 8. Once again there is a region of elastic behavior for sufficiently small values of s, and this terminates at the point C, beyond which the T v. s relation can no longer be retraced reversibly. The corresponding value of the shear stress is T_c, and it depends on both ρ_o and N. Beyond C, T continues to increase as s increases, eventually approaching an asymptotic value T_∞, which is again independent of ρ_o. These features are sketched in Fig. 8a. As indicated in Fig. 8b, and in contrast to the previous case, the shear is now distributed uniformly across the layer, and a measurement of local bulk density would reveal a uniform increase in ρ throughout the material as deformation proceeds. This is consistent with the increasing strength indicated in Fig. 8a, and it manifests itself externally by a substantial normal displacement n of the upper plate, which accompanies the lateral displacement s. Because of this, the phenomenon is referred to as consolidation, and a curve relating T_c to N for a fixed value of ρ_o is called

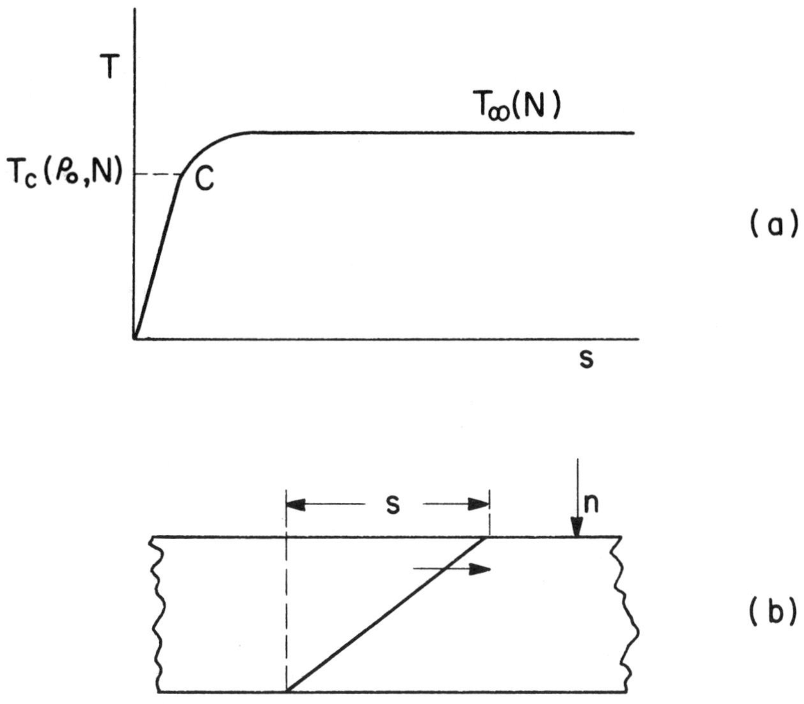

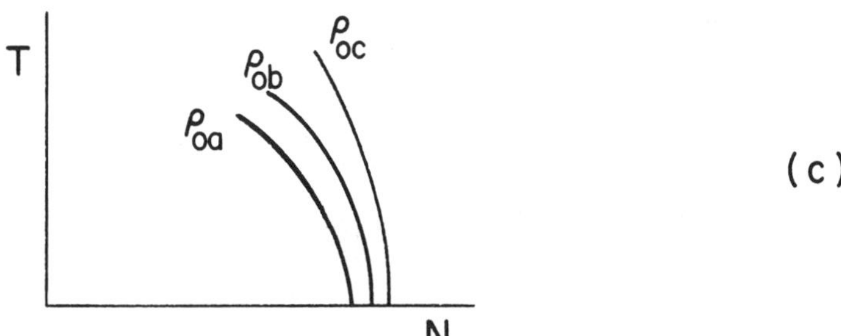

Figure 8. Consolidation in plane shear

a consolidation locus. A set of these loci, for different values of ρ_o, is sketched in Fig. 8c. Note that they intersect the $T_c = 0$ axis for sufficiently large values of N. At such an intersection point the material consolidates under a normal load alone, without any assistance from shear stress.

For most non-cohesive materials consolidation loci are difficult to observe since they correspond to rather small values of ρ_o. Indeed, the consolidation due to gravity alone often represents virtually all that can be achieved without very large normal loads, which may fracture the individual particles. There is no such difficulty in the case of certain cohesive materials, such as clays, and it is from measurements on this type of material that we derive our physical ideas on consolidation.

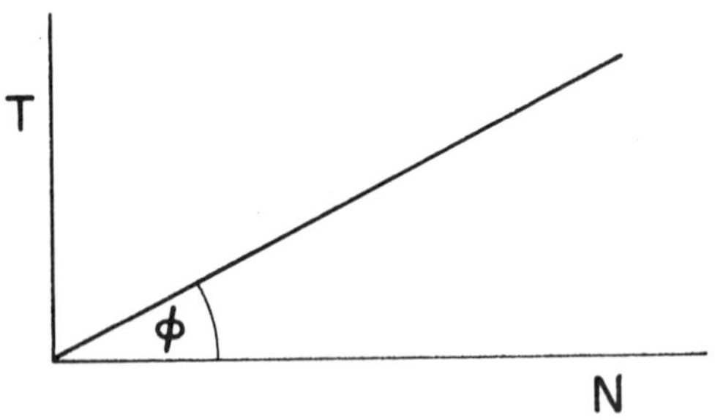

Figure 9. Asymptotic relation between T and N for plane shear

Finally, there is evidence that a plot of the asymptotic shear stress T_∞ against N is very close to a common straight line through the origin, as shown in Fig. 9, for results from both failure and consolidation tests. Thus, after extensive deformation, the granular material approaches the behavior of a Coulomb powder, at least so far as the relation between T and N is concerned. Furthermore, dilation or compaction ceases as the asymptote is approached, and deformation then continues without any further change in bulk density.

The simple Drucker-Prager model introduced earlier clearly cannot account properly for the behavior just described. For a Coulomb yield locus we have seen that the plastic potential flow rule predicts a large and continuing dilation as deformation proceeds, regardless of the initial bulk density or state of stress, whereas, in practice, either dilation or compaction may accompany deformation. Of course, the experiments also show that the failure and consolidation loci are not of the Coulomb form and depend quite strongly on the initial bulk density. Nevertheless, the contradiction remains, since the observed asymptotic relation between T and N, for large deformation, is of the Coulomb form, while deformation then continues without change in volume, in contrast to the Drucker-Prager prediction of rapid dilation [6] [14].

Another inadequacy of the Drucker-Prager model is revealed on calculating the dissipation rate, which is found to be zero for a yield locus of the non-cohesive Coulomb form. This is quite inconsistent with our physical picture of resistance to deformation being associated with friction between particles in contact. It is not difficult to show [19] that the yield locus must be convex if the plastic potential flow rule is to predict a positive, and therefore acceptable, dissipation rate.

The evidence summarized in Figs. 7-9 can be accounted for quite compactly by adopting a viewpoint put forward by the Cambridge school of soil mechanics and extensively discussed by Schofield and Wroth [16]. For a given bulk density ρ_o the failure and consolidation loci are regarded as images in the (T,N)-plane of different arcs of a single yield locus in the plane (σ_1, σ_2) of the principal stresses. This locus has a shape of the sort indicated in Fig. 10, which also shows the directions of the inward normals to the locus at three points, 1, 2 and C. At point 1, corresponding to a low value of the mean stress, the inward normal has a positive projection on the (1,1) line so, according to the plastic potential flow rule, yielding is accompanied by dilation. At point 2, on the other hand, corresponding to

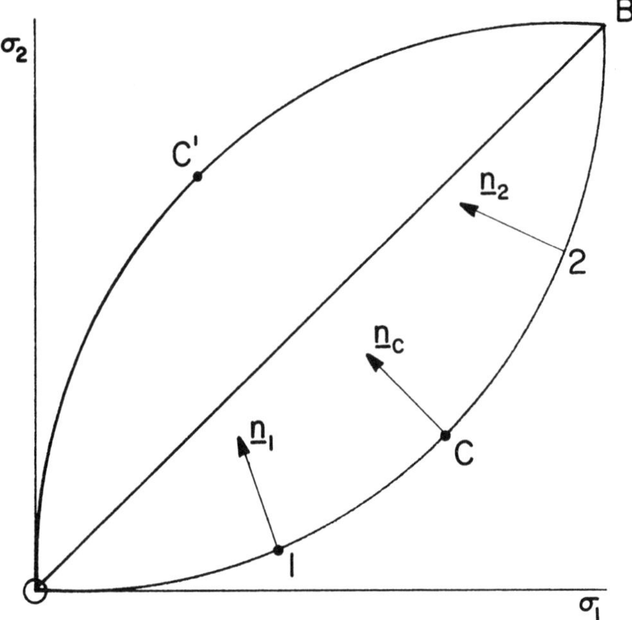

Figure 10. Yield locus, illustrating dilation, compaction and critical states

a high value of the mean stress, the inward normal has a negative projection on (1,1), and yielding is accompanied by compaction. At the pair of points C, C', the inward normal is orthogonal to (1,1), so the material yields without change in volume. These are called <u>critical states</u> by Schofield and Wroth [16].

With the plastic potential flow rule (but not otherwise [20]) this locus can be translated into a corresponding (T,N) yield locus by drawing the set of all Mohr circles corresponding to points on the (σ_1, σ_2)-locus. The envelope of these circles then gives the (T,N) yield locus, and the images of the arcs OC and CB in Fig. 10 form the failure and consolidation loci, respectively. Thus there is a one-one relation between the yield locus in the (σ_1, σ_2)-plane and the failure locus-consolidation locus pair in the (T,N)-plane.

Fig. 10 shows the yield locus for only one value of the bulk density. Similar loci exist for all other values of ρ, as is shown in Fig. 11. The loci are drawn at equal increments in ρ, and their increasingly rapid growth with increasing ρ is typical of a non-cohesive material.

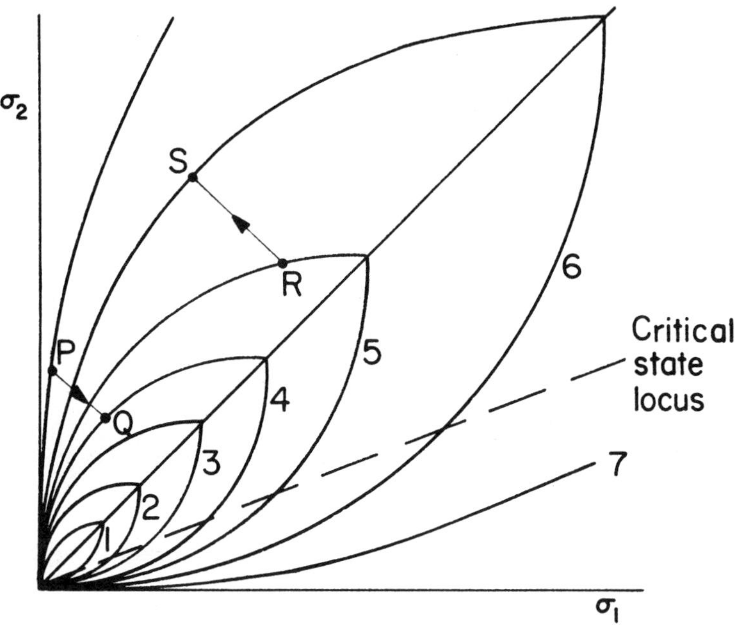

Figure 11. Set of yield loci for various densities

Clearly, if yield is initiated at a point on a yield locus between the critical state and the origin, it will be accompanied by dilation. Then, if the mean stress σ (= $(\sigma_1 + \sigma_2)/2$) is held constant, dilation will continue until a bulk density is reached which corresponds to the critical state at this value of σ. Deformation will then proceed with no further change in density. This sequence of events corresponds to the path PQ in Fig. 11. On the other hand, if yield is initiated at a point beyond the critical state, such as R in Fig. 11, it will be accompanied by compaction. Then, if σ is held constant, the compaction will continue until a critical state is reached, as indicated by the path RS. Thus, we see that this model accounts correctly for the shear-weakening and shear-strengthening behavior associated with yield on the failure and consolidation loci, respectively, as illustrated in Figs. 7 and 8, and also shows why T approaches an asymptote after large deformation.

Finally, to reproduce the observed linear relation between T_∞ and N shown in Fig. 9, we need only postulate that the set of critical states for all the yield loci lie on a pair of straight lines through the origin, with slopes $(1+\sin\phi)/(1-\sin\phi)$ and $(1-\sin\phi)/(1+\sin\phi)$, respectively. These are indicated in Fig. 11. From this point of view, the observed proportionality between T and N after large deformations does not imply that the yield locus degenerates into the Coulomb form, but merely reflects the shape of the locus of critical states.

There is now no difficulty with the dissipation inequality. Because of the convexity of the individual yield loci, on whose form the flow rule is based, the dissipation rate is positive even at the critical states.

We also note that the model provides a true constitutive relation, in that it permits the stress tensor to be found, given the bulk density and the rate of deformation tensor. To see this we note that the orientation of the principal axes of rate of deformation determines the orientation of the principal axes of stress, by the assumption of co-axiality, while the values of the principal rates of deformation determine the direction of a vector in the (σ_1, σ_2)-plane whose components are in the same ratio as the principal rates of deformation. We then seek the point where the yield locus is orthogonal to this vector, and the co-ordinates of this point give the values of the principal stresses, thus completing the specification of the stress tensor.

With yield loci of the form sketched in Figs. 10 and 11 there is a unique stress tensor corresponding to each rate of deformation tensor, but the converse is not true. An ambiguity exists at discontinuities in the slope of the yield locus. For example, any vector pointing into the first quadrant at 0 may represent acceptable principal rates of deformation.

We have already noted that there is a one-one relation between a yield locus in the (σ_1, σ_2)-plane and the corresponding locus representing a plane yield experiment in the (T,N)-plane. Images of the (σ_1, σ_2) yield loci in the (σ, τ)-plane are also useful in certain circumstances, where $\sigma = (\sigma_1 + \sigma_2)/2$, $\tau = (\sigma_1 - \sigma_2)/2$. We shall later formulate equations of motion in terms of this representation of the yield loci.

Finally, there is no difficulty in generalizing these ideas to three-dimensional situations. The yield loci in the (σ_1, σ_2)-plane then become yield surfaces in the $(\sigma_1, \sigma_2, \sigma_3)$-space. These are disposed about the (1,1,1)-line as axis, with cross-sectional shapes which are invariant under rotation through multiples of $2\pi/3$, as discussed earlier, so surfaces of the von Mises or Tresca types are acceptable.

Before moving on to the formulation of equations of motion, we should emphasize that our discussion of constitutive relations has been confined to forces transmitted between particles in permanent rolling and sliding contact at low shear rates. At higher shear rates momentum transfer by collision becomes a significant mechanism in generating stress. In contrast to the effects discussed here, this contribution is strongly dependent on shear rate. It is discussed by S. B. Savage in another contribution to this seminar.

EQUATIONS OF MOTION AND THE CRITICAL STATE APPROXIMATION

Equations of motion for a granular material are obtained by supplementing the equations of continuity and momentum balance with constitutive relations of the type discussed above. These are two in number, namely the coaxiality condition, requiring alignment of the principal axes of stress and rate of deformation, and the flow rule relating the ratio of the principal rates of deformation to the geometry of the yield loci. For the present purpose it is convenient to work with yield loci in the (σ,τ)-plane, where

$$\sigma = (\sigma_{maj} + \sigma_{min})/2 \quad ; \quad \tau = (\sigma_{maj} - \sigma_{min})/2. \tag{6}$$

The set of yield loci is then defined by a functional relation of the form

$$\tau = f(\sigma,\rho). \tag{7}$$

It is also convenient to describe the stress tensor in terms of the Sokolovski variables σ and γ, with τ determined by eqn. (7).

Then, if v_x and v_y are Cartesian components of velocity, it is straightforward to show that the condition of co-axiality can be written

$$\cos 2\gamma \left(\frac{\partial v_x}{\partial y} + \frac{\partial v_y}{\partial x} \right) = \sin 2\gamma \left(\frac{\partial v_y}{\partial y} - \frac{\partial v_x}{\partial x} \right) \tag{8}$$

while the flow rule takes the form

$$\cos 2\gamma \left(\frac{\partial v_y}{\partial y} + \frac{\partial v_x}{\partial x} \right) = \frac{\partial \tau}{\partial \sigma} \left(\frac{\partial v_y}{\partial y} - \frac{\partial v_x}{\partial x} \right) \tag{9}$$

The continuity equation has the familiar form

$$\frac{\partial \rho}{\partial t} + v_x \frac{\partial \rho}{\partial x} + v_y \frac{\partial \rho}{\partial y} = -\rho \left(\frac{\partial v_x}{\partial x} + \frac{\partial v_y}{\partial y} \right) \tag{10}$$

and the x and y-components of momentum balance are

$$\left[1 + \frac{\partial \tau}{\partial \sigma} \cos 2\gamma\right] \frac{\partial \sigma}{\partial x} - 2\tau \sin 2\gamma \frac{\partial \gamma}{\partial x} + \frac{\partial \tau}{\partial \rho} \cos 2\gamma \frac{\partial \rho}{\partial x}$$
$$- \frac{\partial \tau}{\partial \sigma} \sin 2\gamma \frac{\partial \sigma}{\partial y} - 2\tau \cos 2\gamma \frac{\partial \gamma}{\partial y} - \frac{\partial \tau}{\partial \rho} \sin 2\gamma \frac{\partial \rho}{\partial y}$$
$$+ \rho \left[\frac{\partial v_x}{\partial t} + v_x \frac{\partial v_x}{\partial x} + v_y \frac{\partial v_x}{\partial y} \right] = \rho g_x \tag{11}$$

and

$$- \frac{\partial \tau}{\partial \sigma} \sin 2\gamma \frac{\partial \sigma}{\partial x} - 2\tau \cos 2\gamma \frac{\partial \gamma}{\partial x} - \frac{\partial \tau}{\partial \rho} \sin 2\gamma \frac{\partial \rho}{\partial x}$$
$$+ \left[1 - \frac{\partial \tau}{\partial \sigma} \cos 2\gamma\right] \frac{\partial \sigma}{\partial y} + 2\tau \sin 2\gamma \frac{\partial \gamma}{\partial y} - \frac{\partial \tau}{\partial \rho} \cos 2\gamma \frac{\partial \rho}{\partial y}$$
$$+ \rho \left[\frac{\partial v_y}{\partial t} + v_x \frac{\partial v_y}{\partial x} + v_y \frac{\partial v_y}{\partial y} \right] = \rho g_y \tag{12}$$

In the particular case of steady flow, which will be our main concern, (8)-(12) provide a set of five first order, hyperbolic partial differential equations for the five dependent variables v_x, v_y, σ, γ, ρ in terms of the two independent variables x, y. It can be shown that their characteristics are

(i) the streamlines of the velocity field, and

(ii) two pairs of coincident directions making angles $\pm\psi$ with the axis of major principal stress, where

$$\tan \psi = \sqrt{\frac{1 - \partial\tau/\partial\sigma}{1 + \partial\tau/\partial\sigma}}. \tag{13}$$

The critical states are points on the yield locus (7) at which $\partial\tau/\partial\sigma = 0$. Thus $\psi = \pi/4$, so the characteristics (other than the streamlines) are inclined at ±45° to the major principal stress axis when the material is in a critical state.

A useful simplification of the equations of motion is possible in the case of materials like coarse sand, for which the yield loci expand very rapidly with increasing ρ, as indicated in Fig. 11. Then we expect only a narrow range of densities to be found in the flowing material, and consequently any departure from the critical state can quickly be eliminated by the dilation or compaction predicated by the flow rule. It is, therefore, a reasonable approximation to take the bulk density as constant and to neglect departures from the critical state.

Mathematically, then, we assume

(i) $\frac{\partial\tau}{\partial\sigma} \to 0$ (critical state)

(ii) $\frac{\partial\tau}{\partial\rho} \to \infty$ (yield loci very sensitive to density)

(iii) ρ = constant

Conditions (ii) and (iii) reduce the flow rule (9) to a triviality, since both sides of the equation vanish. In the momentum balances, however, a little more care is required, since $\partial\tau/\partial\sigma$ and $\partial\tau/\partial\rho$ always appear as the combinations

$$\frac{\partial\tau}{\partial\sigma}\frac{\partial\sigma}{\partial x} + \frac{\partial\tau}{\partial\rho}\frac{\partial\rho}{\partial x} = \frac{\partial\tau}{\partial x} \quad \text{and} \quad \frac{\partial\tau}{\partial\sigma}\frac{\partial\sigma}{\partial y} + \frac{\partial\tau}{\partial\rho}\frac{\partial\rho}{\partial y} = \frac{\partial\tau}{\partial y}$$

But these are easily evaluated, since the variations in σ and ρ are balanced in such a way as to maintain the material at a critical state, and the critical states all lie on the straight line

$$\tau = \sigma \sin\phi \tag{14}$$

Thus

$$\frac{\partial\tau}{\partial\sigma}\frac{\partial\sigma}{\partial x} + \frac{\partial\tau}{\partial\rho}\frac{\partial\rho}{\partial x} = \frac{\partial\sigma}{\partial x}\sin\phi \ ; \ \frac{\partial\tau}{\partial\sigma}\frac{\partial\sigma}{\partial y} + \frac{\partial\tau}{\partial\rho}\frac{\partial\rho}{\partial y} = \frac{\partial\sigma}{\partial y}\sin\phi . \tag{15}$$

which can be used in equations (11) and (12).

With these simplifications equations (8)-(12) reduce to the following four equations

$$\cos 2\gamma\left(\frac{\partial v_x}{\partial y} + \frac{\partial v_y}{\partial x}\right) = \sin 2\gamma\left(\frac{\partial v_y}{\partial y} - \frac{\partial v_x}{\partial x}\right) \tag{16}$$

$$\frac{\partial v_x}{\partial x} + \frac{\partial v_y}{\partial y} = 0 \tag{17}$$

$$\begin{aligned}[1 + \sin\phi\cos 2\gamma]\frac{\partial\sigma}{\partial x} - 2\sigma\sin\phi\sin 2\gamma\frac{\partial\gamma}{\partial x} - \sin\phi\sin 2\gamma\frac{\partial\sigma}{\partial y}\\ - 2\sigma\sin\phi\cos 2\gamma\frac{\partial\gamma}{\partial y} + \rho\left[\frac{\partial v_x}{\partial t} + v_x\frac{\partial v_x}{\partial x} + v_y\frac{\partial v_x}{\partial y}\right] = \rho g_x\end{aligned} \tag{18}$$

$$\begin{aligned}- \sin\phi\sin 2\gamma\frac{\partial\sigma}{\partial x} - 2\sigma\sin\phi\cos 2\gamma\frac{\partial\gamma}{\partial x} + [1-\sin\phi\cos 2\gamma]\frac{\partial\sigma}{\partial y}\\ + 2\sigma\sin\phi\sin 2\gamma\frac{\partial\gamma}{\partial y} + \rho\left[\frac{\partial v_y}{\partial t} + v_x\frac{\partial v_y}{\partial x} + v_y\frac{\partial v_y}{\partial y}\right] = \rho g_y\end{aligned} \tag{19}$$

The first of these (identical with equation (8)) is the co-axiality condition, the second is the continuity equation, and the third and fourth are the two components of momentum balance. They will be referred to as the critical state approximation to the momentum balance and are identical with the equations of motion for an incompressible, isotropic Coulomb powder.

For steady flows (16)-(19) provide a set of four first order, hyperbolic partial differential equations for the four dependent variables v_x, v_y, σ and γ in terms of the independent variables x and y. Their characteristics form two pairs, the first of which make angles $\pm \pi/4$ with the major principal stress axis and are called <u>velocity characteristics</u>, while the second pair make angles $\pm(\pi/4-\phi/2)$ with the major principal stress axis and are called <u>stress characteristics</u>.

The nature of the limiting process which carries the full equations (8)-(12) into the critical state approximation (16)-(19) calls for some comment. We expect the critical state approximation to be good when the yield loci (7) expand sufficiently rapidly with increasing ρ, that is, when $\partial\tau/\partial\rho$ is sufficiently large. Indeed equations (16)-(19) correspond to the limiting situation in which $\partial\tau/\partial\rho \to \infty$. But the characteristics of the complete equations make angles $\pm \pi/4$ with the major principal stress axis whenever the material is in the critical state, regardless of the value of $\partial\tau/\partial\rho$, while the characteristics of the equations of the critical state approximation make angles $\pm \pi/4$, $\pm(\pi/4-\phi/2)$ with the major principal stress axis. Thus, despite the fact that the complete equations appear to pass continuously to the critical state equations as a limit, there is a discontinuous jump in the structure of the characteristics, and hence in the domain of dependence, at this limit.

Apart from a few attempts to allow for variations in bulk density in an ad hoc manner, by postulating a functional relation between σ and ρ, existing work on the motion of granular materials has been based on the simplified equations (16)-(19). The validity of these equations as approximations to the full equations of motion is questionable mainly when ρ becomes small and, correspondingly, the yield loci in Fig. 11 shrink towards the origin. There is then some evidence that the size of the yield locus becomes less sensitive to density variations, and consequently we are no longer justified in assuming that the material lies close to a critical state.

PHYSICAL FEATURES AND MATHEMATICAL FORMULATION OF THE WEDGE-SHAPED HOPPER PROBLEM

So far we have been concerned with the formulation of equations of motion for a granular material. The solution of these equations also poses interesting mathematical and physical problems. Some of these are illustrated by one of the simplest non-trivial problems, namely the flow of granular material under gravity out of a wedge-shaped hopper formed by plane walls equally inclined to the vertical, as indicated in Fig. 12. First we will describe the salient physical features of this flow and contrast them with the corresponding behavior of a liquid flowing from the same vessel.

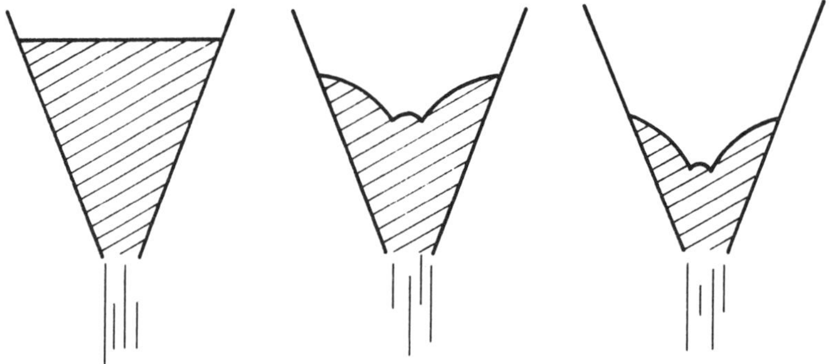

Figure 12. Stages in the discharge of a wedge-shaped hopper

Figure 12 indicates the gross configuration of the surface of the material at three successive times during discharge, and it is seen that a depression rapidly forms at the axis of symmetry and deepens as flow continues, even when the depth of the fill is very large compared with the width of the exit slot. Close examination of the motion of the particles below the surface (through a transparent end wall, for example) reveals that it is very complex, with thin "shock-like" layers separating blocks of material which slide with little distortion. In contrast, for a liquid, the free surface remains almost flat until the depth becomes a relatively small multiple of the slot width, and the velocity field below the surface is quite smooth.

For the granular material the complex motion high above the exit slot is strongly dependent on the surface configuration and on stresses applied to the surface, but on descending the hopper all these variants degenerate into a common motion, determined only by the nature of the material, the roughness and angle of inclination of the walls, and the width of the exit slot. In particular, the discharge rate is found to be independent of the depth of fill, provided this is large compared with the width of the slot, and it is not influenced by the shape of the upper surface or by stresses applied to this surface. For a liquid, on the other hand, the discharge rate increases with increasing head in the vessel, even when this is large compared with the width of the exit slot, and the flow rate can also be changed substantially by applying pressure to the free surface.

The "decoupling" of the motion near the exit slot from conditions higher in the hopper is one of the most striking features of the behavior of a granular material in this system, and it suggests the possibility of a corresponding mathematical decoupling, permitting the discharge rate and the motion near the exit to be calculated without having to analyze the complex motion higher in the hopper. It is fortunate that such a possibility presents itself since, as noted earlier, the motion near the surface evolves in time as the surface descends, so there is no clear sense in which it might be approximated by a steady motion. Meanwhile, however, the motion near the exit remains essentially unchanged, so there is good reason to expect that it can be approximated as the solution of a steady flow problem.

The mathematical reason for the physical decoupling of the motions high and low in the hopper will be traced below, but first we formulate the equations of motion in a suitable dimensionless form. Using polar co-ordinates with origin at the intersection of the walls and $\theta = 0$ on the median line of the hopper, and defining the following dimensionless variables:

$$\beta = \sigma/\rho g L, \quad u = v_r/\sqrt{gL}, \quad v = v_\theta/\sqrt{gL}, \quad \xi = r/L \qquad (20)$$

where L is a characteristic length, which will be selected later, the equations of motion corresponding to the critical state approximation (16)-(19) are

$$2\tan 2\gamma \frac{\partial u}{\partial \xi} - \left(\frac{\partial v}{\partial \xi} + \frac{1}{\xi}\frac{\partial u}{\partial \theta} - \frac{v}{\xi}\right) = 0 \tag{21}$$

$$\frac{\partial u}{\partial \xi} + \frac{1}{\xi}\frac{\partial v}{\partial \theta} + \frac{u}{\xi} = 0 \tag{22}$$

$$u\frac{\partial u}{\partial \xi} + \frac{v}{\xi}\frac{\partial u}{\partial \theta} - \frac{v^2}{\xi} + [1-\sin\phi\cos 2\gamma]\frac{\partial \beta}{\partial \xi} + 2\beta\sin\phi\sin 2\gamma\frac{\partial \gamma}{\partial \xi}$$
$$-\sin\phi\sin 2\gamma\frac{1}{\xi}\frac{\partial \beta}{\partial \theta} - \frac{2\beta}{\xi}\sin\phi\cos 2\gamma\left(\frac{\partial \gamma}{\partial \theta} + 1\right) + \cos\theta = 0 \tag{23}$$

$$u\frac{\partial v}{\partial \xi} + \frac{v}{\xi}\frac{\partial v}{\partial \theta} + \frac{uv}{\xi} - \sin\phi\sin 2\gamma\frac{\partial \beta}{\partial \xi} - 2\beta\sin\phi\cos 2\gamma\frac{\partial \gamma}{\partial \xi}$$
$$+ [1+\sin\phi\cos 2\gamma]\frac{1}{\xi}\frac{\partial \beta}{\partial \theta} - \frac{2\beta}{\xi}\sin\phi\sin 2\gamma\left(\frac{\partial \gamma}{\partial \theta} + 1\right) - \sin\theta = 0 \tag{24}$$

These are subject to boundary conditions at the hopper walls, where v clearly vanishes. Also, as a result of sliding friction, the ratio of the tangential and the normal stress on the wall is given by $T/N = \tan\delta$, where δ is the angle of friction between the granular material and the wall. It can be shown that this condition determines a value for γ at the wall, namely

$$\gamma_w = \frac{1}{2}\left[\delta + \sin^{-1}\left(\frac{\sin\delta}{\sin\phi}\right)\right] \tag{25}$$

with

$$0 \le \sin^{-1}\left(\frac{\sin\delta}{\sin\phi}\right) \le \pi/2$$

Other boundary conditions are associated with the upper free surface of the material and the motion through the exit slot. Clearly the upper surface must be a traction-free surface and, for a non-cohesive material, this implies that $\sigma = 0$. However, this surface condition is one which will not concern us directly if we are successful in decoupling the motion near the exit from conditions near the surface. The boundary condition at exit is less obvious. In the existing literature it has been assumed that the exit slot is spanned by a traction-free surface (which is synonymous with a surface $\sigma = 0$ for non-cohesive material), below which the particles lose contact with each other and fall freely under gravity. The shape of this surface is then to be determined as part of the problem. Though we shall question this form of exit condition later, it will be adopted for the time being.

STRUCTURE OF THE SOLUTION OF THE HOPPER PROBLEM

The nature of the solution of the hopper problem is elegantly revealed by an approximation introduced by Savage [21], which permits the equations of motion to be solved explicitly. Savage considered the case of smooth walls ($\delta = 0$), for which eqn. (25) shows that $\gamma_w = 0$, and he also replaced the gravitational field, which acts vertically, by a body force field of equal magnitude directed radially inward towards the intersection of the hopper walls. For a small hopper angle θ_w this does not distort the true gravitational field very seriously, but the combination of smooth walls and radially directed gravity admits a solution which depends only on the radial co-ordinate ξ.

Specifically, the continuity equation (22) is now satisfied by $u = -A/\xi$, $v = 0$, for any constant value of A. Equations (21) and (24) then reduce to trivialities if we take $\gamma = 0$, and the radial component of momentum balance, eqn. (23), becomes

$$\frac{d\beta}{d\xi} - (K-1)\,\frac{\beta}{\xi} = \frac{K+1}{2}\left(\frac{A^2}{\xi^2} - 1\right) \tag{26}$$

where $K = (1 + \sin\phi)/(1 - \sin\phi)$. This has the general solution

$$\beta = \frac{1}{2}\left[\left(\frac{K+1}{K-2}\right)\xi - \frac{A^2}{\xi^2}\right] - I\,\xi^{K-1} \tag{27}$$

where I is an integration constant.

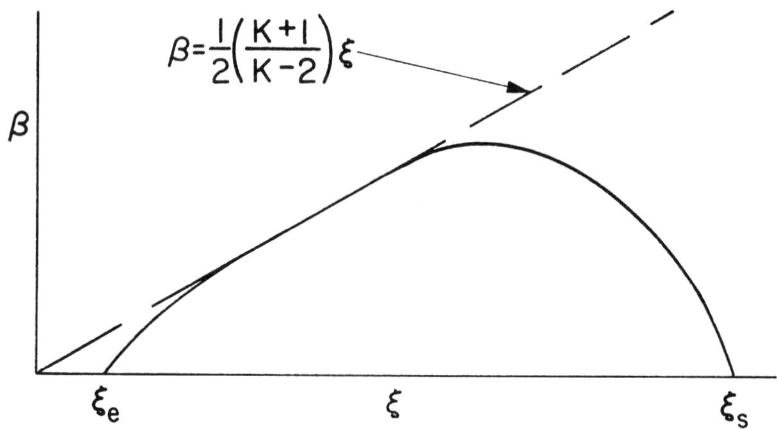

Figure 13. Savage's solution for smooth walls and radial gravity

The form of β, for small values of A and I, is sketched in Fig. 13. It is seen that there is an interval in which β lies close to $\frac{1}{2}$ (K+1/K-2)ξ, and this is called the radial stress field, since β is proportional to radial distance. In this region the inertial terms in the momentum balance, represented by A^2/ξ^2 in (26), are negligible. However, as ξ decreases, A^2/ξ^2 increases, and the effect of this is to displace the solution below the radial stress field so that it crosses the axis $\beta = 0$ at some point $\xi = \xi_e > 0$. ξ_e then locates the position of the exit slot, and the section of the traction-free surface which spans this slot is the circular arc $\xi = \xi_e$. Clearly ξ_e depends on A, while A also determines the value of the flow rate through the channel, so the relation between the discharge rate and the width of the exit slot follows.

For large values of ξ, β is displaced below the radial stress field by the influence of the term I ξ^{K-1} in (27). Thus the β v. ξ curve crosses the axis $\beta = 0$ again at some point $\xi = \xi_s$, which locates a possible position for the free upper surface of the fill in the hopper. This simple solution is valid only for an upper surface whose section is a circular arc, of course.

When the fill is deep compared with the slot width, $\xi_s >> \xi_e$ and the reason for the decoupling between conditions near the slot and conditions near the surface is now clear from the form of the solution. Whatever the value of ξ_s (provided it is sufficiently large) β converges to the same radial stress field as ξ decreases below ξ_s. When ξ_s is large enough, this convergence is essentially complete before the inertial terms become significant, so the form of the solution near $\xi = \xi_e$ is uninfluenced by the value of ξ_s. Thus, the decoupling depends on the existence of a solution of the inertia-free equations, namely the radial stress field, which is approached asymptotically on moving up from the exit and on moving down from the surface, so that it provides a "buffer" between these two regions.

The question which remains to be addressed is whether a comparable solution exists for the complete problem, with rough hopper walls and gravity directed vertically. Such a solution was, in fact, found quite early by Jenike [22], who noted that, if inertial terms are omitted, equations (23) and (24) admit an exact solution in the form of a radial stress field:

$$\beta = b_o(\theta)\xi \quad , \quad \gamma = g_1(\theta) \tag{28}$$

where $b_o(\theta)$ and $g_o(\theta)$ must satisfy the following ordinary differential equations

$$\frac{d}{d\theta}[-b_o \sin\phi \sin 2g_o] + b_o(1-3\sin\phi\cos 2g_o) + \cos\theta = 0 \tag{29}$$

$$\frac{d}{d\theta}[b_o(1+\sin\phi\cos 2g_o)] - 3b_o\sin\phi\sin 2g_o - \sin\theta = 0 \tag{30}$$

and the boundary conditions

$$g_o(0) = 0, \quad g_o(\theta_w) = \gamma_w. \tag{31}$$

There is also a corresponding solution of the co-axiality and continuity equations (21) and (22), namely

$$u = f_o(\theta)/\xi \quad , \quad v = 0 \tag{32}$$

with $f_o(\theta)$ given by

$$f_o(\theta) = f_o(0)\ \exp[-2\int_0^{\theta} \tan\{2g_o(\theta')\}d\theta'] \tag{33}$$

and this will be called the radial velocity field.

The value chosen for $f_o(0)$ is directly related to the length scale parameter L, which has so far been left undefined. In fact, it can be shown that the convenient choice $f_o(0) = -1$ corresponds to

$$L = \lim_{r\to\infty} \left[\frac{(rv_r)^2}{g}\right]^{1/3}_{\theta=0}$$

There remains the question whether this solution is approached asymptotically on moving down the hopper, and this has not yet been resolved with complete certainty. Some computational solutions of the inertia-free momentum balances obtained by Johanson [22], starting from a traction-free upper surface, appear to converge towards the radial stress field as ξ decreases, and similar computions of both stress and velocity fields by Kaza [20] also approach the radial fields (28) and (32) from various initial conditions. Though the eigenvalues of the linearized equations describing small departures from the radial fields have not been found, Kaza [20] has shown that the eigenvalues of certain simpler but closely related equations have signs consistent with convergence to the radial fields as ξ decreases. If we accept this evidence, then boundary conditions for the complete equations, including inertial terms, can be generated by requiring their solutions to approach the radial fields (28), (32) as $\xi \to \infty$. These should define the solution we seek, which is common to all surface conditions with a deep fill of granular material.

SOLUTION OF THE EQUATIONS FOR HOPPER FLOW

Since equations (21)-(24) are hyperbolic, integration down the hopper from initial conditions corresponding to the radial fields is straightforward, in principle. A numerical procedure based on the method of characteristics is quite successful in generating the solution well into

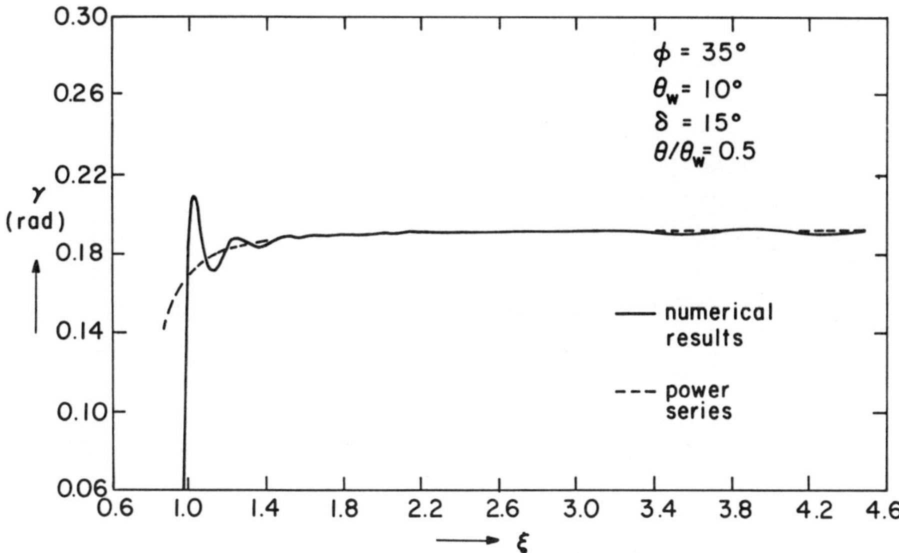

Figure 14. γ v. ξ at $\theta = \theta_w/2$ from numerical integration

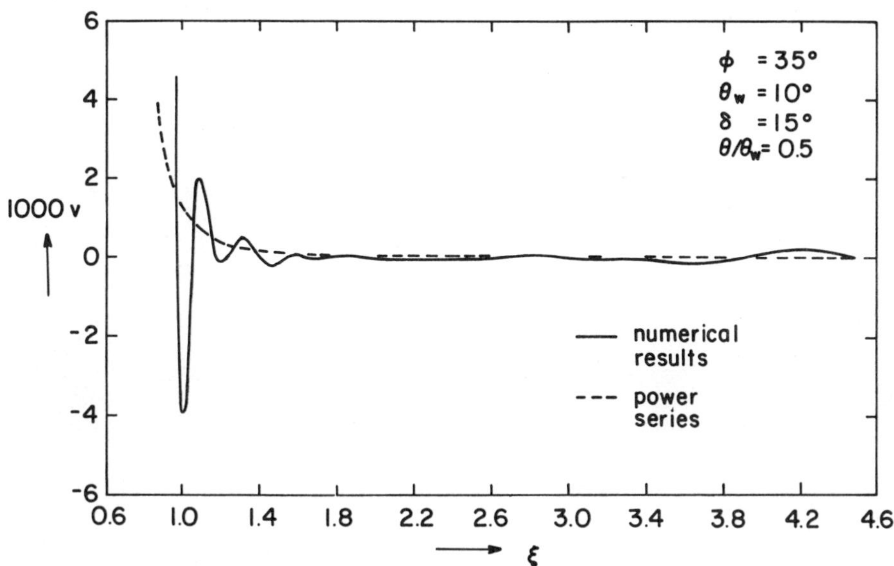

Figure 15. v v. ξ at $\theta = \theta_w/2$ from numerical integration

the region where inertial terms become important, but it fails before reaching the traction-free curve $\beta = 0$ for a reason which becomes obvious on examining equations (23) and (24). In these equations the derivatives $\partial\gamma/\partial\xi$ and $\partial\gamma/\partial\theta$ are multiplied by β wherever they appear, so $\beta = 0$ is a singular curve. The consequence of this is seen in Figs. 14 and 15, which show the behavior of γ and v, found from the numerical integration, as functions of ξ at $\theta = \theta_w/2$. In each case diverging oscillations appear on approaching the singular curve, so the numerical solution cannot be used to locate this curve. Since the position of the singular curve determines the width of the exit slot, it follows that the discharge rate - slot width relation cannot be obtained, so the method is of little value.

Three procedures have been published for obtaining approximate solutions which can be used to locate the curve $\beta = 0$, despite the singularity there. The first, due to Brennen and Pearce (23), uses a trial solution in the form of a series in ascending powers of θ. If the equations are expanded only up to terms of $0(\theta)$, the series expansions of the dependent variables can be truncated as follows:

$$u = u_o(\xi) + u_2(\xi)\theta^2 \quad ; \quad v = v_1(\xi)\theta$$

$$\beta = \beta_o(\xi) + \beta_2(\xi)\theta^2 \quad : \quad \gamma = \gamma_1(\xi)\theta$$

The boundary conditions $v = 0$, $\gamma = \gamma_w$ at $\theta = \theta_w$ then demand that $v_1 = 0$, $\gamma_1 = \gamma_w/\theta_w$, and to $0(1)$ the radial component of the momentum balance becomes

$$(1-\sin\phi)d\beta_o/d\xi - 2\sin\phi(1+\gamma_w/\theta_w)\beta_o/\xi + 1 + u_o\frac{du_o}{d\xi} = 0$$

while the continuity equation gives

$$\frac{du_o}{d\xi} + u_o/\xi = 0$$

These ordinary differential equations can easily be solved for β_o and u_o, then β_2 and u_2 can be found from the $0(\theta)$ approximation to the co-axiality condition and the circumferential component of momentum balance.

To this lowest order the solution is very simple, since the above equations have the same structure as those describing Savage's approximation. However, terms of the next higher order in θ give ordinary differential equations with a singularity where $\beta = 0$, which is the same as the difficulty presented by the complete equations. Thus it is not a straightforward matter to generate successive approximations in order to judge convergence.

Savage and Sayed [24] have proposed an alternative approximation, in which the partial differential equations are reduced to ordinary differential equations by integrating the momentum balances across the width of the hopper. This resembles the method of integral balances in fluid mechanics and, as in the corresponding fluid mechanical problems, it is necessary to assume that the profile of u across the hopper is "reasonable", in the sense that $\overline{u^2} \approx (\bar{u})^2$ where the bar denotes a cross-sectional average. The procedure makes no attempt to generate a sequence of successive approximations, so again there is no real basis for judging its success.

Discharge rates calculated from the Brennen-Pearce solution are compared with experimental measurements of Sullivan [25] in Figs. 16 and 17. The discharge rates are expressed as a characteristic velocity V_D, defined by

$$V_D = \dot{M}/\rho BD\sqrt{gD} \tag{34}$$

where $\dot{M}$ is the mass discharge rate, while D and B denote the width and length of the exit slot, respectively. It is seen that the theory overestimates the observed flows, particularly at small values of θ, where the theoretical approximation would be expected to be best.

A third procedure, recently presented by Kaza and Jackson [26], generates a sequence of approximations starting from the radial stress and velocity fields. The method is illustrated by Table 1, the first line of which represents the radial stress and velocity fields. On substituting the radial velocity field into the momentum balances (23) and (24) a pair of equations in the two variables β

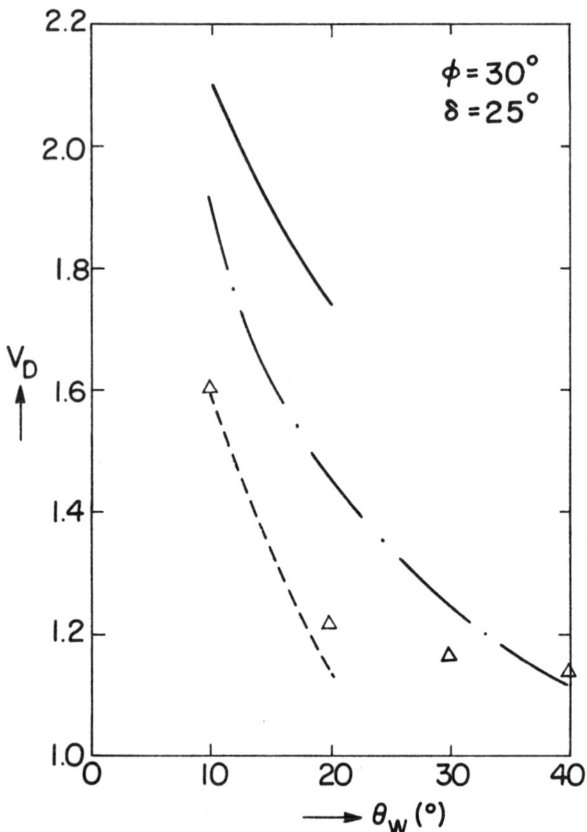

Figure 16. Comparison of theoretical predictions with experimental results of Sullivan (25)

— - — - Theory of Brennen and Pearce

——— Present theory - traction free surface

- - - - - - Present theory - surface of vertical free fall

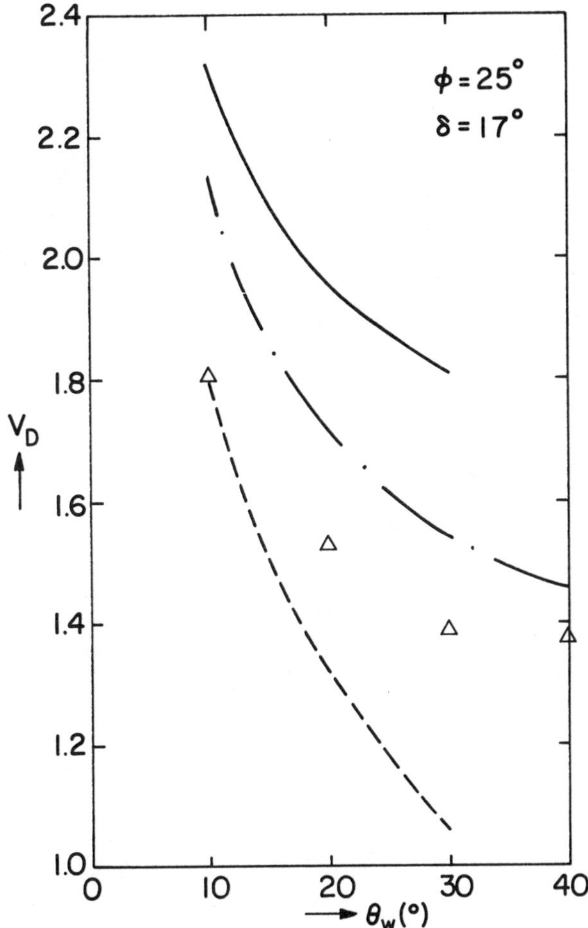

Figure 17. (See Fig. 16 above)

and γ results. A solution of these is sought in the form

$$\beta = \xi b_o(\theta) + \beta_1 \quad ; \quad \gamma = g_o(\theta) + \gamma_1$$

and the equations are linearized in the perturbations β_1

and γ_1. The resulting linear equations are then found to be separable in θ and ξ, and their solution is of the form

$$\beta_1 = b_1(\theta)/\xi^2 \quad , \quad \gamma_1 = g_1(\theta)/\xi^3$$

where b_1 and g_1 can be obtained by solving a two-point boundary value problem for a pair of ordinary differential equations. The second approximation to the stress fields, found in this way, appears as the second entry in the left-hand column of Table 1, and it is used in the co-axiality condition and continuity equation, (21) and (22), to give a pair of equations for u and v. A solution of these is then sought in the form

$$u = f_o(\theta)/\xi + u_1 \quad , \quad v = v_1$$

and the equations are linearized in u_1 and v_1. The resulting linear equations are again separable in θ and ξ, and their solutions are of the form

$$u_1 = f_1(\theta)/\xi^4 \quad , \quad v_1 = h_1(\theta)/\xi^4$$

where f_1 and h_1 can be found by solving a two-point boundary value problem for a pair of ordinary differential equations. Thus, a second approximation to the velocity fields is found.

This process can be continued, as indicated in Table 1, alternating between solutions of linearized momentum balances and linearized co-axiality and continuity equations. At every stage separable solutions can be found, and the process generates an approximate solution as a set of expansions in inverse cubic powers of ξ:

$$\begin{aligned} \beta &= b_o(\theta)\xi + b_1(\theta)/\xi^2 + b_2(\theta)/\xi^5 + \\ \gamma &= g_o(\theta) + g_1(\theta)/\xi^3 + g_2(\theta)/\xi^6 + \\ u &= f_o(\theta) + f_1(\theta)/\xi^4 + f_2(\theta)/\xi^7 + \\ v &= 0 + h_1(\theta)/\xi^4 + h_2(\theta)/\xi^7 + \end{aligned} \tag{35}$$

Table 1. Perturbation Scheme

Momentum balances		Isotropy and continuity equations
$\beta = \xi b_o(\theta)$ $\gamma = g_o(\theta)$	→	$u = f_o(\theta)/\xi$ $v = 0$
$\beta = \xi b_o(\theta) + \beta_1$ $\gamma = g_o(\theta) + \gamma_1$	←	$u = f_o(\theta)/\xi + u_1$ $v = 0 + v_1$
Linearize. Separable solns. $\beta = \xi b_o(\theta) + b_1(\theta)/\xi^2$ $\gamma = g_o(\theta) + g_1(\theta)/\xi^3$	→	Linearize. Separable solns. $u = f_o(\theta)/\xi + f_1(\theta)/\xi^4$ $v = 0 + h_1(\theta)/\xi^4$
$\beta = \xi b_o(\theta) + b_1(\theta)/\xi^2 + \beta_2$ $\gamma = g_o(\theta) + g_1(\theta)/\xi^3 + \gamma_2$	←	$u = f_o(\theta)/\xi + f_1(\theta)/\xi^4 + u_2$ $v = 0 + h_1(\theta)/\xi^4 + v_2$
Linearize. Separable solns. $\beta = \xi b_o(\theta) + b_1(\theta)/\xi^2 + b_2(\theta)/\xi^5$ $\gamma = g_o(\theta) + g_1(\theta)/\xi^3 + g_2(\theta)/\xi^6$	→	Linearize. Separable solns. $u = f_o(\theta)/\xi + f_1(\theta)/\xi^4 + f_2(\theta)/\xi^7$ $v = 0 + h_1(\theta)/\xi^4 + h_2(\theta)/\xi^7$

These series clearly must diverge for small enough values of ξ, but is is sufficient for the present purpose that they should converge throughout the region in which $\beta \geq 0$, so that the curve $\beta = 0$ can be located. Though there is no formal proof of convergence, calculation of the first few terms suggests convergence in this region for small enough values of θ_w, though for large θ_w there is every indication of divergence before β reaches zero. Thus this procedure seems to be feasible in hoppers of sufficiently small angle.

Kaza and Jackson [26] truncated their solution after four correction terms and found that the corresponding residues for the differential equations, computed near the curve $\beta = 0$, were significantly smaller than the residues generated by the Brennen-Pearce approximation, so it is likely that theirs is the more accurate solution for hoppers of small angle.

Once the approximate solution has been found the discharge rate can be calculated. Indeed, the traction free curve $\beta = 0$ is located using the first of equations (35), which gives

$$b_o(\theta)\xi + b_1(\theta)/\xi^2 + b_2(\theta)/\xi^5 + \ldots = 0$$

Setting $\theta = \theta_w$ this can then be solved for ξ, and the corresponding value ξ_e locates the bottom edge of the hopper wall. It is not difficult to show that the characteristic discharge velocity V_D, defined by eqn. (34), is then given by

$$V_D = \frac{\theta_w}{\sin\theta_w} \frac{|\bar{f}_o|}{\sqrt{2\xi_e^3 \sin\theta_w}} \tag{36}$$

where

$$\bar{f}_o = \frac{1}{\theta_w} \int_o^{\theta_w} f_o(\theta)\, d\theta$$

Theoretical discharge velocities based on Kaza and Jackson's solution are shown in Figs. 16 and 17 for comparison with the results of Brennen and Pearce and with the experimental measurements. Despite the fact that Kaza and Jackson's is presumably the more accurate of the two solutions, its predicted discharge rates are further from the experimental results. Thus, it is unlikely that the significant discrepancy between theory and experiment can be eliminated by better approximation techniques for solving the equations of motion, but rather it points to a physical shortcoming in the formulation of the problem.

A RE-EXAMINATION OF THE EXIT BOUNDARY CONDITION

In seeking a physical reason for the discrepancy between theory and experiment seen in Figs. 16 and 17 we are led to question the exit boundary condition. This demands that a region above, with constant bulk density, should be separated by a traction-free surface from a region below, where the bulk density decreases as the particles accelerate freely under gravity. On moving down to the traction free surface from above, β decreases continuously from positive values to zero at the surface. Surely, then, if the motion were to continue with constant bulk density, β would have to pass to negative values below the traction-free surface. In a non-cohesive material this is impossible, so we would expect the bulk density to <u>increase</u> if β retained the constant value zero. But this contradicts the physical picture, sketched above, of the behavior in the neighborhood of the traction-free surface.

It is not difficult to formalize this argument and show that downflow through a traction-free surface, with constant bulk density above and decreasing bulk density below, is impossible unless the normal derivative of β vanishes at the surface. Since it certainly does not vanish for the solution found above, the traction-free surface is not an acceptable terminal condition for the hopper.

The resolution of this difficulty is not yet by any means complete, but an alternative boundary condition is suggested by photographs of the motion near the exit slot.

Fig. 18 reproduces such a photograph from Bcsley et al., [27] and it is seen that the particles fall almost vertically below the exit slot, with a sharp change in the direction of motion where they lose contact with the hopper walls. Since this change in velocity must be accompanied by a stress jump, β cannot approach zero on moving down to the end of the hopper wall, as it would if the slot were spanned by a traction-free surface.

Figure 18. Photograph of discharge from a steep-walled, wedge-shaped hopper. From Bosley et al. (27)

Accepting this photographic evidence we should explore the consequences of assuming that the slot is spanned by a surface below which the particles fall vertically under gravity alone [28]. This must be located so that the stress jump across it matches the jump in momentum flux associated with the sudden change in the direction of motion. With this terminal condition for the hopper, which we will call a surface of vertical free fall, the bulk density of the vertically falling material will certainly decrease as it accelerates under gravity, so the difficulty associated with the traction-free surface does not arise.

Suppose the orientation of a surface of vertical free fall is such that its upward normal subtends an angle ψ with the radial direction, measured in a clockwise sense, and let $\tilde{\rho}$ denote the ratio of the bulk density of the material immediately below the surface to that of the material immediately above. Then continuity of particle flux requires that

$$u\cos\psi + v\sin\psi = -\tilde{\rho}w\cos(\theta+\psi) \tag{37}$$

where w is the speed of fall of the particles immediately below the surface. Furthermore, if β and γ denote the values of the Sokolovski variables immediately above the surface, the jump relations for the normal and tangential components of stress and momentum flux are

$$\beta[1-\sin\phi\cos2(\gamma-\psi)] = -(u\cos\psi+v\sin\psi)[w\cos(\theta+\psi) + u\cos\psi+v\sin\psi] \tag{38}$$

and

$$\beta\sin\phi\sin2(\gamma-\psi) = -(u\cos\psi+v\sin\psi)[w\sin(\theta+\psi) + u\sin\psi-v\cos\psi] \tag{39}$$

respectively. Since β, γ, u and v are known from the solution within the hopper, (37)-(39) provide three equations for $\tilde{\rho}$, w and ψ. In particular, ψ determines the orientation of the surface of vertical free fall, since

$$\tan\psi = -\frac{1}{\xi}\frac{d\xi}{d\theta} \tag{40}$$

Eliminating $\tilde{\rho}$, w and ψ between equations (37)-(40) then gives the following differential equation for the section of a surface of vertical free fall

$$\frac{1}{\xi}\frac{d\xi}{d\theta} = \frac{\beta[\sin\theta-\sin\phi\sin(2\gamma+\theta)] + u[u\sin\theta+v\cos\theta]}{\beta[\cos\theta+\sin\phi\cos(2\gamma+\theta)] + v[u\sin\theta+v\cos\theta]} \tag{41}$$

A one-parameter set of possible surfaces can be generated by integrating (41) from initial conditions $\xi = \xi_o$ at $\theta = 0$, where ξ_o is any value of ξ for which $\beta(\xi_o,0) \geq 0$.

To each member of this set there corresponds a value of $\xi_e = \xi(\theta_w)$ and hence, through equation (36), a value for the discharge rate V_D. Thus, the possible values for V_D occupy an interval whose upper bound corresponds to the extreme surface of vertical free fall, for which $\beta(\xi_o, 0) = 0$. This upper bound is plotted on Figs. 16 and 17, and it is seen to give values of V_D substantially lower than those predicted from the traction-free surface. Indeed, the agreement with experiment is now very good for small values of θ_w, where the solution within the hopper is expected to be accurate.

Though encouraging, these considerations by no means resolve the question of the exit boundary condition. They do not determine a unique value for the discharge rate and they do not, in themselves, provide any physical reason for selecting the upper bound of the predicted range of discharge rates. There is also a difficulty of internal consistency, which can be seen by considering the shape of the extreme surface of vertical free fall, corresponding to the maximum discharge rate. This is shown in Fig. 19 and it is seen to have a cusp on the median line, at which point it touches the curve $\beta = 0$. The same diagram shows the boundary of the domain of dependence for the hopper equations (21)-(24) associated with the wall boundary conditions (25), and the extreme surface of vertical free fall is seen to lie outside this domain. At first sight this seems impossible, since the surface of vertical free fall was derived from the known solution of eqns. (21)-(24) using the jump conditions (37)-(39). However, the solution used was the power series (35), and this was obtained by applying the boundary condition $\gamma = \gamma_w$ and $\theta = \theta_w$ for <u>all values</u> of ξ, not merely those corresponding to the walls of the hopper. Thus, the solution used to construct the surface of vertical free fall shown in Fig. 19 satisfies an <u>extrapolation</u> of the wall boundary condition beyond the end of the physical walls. This is clearly inappropriate.

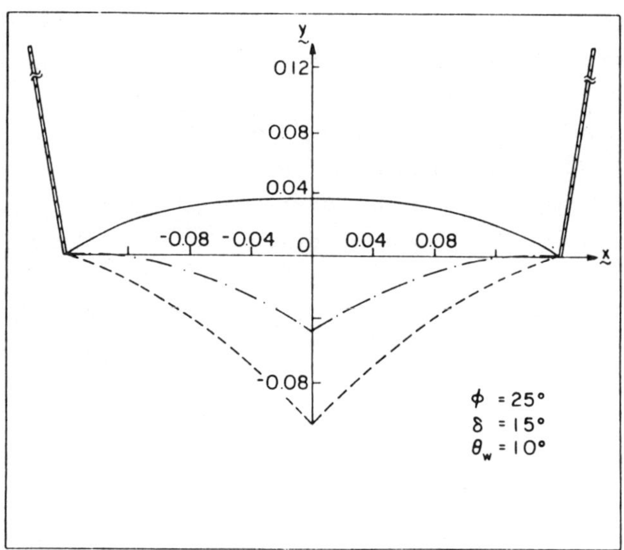

Figure 19. Traction-free surface and surface of vertical free fall spanning the exit slot, in relation to domain of dependence for wall boundary conditions.

——————— Traction-free surface

— — — — Surface of vertical free fall

—·—·—·— Boundary of domain of dependence

CONCLUSION

In the above we have discussed some problems associated with the formulation and solution of differential equations of motion for a continuum model of a granular material. The intention has been to illustrate the sort of difficulties which must be faced, rather than attempt an exhaustive survey of existing work or present a definitive solution of any one problem.

Indeed, one of the simplest non-trivial problems, namely flow from a wedge-shaped hopper, presents a formidable array of difficulties, mostly unresolved or incompletely resolved at the moment. We have examined only the motion near the exit slot, neglected all but frictional stresses, which is questionable, and invoked the critical state approximation in just the place where it is least likely to be valid. Even so we have not found a completely satisfactory way of effecting the transition from motion within the hopper to free fall below.

The whole area of creeping flow, where inertial terms in the equations of motion can be neglected, has scarcely been touched upon. This includes the fascinating complexities of kinematics and stress distribution well above the exit slot in the hopper problem, and similar difficulties must also be faced in seeking satisfactory test methods to characterize the constitutive relations for contact stresses.

In many practical situations the flow of a granular material is strongly influenced by drag forces resulting from relative motion of the solid material and the fluid (liquid or gas) which occupies its interstices. This is known to influence the discharge rate of fine particles from a hopper, for example, and is a dominant factor in determining the behavior of systems such as fluidized beds, pneumatic transport lines, and standpipes.

The field of granular material flow is therefore large, and is still in an early stage of development, but from what is already known it is certain to be a fertile source of mathematical problems.

REFERENCES

1. Coulomb, C. A., Essai sur une application des règles de maximis et minimis à quelques problèmes de statique, relatifs à l'architecture, Memoires de Mathématique de l'Academie Royale des Sciences (Paris) 7 (1776), 343-382.

2. Sokolovski, V. V., Statics of Granular Media, Pergamon Press, 1965.

3. Mandel, J., Sur les lignes de glissement et le calcul des déplacements dans la déformation plastique, C.r. hebd. Séanc. l'Acad. Sci. Paris 225 (1947), 1272.

4. De Josselin de Jong, G., Statics and kinematics in the failable zone of granular material, Thesis, University of Delft, 1959.

5. Reynolds, O., On the dilatancy of media composed of rigid particles in contact, Phil. Mag. 20 (1885), 469-481.

6. Drucker, D. C. and Prager, W., Soil mechanics and plastic analysis or limit design, Q. Appl. Math. 10 (1952), 157-165.

7. Spencer, A. J. M., A theory of the kinematics of ideal soils under plane strain conditions, J. Mech. Phys. Solids 12 (1964), 337-351.

8. Mandl, G. and Luque, R. F., Fully developed plastic shear flow of granular materials, Géotech. 20 (1970), 277-307.

9. De Josselin de Jong, G., The double-sliding free rotating model for granular assemblies, Géotech. 21 (1971), 155-163.

10. De Josselin de Jong, G., Mathematical elaboration of the double-sliding free rotating model, Arch. Mech. 29 (1977), 561-591.

11. Mehrabadi, M. M. and Cowin, S. C., Initial planar deformation of dilatant granular materials, J. Mech. Phys. Solids 26 (1978), 269-284.

12. Mehrabadi, M. M. and Cowin, S. C., On the double-sliding, free rotating model for the deformation of granular materials, J. Mech. Phys. Solids 29 (1981), 269-282.

13. Shield, R. T., On Coulomb's law of failure in soils, J. Mech. Phys. Solids 4 (1955), 10-16.

14. Jenike, A. W. and Shield, R. T., On the plastic flow of Coulomb solids beyond failure, J. Appl. Mech. 26 (1959), 599-602.

15. Roscoe, K. H., Schofield, A. N. and Wroth, C. P., On the yielding of soils, Géotech. 8 (1958), 22-53.

16. Schofield, A. N. and Wroth, C. P., Critical State Soil Mechanics, McGraw-Hill Publishing Company, 1968.

17. Roscoe, K. H., The influence of strains in soil mechanics, Géotech. 20 (1970), 129-170.

18. Drescher, A., An experimental investigation of flow rules for granular materials using optically sensitive glass particles, Géotech. 26 (1976), 591-601.

19. Drucker, D. C., A more fundamental approach to plastic stress-strain relations, Proc. 1st U.S. Nat. Congr. Appl. Mech., 1951, 487-491.

20. Kaza, K. R., PhD Thesis, University of Houston, 1982.

21. Savage, S. B., The mass flow of granular materials from coupled velocity-stress fields, Brit. J. Appl. Phys. 16 (1965), 1885-1888.

22. Johanson, J. R., Stress and velocity fields in the gravity flow of bulk solids, J. Appl. Mech. 31 (1964), 499-506.

23. Brennen, C. and Pearce, J. C., Granular material flow in two-dimensional hoppers, J. Appl. Mech. 45 (1978), 43-50.

24. Savage, S. B. and Sayed, M., Gravity flow of cohesionless granular materials in wedge-shaped hoppers, in Mechanics Applied to the Transport of Bulk Materials, (S. C. Cowin ed.), American Society of Mechanical Engineers, New York, 1979, 1-24.

25. Sullivan, W. N., PhD Thesis, California Institute of Technology, 1972.

26. Kaza, K. R. and Jackson, R., The rate of discharge of granular material from a wedge-shaped mass flow hopper, Powder Tech. (to appear).

27. Bosley, J., Schofield, C. and Shook, C. A., An experimental study of granule discharge from model hoppers, Trans. Inst. Chem. Eng. 47 (1969), 147-153.

28. Kaza, K. R. and Jackson, R., A problem in the flow of granular materials, Ninth U.S. National Congress of Applied Mechanics, 1982.

The author's work in this field is supported by National Science Foundation Grant CPE-8011268.

Roy Jackson
Department of Chemical Engineering
University of Houston
Houston, TX 77004

Granular Flows at High Shear Rates

S. B. Savage

1. INTRODUCTION

Rapid granular flows belong to one of the limiting types of dispersed two-phase flows of mixtures of particulate solids and fluids, namely, the limit corresponding to the case of flows at high solids concentrations and high deformation rates. Such flows are characterized by strong and direct interactions between individual particles. In contrast with the flows of dilute suspensions or fluidized beds, the fluid phase generally plays a minor role in the dynamics of granular flows. This is particularly so when the ratio of fluid to solid mass densities is very small, as in the case of a gaseous interstitial fluid. Granular flows occur in a large number of wide ranging engineering and physical processes. Interest in these flows from an engineering standpoint is centered around problems of materials handling and transportation, and mineral and powder processing. Examples of granular flows in the context of geophysics are rock falls, debris flows, sediment transport in rivers, sub-aqueous grain flows, snow avalanches and drift of pack ice. A review of experimental and theoretical work on granular flows with particular emphasis on chemical engineering applications has been provided recently by Nedderman, Houlsby, Savage and Tuzun [1,2,3].

Although the engineering applications of granular flows are widespread and the field observations of geophysical granular flows have been well documented, the mechanics of such flows is poorly understood. One of the central problems in the theory of granular flows is the determination of the constitutive equations for the fluxes of mass, momentum and energy.

ISBN 0-12-493120-0

Knowledge of these constitutive equations would permit the calculation of the continuum bulk behaviour in various flow situations. For the slower, quasi-static types of flows, considerable work has been done in the context of soil mechanics. Although the issue is still far from settled, there are several theories for quasi-static flows which have achieved some success in describing the deformation fields under limited ranges of conditions [4]. For the case of rapid flows, things are far less satisfactory. There have been comparatively few theoretical studies carried out, and even fewer experimental investigations to obtain data which can be used for the formulation as well as the verification of constitutive theories.

In the present paper I shall describe some ongoing experimental and theoretical studies of rapid granular flows carried out in collaboration with graduate students M. Sayed and C. Lun, and colleagues, D.J. Jeffrey and J.T. Jenkins. I shall begin with a brief report of some typical experimental results for stresses developed by rapidly sheared dry granular material. Recent work on a microstructural kinetic theory for granular materials is then outlined. The theory is applied to two simple boundary value problems, a simple shear flow and a fully developed free surface flow down a rough inclined plane. The predictions of the theory are compared with laboratory measurements of stresses and velocity fields.

2. EXPERIMENTAL MEASUREMENTS OF STRESSES

Sayed [5] (see also [6]) has recently obtained a considerable amount of experimental data on stresses developed by sheared granular materials by using an annular shear cell that I began to develop about four years ago [7]. To emphasize the scarcity of experimental data, I point out that these experiments are, to my knowledge, the only attempts so far to measure the stresses generated during the rapid shear of dry granular materials as functions of both bulk density and shear-rate.

2.1 Annular Shear Cell Apparatus

The annular shear cell apparatus (Figure 1) consists of two concentric circular disk assemblies mounted on a fixed vertical shaft. The bottom disk assembly was driven via a belt and pulley by a variable speed D.C. motor. The upper disk assembly was restrained from rotating by a torque arm but is free to move up and down vertically. Granular material was contained within an annular trough in the lower disk assembly and capped by an annular lip on the top disk. The vertical walls of the trough were made as smooth as possible to permit the granular material to slip there whereas the bottom of the trough and the lip of the annular ring were lined with very coarse sandpaper to develop no-slip conditions there when the granular material was sheared by rotating the lower disk assembly. A displacement transducer was mounted to measure the displacement of the top disk and hence the overall bulk density or solids concentration ν (volume of solids per unit volume). Various loads could be applied to the upper disk through a system of weights and counterweights in order to vary the applied shear and normal stress. The apparatus can therefore determine the shear stress and normal stress developed during the shear of dry granular materials as functions of strain-rate and solids fraction ν.

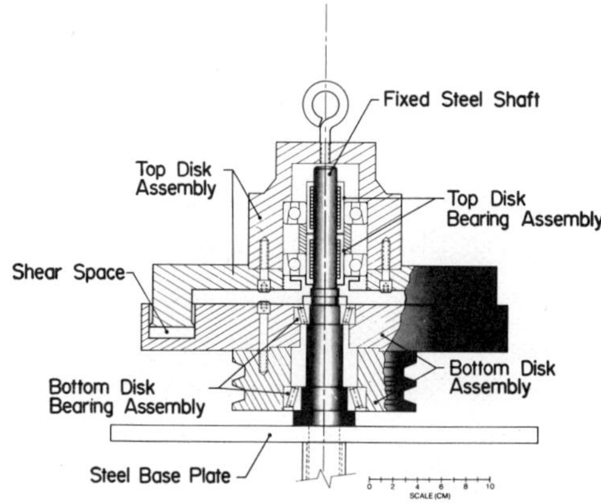

Figure 1 Diagram of annular shear cell

Tests have been performed with several types of dry granular materials. A number of experiments used essentially monosized spherical particles of glass ballotini and polystyrene having specific gravities of 2.97 and 1.095 respectively. To study the diameter dependence of the stresses, some tests were carried out with particles of different nominal diameters, σ, ranging from about 0.5 to 2 mm. A few tests have been made with binary mixtures of particles of different sizes to examine size distribution effects. To study the effects of particle angularity, experiments were performed using crushed walnut shells having a nominal volumetric 'diameter' of 1.2 mm and a specific gravity of 1.18.

Because of centrifugal effects, the normal and shear stresses over the top lip are not uniform, but increase with radius. In order to determine the average shear stress from the torque measurements, we had to make small 'corrections' for the outward displacement of the radius of action of the effective shear force arising from the non-uniform stress distribution.

In the experiments we could only determine the apparent shear rate, $(\bar{u}/H)$, which is the linear velocity of the lower disk assembly at mid-annulus radius, $\bar{u}$, divided by H, the height of the granular material in the trough. It was important to insure that shear occurred over the full depth of flow in order that the apparent shear rate be representative of the actual shear rate. If H was too large, a plug flow region formed near the bottom of the trough. The typical shear layer thickness was determined by putting a column of dyed beads among the others when initially filling the trough, then shearing the mass of beads and carefully removing the beads with a vacuum cleaner near the area of the initial dyed column placement until the undisturbed portion of the column was reached. It was usually found that the sheared layer was about 10 particle diameters thick. All the final stress, strain-rate tests were performed at values of H such that the entire material in the trough was sheared.

2.2 Experimental Results

Some typical experimental results are shown in Figures 2, 3 and 4. Figure 2 is a plot of the normal stress τ_{22}, made nondimensional by dividing by the product of individual particle mass density ρ_p, gravitational acceleration g, and particle diameter σ, versus the nondimensional shear-rate $(\sigma/g)^{\frac{1}{2}}\ \bar{u}/H$. The data are for nearly uniform 1 mm diameter polystyrene beads having a specific gravity of 1.095. Each set of data points corresponds to a different value of the solids fraction ν. At the lower concentrations, the nondimensional stress versus shear-rate plotted on log-log paper has a slope of 2, i.e., the stress varies as the square of the shear-rate. There is a very strong increase in the stress with a relatively slight increase in concentration. At the higher concentrations the curves tend to flatten out, with a slope of less than 2 as the shear-rate is decreased.

Figure 3 shows the same type of data for a bimodal mixture of polystyrene spheres; 30% by weight of 0.55 mm beads and 70% by weight of 1.65 beads. These tests exhibit a behaviour similar to that shown in Figure 2, a slope close to 2 at the lower concentrations and a somewhat more pronounced flattening out with increase in concentration and decrease in shear-rate.

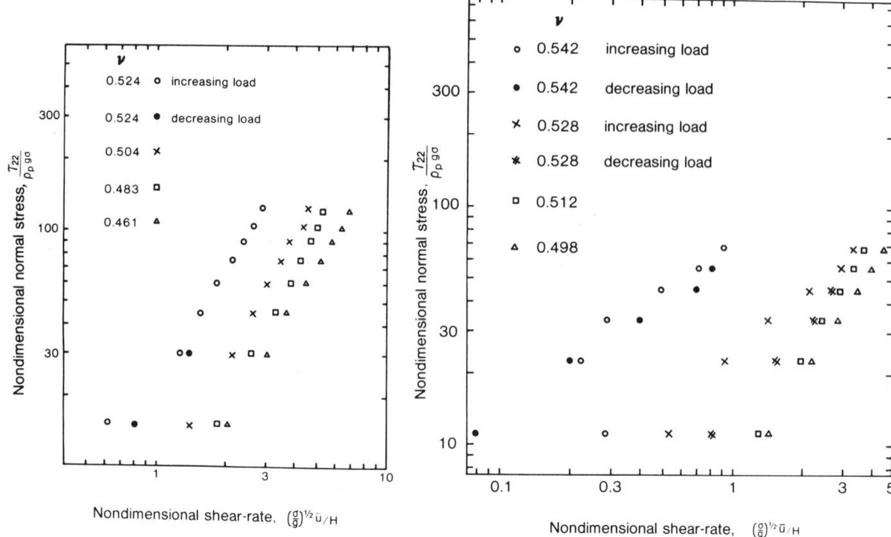

Figure 2 Shear cell data for 1 mm polystyrene beads

Figure 3 Shear cell data for bimodal mixture of polystyrene beads (30% by weight of 0.55 mm spheres and 70% 1.65 mm spheres)

The curves of shear stress versus shear-rate are quite similar to those just shown for the normal stresses. Figure 4 presents data for 1.32 mm polystyrene spheres in another way; in the form of the ratio of shear stress to normal stress ($\tau_{12}/\tau_{22} = \tan \phi_D$, where ϕ_D is the 'dynamic' friction angle) plotted versus the nondimensional apparent shear-rate for various concentrations. Note that the dynamic friction angle ϕ_D is not greatly different from the quasi-static internal friction angle ϕ and that ϕ_D is only weakly dependent upon shear-rate and concentration. With decreasing ν there is a small but noticeable increase in ϕ_D.

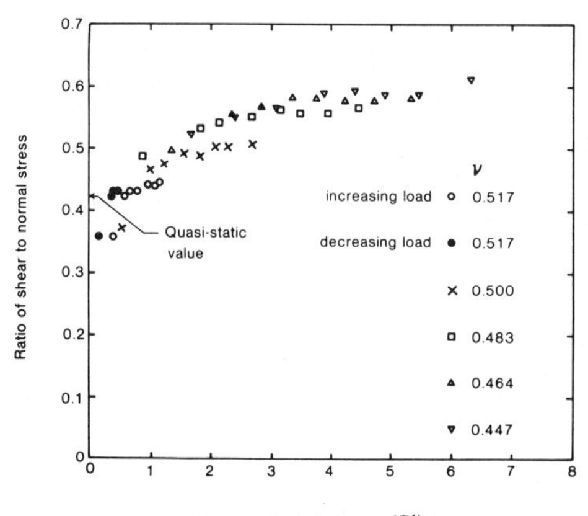

Figure 4 Dynamic friction angle for 1.32 mm polystyrene beads

2.3 Discussion of Results

The data at the lower concentrations are consistent with the behaviour predicted by Bagnold [8] some 30 years ago for the stresses developed during the shear of particulate solids. Bagnold defined a 'grain-inertia' flow regime in which stresses were generated as a result of glancing collisions as the particles of one layer overtook those of an adjacent layer. The collisions give rise to a 'dispersive' normal stress as well as a shear stress. Since both the momentum transferred during a single collision and the rate at which collisions occur are proportional to the relative velocity of the two layers, the stresses are proportional to the square of the shear-rate. In a somewhat more detailed analysis, Bagnold determined expressions for the stresses in a mass of identical spherical particles subjected to a simple mean shear (du_1/dx_2). He found the normal stress in the x_2-direction to be

$$\tau_{22} = f(\nu)\ \rho_p\ \sigma^2\ \left(\frac{du_1}{dx_2}\right)^2 \cos\phi_D \qquad (2.1)$$

and the shear stress to be

$$\tau_{12} = \tau_{22} \tan\phi_D \qquad (2.2)$$

where $f(\nu)$ is some unknown function of solids fraction ν. The proportionality between normal and shear stress (2.2) for this dynamic flow is reminiscent of the Mohr-Coulomb criterion for the quasi-static deformation of a cohesionless granular material [9]. At lower values of the solids fraction ν, the experimental normal stresses shown in Figures 2 and 3 vary approximately as the square of the shear-rate as predicted by equation (2.1). As is seen from Figure 4 the shear stresses have approximately the same dependence upon shear-rate, in agreement with equation (2.2). Tests with materials of different sizes and densities indicated that the stresses at lower concentrations were also dependent upon particle mass density and the square of particle diameter as given by (2.1).

However, with increasing concentration the data depart from the behaviour given by (2.1) and (2.2). A considerable amount of test data has been acquired using the annular shear cell; all this data show the same trends apparent in Fgiure 2, 3 and 4. By making a least square fit to each set of data at a particular value of solids fraction ν it is possible to separate the total measured normal stress (or the shear stress) into two parts, for example

$$\tau_{total} = \tau(\nu)_{Coulomb} + \mu(\nu)\, \rho_p\, \sigma^2 \left(\frac{du_1}{dx_2}\right)^2 \qquad (2.3)$$

Rate-independent, dry friction part — Rate-dependent, viscous part

We may regard the 'dynamic' or 'viscous' contribution as being due to momentum transfer during nearly instantaneous collisions between particles as described by Bagnold [8] for his 'grain-inertia' regime. The rate-independent part results from dry Coulomb friction and particle overriding during enduring contacts between particles. The stress contribution $\tau_{Coulomb}$ and the coefficient μ are monotonic increasing functions of ν. At high concentrations and low shear-rates the first term on the right hand side of (2.3), $\tau_{Coulomb}$, is dominant. At low concentrations and high shear rates, the $\tau_{Coulomb}$ term is negligible and the collisional transfer term prevails.

2.4 Stress Generation in Various Flow Regimes

A tentative physical description of how the stresses are generated in the transition from the rate-dependent 'grain-inertia' regime to the rate-independent type of flow is now given. Consider the example of a shear flow (du_1/dx_2) developed in a granular material contained between two rough plates aligned parallel to the x_1-direction, one plate stationary and one moving. At the very lowest concentrations and highest shear-rates, momentum transfer occurs as a result of particle translation from one shear layer to another as in a dilute gas.

At somewhat higher concentrations, translation between layers is less likely and momentum transport occurs primarily by interparticle collisions as in Bagnold's [8] 'grain-inertia' regime. Because of the collisions the particles acquire random fluctuating spins and translational fluctuating velocities in addition to their mean translational velocities. The particles are inelastic and frictional, and hence energy is dissipated into heat with each collision. The mean shear drives the fluctuations and for this simple flow an equilibrium, as far as the fluctuations are concerned, is reached when the shear work $\tau_{12}u_{1,2}$ equals the rate of dissipation per unit volume arising because of the particles surface friction and inelasticity (neglecting any flux of energy through the bounding pate walls). Of course, the individual particles will continue to get hotter and hotter unless heat is carried away by the interstitial fluid. This leads us to think of two kinds of 'temperatures', 'granular' temperatures associated with the motions of the grains making up the granular material, and the usual kind associated with the motion of the molecules which make up each individual grain.

With increasing concentration and decreasing shear rate, clusters or groups of particles form and break-up intermittently. With further increase in concentration, the clusters grow in size such that they span the full width of the shear region. Such columns of particles in contact, such as to form force networks lined up along the principal stress directions, have been observed in photoelastic studies of quasi-static flow [10]. It is proposed that the deformation of these columns adds a quasi-static component to the stress. At the highest concentrations almost all the particles are in contact, the shear rates are low enough that the inertia forces associated with particle overriding make a negligibly small contribution to the stresses, and we end up with rate-independent stresses.

3. DENSE GAS TYPE THEORY FOR INELASTIC GRANULES

I shall now briefly outline a theory for granular flows in the 'grain inertia' regime that is presently being developed in collaboration with colleagues D.J. Jeffrey and J.T. Jenkins and one of my graduate students C. Lun [11, 12, 13]. This theory makes use of ideas developed previously for the kinetic theory of dense gases. However, there are important differences between the flows of granular materials and flows of dense gases. While it may be justifiable in certain cases to model the molecules of a dense gas as perfectly elastic spheres which conserve energy during collisions (as in the most elementary dense gas theories), we must consider the particles in a granular flow as inelastic, perhaps rough, particles which dissipate energy during each collision. The energy dissipation is an essential feature which should be incorporated in any complete granular flow theory. For a given mean deformation field, we can form one of the important flow parameters by taking a typical characteristic mean shear velocity and dividing by $\langle C^2\rangle^{\frac{1}{2}}$, the root mean square of the velocity fluctuations arising from collisions. (In the kinetic theory of gases $\langle C^2\rangle$ is proportional to three times the temperature.) For the case of a simple shear flow (du_1/dx_2), this parameter can be defined as

$$R = \sigma|du_1/dx_2|/\langle C^2\rangle^{\frac{1}{2}} \tag{3.1}$$

Typically for a gas R << 1, whereas for a granular flow the evidence from several theories [12,13,14,15,16] and numerical experiments [17] indicates that $R \simeq O(1)$. This difference arises because of the significant dissipation present in the granular flows. In the case of granular flows, the mean shear drives the velocity fluctuations; the mean shear _must_ be present to maintain them, unlike the case of a gas at the molecular level. Significant spatial and collisional anisotropies can be associated with the large value of R which characterizes granular flows.

For the present we consider the granular material to be composed of smooth, identical, inelastic spherical particles of diameter σ and mass m (calculations to consider more general and more complex materials are in progress). Analysing the dynamics of binary collisions between two particles at $\underline{r}_1$ and $\underline{r}_2$ moving with velocities of $\underline{c}_1$ and $\underline{c}_2$ respectively, it is found that the change in kinetic energy during a single collision is

$$\Delta E = \frac{m}{4}(e^2 - 1)(\underline{k}\cdot\underline{c}_{12})^2 \qquad (3.2)$$

where e is the coefficient of restitution, $\underline{k}$ is the unit vector from particle 1 to particle 2 and $\underline{c}_{12} = \underline{c}_1 - \underline{c}_2$.

The ensemble average of some function ψ is defined as

$$\langle\psi\rangle = \frac{1}{n}\int \psi\, f^{(1)}(\underline{r}_1, \underline{c}_1; t)\, d\underline{c}_1 \qquad (3.3)$$

where n is the number density and $f^{(1)}(\underline{r}_1,\underline{c}_1; t)$ is the usual single particle velocity distribution function from kinetic gas theory.

It is possible to derive the Maxwell-Chapman Transport Equation for this dissipative system that is the same as the usual form [18]

$$\frac{\partial}{\partial t}\langle n\psi\rangle = n\langle D\psi\rangle - \nabla\cdot\langle n\underline{c}\psi\rangle - \nabla\cdot\Theta + \chi \qquad (3.4)$$

where $D\psi = \frac{d\underline{c}}{dt}\cdot\frac{\partial\psi}{\partial\underline{c}} = \frac{1}{m}\underline{F}\cdot\frac{\partial\psi}{\partial\underline{c}}$

and $\underline{F}$ is the external force acting on a particle. The collisional flux term

$$\Theta(\psi) = -\frac{1}{2}\sigma \iiint_{\underline{c}_{12}\cdot\underline{k}>0} \underline{k}\,(\psi_1' - \psi_1) f^{(2)}(\underline{r}_1,\underline{c}_1,\underline{r}_2,\underline{c}_2)\,\sigma^2(\underline{c}_{12}\cdot\underline{k})\,d\underline{k}\,d\underline{c}_1\,d\underline{c}_2 \qquad (3.5)$$

and the collisional source term

$$\chi(\psi) = \frac{1}{2} \underset{\underline{c}_{12}\cdot\underline{k}>0}{\iiint} (\psi_1'+\psi_2'-\psi_1-\psi_2) f^{(2)}(\underline{r}_1,\underline{c}_1,\underline{r}_2,\underline{c}_2)\sigma^2(\underline{c}_{12}\cdot\underline{k})d\underline{k}d\underline{c}_1 d\underline{c}_2 \quad (3.5)$$

where $f^{(2)}(\underline{r}_1,\underline{c}_1,\underline{r}_2,\underline{c}_2)$ is the complete pair distribution function defined such that $f^{(2)}(\underline{r}_1,\underline{c}_1,\underline{r}_2,\underline{c}_2)\ \delta\underline{c}_1\delta\underline{c}_2\delta\underline{r}_1\delta\underline{r}_2$ is the probability of finding a pair of particles in the volume element $\delta\underline{r}_1$, $\delta\underline{r}_2$ centered on the points $\underline{r}_1$, $\underline{r}_2$ and having velocities within the ranges $\underline{c}_1$ and $\underline{c}_1 + \delta\underline{c}_1$ and $\underline{c}_2$ and $\underline{c}_2 + \delta\underline{c}_2$. Integration over values of $\underline{c}_1$ and $\underline{c}_2$ such that $\underline{c}_{12}\cdot\underline{k}>0$ picks out those particles that are just about to collide.

By taking ψ to be m, $m\underline{c}$ and $\frac{1}{2}mc^2$ in equations (3.3) to (3.6) we generate the usual hydrodynamic equations

$$\frac{d\rho}{dt} = -\rho\nabla\cdot\underline{u} \quad (3.7)$$

$$\rho\frac{d\underline{u}}{dt} = \rho\underline{b} - \nabla\cdot\underline{p} \quad (3.8)$$

$$\frac{3}{2}\rho\frac{dT}{dt} = -\ \underline{p}:\nabla\underline{u} - \nabla\underline{q} - \gamma \quad (3.9)$$

where $\rho = mn = \nu\rho_p$ is the bulk mass density, $\underline{u} = \langle\underline{c}\rangle$ is the bulk velocity, $3T/2 = \langle C^2\rangle/2$ is the fluctuation specific kinetic energy and $\underline{C} = \underline{c} - \underline{u}$, $\underline{p}$ is the pressure tensor composed of a kinetic part $\underline{p}_k = \rho\langle\underline{C}\underline{C}\rangle$ and a collisional part $\underline{p}_c$, q is the flux of fluctuation energy composed of a kinetic part $\underline{q}_k = \rho\langle\underline{C}C^2\rangle/2$ and a collisional part $\underline{q}_c$, and γ is the collisional rate of dissipation per unit volume. The collisional terms $\underline{p}_c$ and $\underline{q}_c$ are given by (3.5), after taking ψ to be respecitvely $m\underline{C}$ and $\frac{1}{2}mC^2$ in (3.5). Putting $\psi = \frac{1}{2}mc^2$ in equation (3.6) yields the rate of dissipation $\gamma = -\ \chi(\psi)$.

For moderate concentrations it is permissible to neglect the kinetic contributions in comparison with the collisional ones, thus we assume $\underline{p}_k \ll \underline{p}_c$ and $\underline{q}_k \ll \underline{q}_c$ for the present treatment. The explicit evaluations of the collisional integrals for $\underline{p}_c$, q_c and γ require the specific form for the pair distribution function $f^{(2)}(\underline{r}_1,\underline{c}_1,\underline{r}_2,\underline{c}_2)$. As a first approximation we assume that the single particle velocity distribution function is Maxwellian about the mean velocity, for example

$$f^{(1)}(\underline{r}_1,\underline{c}_1) = n_1(2\pi T_1)^{-3/2} \exp\{-(\underline{c}_1 - \underline{u}_1)^2/2T_1\} \quad (3.10)$$

and that

$$f^{(2)} = g_o(\nu)[1 - \frac{\alpha\sigma}{\sqrt{\pi T}} \underline{\underline{kk}}:\nabla\underline{u}]\ f^{(1)}(\underline{r}_1,\underline{c}_1)\ f^{(1)}(\underline{r}_2,\underline{c}_2) \tag{3.11}$$

where $g_o(\nu)$ is the radial distribution function and α is a numerical coefficient. Part of the collisional anisotropy arising from the presence of the mean shear is incorporated in $f^{(2)}$ by taking the Maxwellians about the local mean velocity and part is incorporated in the term in square brackets in (3.11). We take the radial distribution function $g_o(\nu)$ to be that proposed by Carnahan and Starling [19] on the basis of numerical simulations

$$g_o(\nu) = \frac{1}{(1-\nu)} + \frac{3\nu}{2(1-\nu)^2} + \frac{\nu^2}{2(1-\nu)^3} \tag{3.12}$$

A linearization of a similar ad hoc expression for $f^{(2)}$ proposed by Savage and Jeffrey [11] gives $\alpha = 1$.

Using expressions (3.10) to (3.12) in the collisional integrals, expanding the Maxwellians about the point of contact and taking $\sigma\underline{\underline{kk}}:\nabla\underline{u}/T^{\frac{1}{2}} << 1$ yields the constitutive equations

$$\underline{q}_c = - \kappa\nabla T \tag{3.13}$$

$$\gamma = \frac{6(1-e)\kappa}{\sigma^2}\ [T + (\frac{\pi}{4} - \frac{\alpha}{3})\ \sigma(\frac{T}{\pi})^{\frac{1}{2}}\ \text{tr}\ \underline{D}\ ] \tag{3.14}$$

$$\underline{p}_c = [\frac{\kappa}{\sigma}\ (\pi T)^{\frac{1}{2}} - \frac{\kappa}{5}\ (2+\alpha)\ \text{tr}\ \underline{D}]\ \underline{I} - \frac{2\kappa}{5}\ (2+\alpha)\ \underline{D} \tag{3.15}$$

where $\kappa = 2\nu^2 g_o(1+e)\rho_p\sigma(T/\pi)^{\frac{1}{2}}$ (3.16)

and $\underline{D} = \frac{1}{2}\ (u_{i,j} + u_{j,i})$ (3.17)

Using the constitutive equations for $\underline{p}_c$, $\underline{q}_c$ and γ in the conservation equations (3.7) to (3.9), after neglecting the kinetic terms, yields five equations for the five unknowns ρ(or ν), $\underline{u}$, and T (or $<C^2>$).

Note that we did not attempt to determine the distribution functions $f^{(1)}$, $f^{(2)}$ etc. by solving the Boltzmann equation or some approximation to it as is done, for example, in the Chapman-Enskog approach [20]. For the present purposes we are not greatly interested in the distribution functions themselves, but more in the averaged properties such as ν, $\underline{u}$ and T that are of concern in flow problems. We have used simple, plausible assumptions for the form of the distribution functions in which ν, $\underline{u}$ and T were regarded as parameters and used these forms to determine the constitutive equations which can be used in the overall conservation equations for the solution of boundary value problems. This approach is analogous, in a sense, to the momentum integral approach used in boundary layer theory. It is probably the simplest kind of analysis that we can expect to be capable of predicting the gross features of rapid granular shear flows at moderate concentrations.

4. APPLICATION OF THE THEORY TO SOME SIMPLE FLOWS

It is interesting to see what the theory just described yields for two relatively straightforward examples of shear flows. Experimental data exist for both cases to be discussed and we can make comparisons with the theoretical predictions.

4.1 Simple Shear Flow

4.1.1 Analysis. Consider the case of a simple flow $u_1(x_2)$ in which there are no gradients of T. Such a situation might approximate the granular Couette flow in the annular shear cell described in Section 2, assuming that the flux of fluctuation energy from the granular material into the shear cell boundary walls is negligible (this is analogous to the boundary condition of an insulated wall in a thermal problem). For this simple flow, equation (3.9) for the conservation of specific fluctuation kinetic energy reduces to

$$p_{12}\, u_{1,2} + \gamma = 0 \tag{4.1}$$

By using the constitutive equations (3.14) to (3.17) in (4.1) and neglecting $\underline{p}_k$, the kinetic contribution to the stress tensor, we obtain

$$p_{12} \simeq p_{c_{12}} = -\frac{6}{5}(3\pi)^{-\frac{1}{2}}(1+e)\,\nu g_o\,\rho <C^2>^{\frac{1}{2}}\, u_{1,2}\,\sigma \tag{4.2}$$

$$p_{22} \simeq p_{c_{22}} = p_{c_{11}} = p_{c_{33}} = \frac{2}{3}(1+e)\,\nu g_o \rho <C^2> \tag{4.3}$$

$$R = \frac{\sigma|u_{1,2}|}{<C^2>^{\frac{1}{2}}} = [\frac{10}{3}(1-e)]^{\frac{1}{2}} = \text{const.} \tag{4.4}$$

$$\frac{|p_{c_{12}}|}{p_{c_{22}}} = \tan\phi_D = \frac{9}{5}(3\pi)^{-\frac{1}{2}} R = \text{const.} \tag{4.5}$$

Substituting (4.4) into (4.2) and (4.3) shows that both the shear and normal stresses vary as the square of the shear rate, as observed in the annual shear cell experiments. The functional dependence of the stresses upon ρ_p and σ is the same as given in equations (2.1) and (2.2) from Bagnold's simple analysis, but we now have obtained the explicit dependence upon ν.

4.1.2 Comparison with experimental results. We can compare equations (4.2) to (4.5) with the annular shear cell results if we assume no slip at the rough boundaries such that the apparent shear rate equals $u_{1,2}$. Figure 5 shows the nondimensional shear stress $p_{12}/[\rho_p \sigma^2 u_{1,2}^2]$ from equation (4.2) plotted versus ν for typical values of the coefficient of restitution of 0.8 and 0.9. The experimental data points for spherical glass and polystyrene beads correspond to the rate-dependent dynamic contributions to the total stress appearing in equation (2.3) and determined by the least square fits to the data as described in Section 2. The experimental 'dynamic' stresses shown in Figure 5 were generally nearly equal to the measured total stresses. The diameters of the glass and polystyrene beads were 1.8 and 1.0 mm respectively and the specific gravities were 2.97 and 1.095. The predicted stresses are of the right order of magnitude and reasonably close to the measured values.

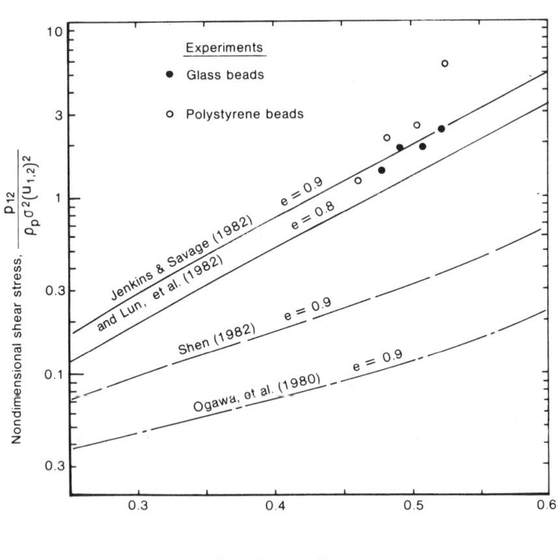

Figure 5 Comparisons of predicted and experimental shear stresses

Also shown on Figure 5 are three other theoretical predictions. The predictions of the analysis of Lun, et al. [13] (which is based upon a more exact evaluation of the collision integrals than the linearizations described above) are almost identical (for this example) to the results given by equation (4.2). The analyses of Ogawa, et al. [14] and Shen and Ackermann [15,16], which treat the collisional dynamics in a more simplified way than just described, predict stresses which are roughly between one and two orders of magnitude lower than observed. The theories of Shen and Ackermann and Ogawa, et al. can treat rough as well as inelastic particles. All the curves shown on Figure 5 are for smooth particles. Putting the particle surface coefficient of friction to be non-zero in the theories of Shen and Ackermann and Ogawa, et al. further reduces the predicted stresses.

4.2 Flow Down a Rough Inclined Chute

4.2.1 Analysis. For this second example, let us consider the fully developed two-dimensional, free surface flow of granular material down a rough plane inclined at an angle ζ to the horizontal as shown in Figure 6. The only variation in flow properties are in the x_2-direction, normal to the incline. The linear momentum equations yield

$$p_{22} = g \cos \zeta \int_0^{x_2} \rho dx_2 \tag{4.6}$$

$$p_{12} = g \sin \zeta \int_0^{x_2} \rho dx_2 \tag{4.7}$$

and

$$\frac{p_{12}}{p_{22}} = \tan \zeta = \text{const.} \tag{4.8}$$

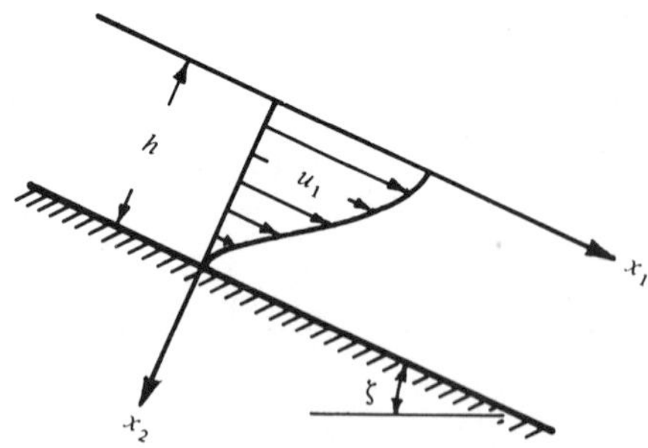

Figure 6 Two-dimensional gravity flow of granular material down an inclined plane

Again we neglect the kinetic terms p_k and q_k in comparison with the collisional contributions. We make the approximation that

$$\int_0^{x_2} \rho dx_2 \simeq \bar{\rho} x_2 = \rho_p \bar{\nu} x_2 \tag{4.9}$$

where $\bar{\rho}$ and $\bar{\nu}$ are average values of ρ and ν over the depth h. Putting $\alpha = 1$, we find

$$\gamma = 6\,(1-e)\,\frac{\kappa}{\sigma^2}\,T \tag{4.10}$$

$$p_{c_{22}} = \frac{\kappa}{\sigma}(\pi T)^{\frac{1}{2}} \simeq g \cos\zeta\, \rho_p\, \bar{\nu} x_2 \tag{4.11}$$

$$p_{c_{21}} = -\frac{6\kappa}{5}\,\frac{u_{1,2}}{2} \simeq g \sin\zeta\, \rho_p\, \bar{\nu} x_2 \tag{4.12}$$

and

$$\frac{p_{12}}{p_{22}} = \frac{p_{c_{12}}}{p_{c_{22}}} = -\frac{3}{5}\left(\frac{3}{\pi}\right)^{\frac{1}{2}} \frac{\sigma u_{1,2}}{\langle C^2\rangle^{\frac{1}{2}}} \tag{4.13}$$

Hence, taking $u_{1,2}$ to be negative

$$R = \frac{5}{3}\left(\frac{\pi}{3}\right)^{\frac{1}{2}} \tan\zeta \tag{4.14}$$

The fluctuation energy equation (3.9) reduces to

$$\frac{3}{2}\rho\frac{dT}{dt} = 0 = -p_{c_{12}}\, u_{1,2} - (-\kappa\, T_{,2})_{,2} - \gamma \tag{4.15}$$

where there is a balance between the three terms on the right hand side, which represent respectively the shear work, the energy flux gradient and the rate of dissipation. Using (4.10) to (4.14) in (4.15) we obtain

$$\frac{d}{dx_2}\left(x_2 \frac{df}{dx_2}\right) + \frac{\beta x_2 f}{\sigma^2} = 0 \tag{4.16}$$

where $f = T^{\frac{1}{2}}$ (4.17)

$\beta = 9R^2/10 - 3(1-e)$ (4.18)

and $\beta \gtrless 0$ if $\tan^2\zeta \gtrless 18(1-e)/5\pi$ (4.19)

Equation (4.16) has the form of Bessel's equation of zeroth order and is subject to the boundary conditions that there is no flux of fluctuation energy at the free surface, i.e. $dT/dx_2 = 0$ at $x_2 = 0$ and a specified flux at the bed $x_2 = h$. The general case will be described elsewhere and we consider here only the case of zero energy flux into the bed. This corresponds to

$$\beta = 0$$

$$\tan\zeta = \tan\zeta_0 = \left[\frac{18(1-e)}{5\pi}\right]^{\frac{1}{2}} \tag{4.20}$$

and $T = \text{const.}$

We merely note that fully developed flows with an energy flux into the bed correspond to $\beta > 0$ and $\zeta > \zeta_0$, whereas an input energy from the bed into the granular material corresponds to $\beta < 0$ and $\zeta < \zeta_0$. Vibrating the bed is one way to put fluctuation energy into the material and this technique is commonly used in materials handling devices to make granular materials flow at slopes considerably less than their natural angle of repose.

From (4.8) and (4.13), with $T = \text{const.}$

$$\frac{du_1}{dx_2} = -\frac{5}{3}\frac{(\pi T)^{\frac{1}{2}}}{\sigma}\tan\zeta_0 = \text{const.} \tag{4.21}$$

Assuming the no-slip condition at the bed, we obtain the velocity

$$u_1 = \frac{5}{3}\frac{(\pi T)^{\frac{1}{2}}}{\sigma}\tan\zeta_0\,[h - x_2] \tag{4.22}$$

Applying (4.11) at the bed using (3.16) yields the fluctuation specific kinetic energy or the 'granular temperature' T as

$$T = \frac{g\cos\zeta_0\,\bar{\nu}\,h}{2\nu_b^2\,g_0(\nu_b)(1+e)} = \text{const.} \tag{4.23}$$

where ν_b is the value of ν at the bed $x_2 = h$. From (4.11) and (3.16) we can also obtain an expression capable of yielding $\nu(x_2)$, thus

$$\frac{\nu^2 g_0(\nu)}{\nu_b^2 g_0(\nu_b)} = \frac{x_2}{h} \tag{4.24}$$

After defining

$$\bar{\nu} = \frac{1}{h} \int_0^h \nu dx_2 \tag{4.25}$$

and specifying a value of ν_b and the physical properties of the granular materials, equations (4.22), (4.23) and (4.25) give the required velocity, granular temperature and solids concentration profiles.

4.2.2 Comparison with experimental results. By using fiber-optic probes, Ishida and Shirai [21] have measured the velocity distributions developed by glass beads (diameters ranging between 0.35 and 0.50 mm) flowing down an inclined chute which had its bed roughened with sandpaper. Some typical velocity distributions for different bed inclindations ζ are shown in Figure 7. At the higher bed inclination angle of $\zeta = 30^0$, the velocity profile has the triangular form predicted by (4.22). With decreasing ζ the magnitudes of the velocities decrease and the profile shape becomes more and more concave.

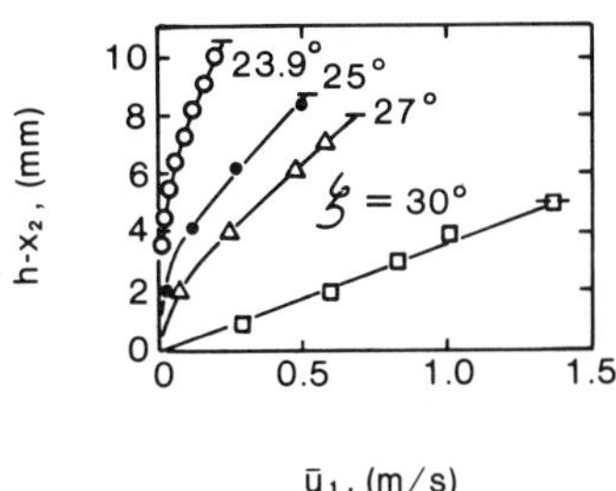

Figure 7 Velocity profiles for flow of glass beads down inclined channel (after Ishida & Shirai [21])

To make a quantitative comparison between (4.22) and the data of Ishida and Shirai, we choose reasonable values for e, ν_b and h of say

$e = 0.8$

$\nu = 0.5$

$h = 6$ mm

and take $\sigma = (0.35 + 0.5)/2 = 0.425$ mm.

From (4.20) we find that the bed inclination for fully developed flow with a triangular velocity profile is

$$\zeta_0 = 25.6^0$$

While this is somewhat lower than the experimental ζ for a triangular profile as shown in Figure 7, it should be noted that the analysis is for an infinitely wide channel, whereas the experiments were carried out in a channel of finite width. Because of friction on the side walls a larger ζ would be required to develop a flow similar to what might occur if the side walls were perfectly smooth or infinitely far apart.

The predicted profiles for velocity u_1, granular temperature T and solids fraction ν are shown in Figure 8. The magnitudes of the predicted velocities are similar to the experimental triangular velocity profile of Ishida and Shirai shown in Figure 7. However, for given material properties the analysis predicts only one value of ζ for fully developed flow (and no fluctuation energy flux through the bed), but the experiments show a range of bed inclination angles over which steady non-accelerating flows occur (see also [22]). It is likely that at the lower values of ζ, enduring contacts between particles occur and particle surface friction becomes increasingly important. At these lower slopes, the rate-independent contribution to the stress tensor discussed in Section 2, cannot be neglected as has been done in the theoretical development.

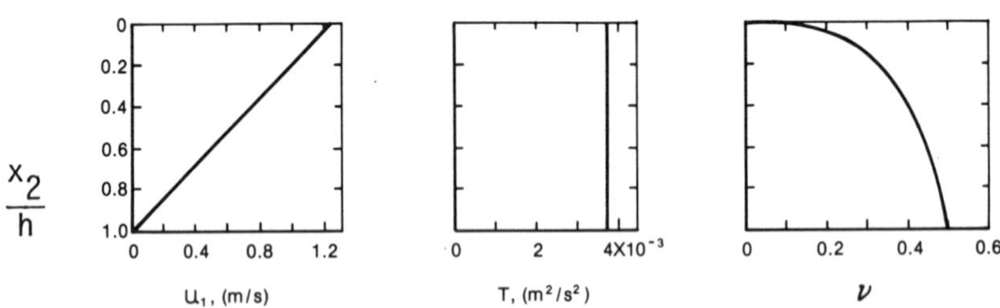

Figure 8 Predicted profiles for velocity, granular temperature and solids fraction during flow down an inclined chute

5. CONCLUSION

The present paper has described some recent theoretical and experimental work on rapid granular flows. While the results of the annular shear cell experiments help to clarify some of the flow phenomena that occur, there is a need for further studies using other experimental devices to corroborate the present results and to acquire a larger store of experimental data to aid in the formulation of constitutive theories.

A rudimentary kinetic theory for granular flow accounting for dissipative binary collisions between smooth particles has been presented. At moderate solids concentrations, this simple theory gives reasonable predictions of the flow behaviour and further developments of the theory seem fruitful. At higher concentrations the assumption of binary collisions breaks down. Extensions and generalizations of the theory to account for enduring contact between larger numbers of particles and interparticle surface friction effects should be studied.

REFERENCES

1. Nedderman, R.M., U. Tuzun, S.B. Savage and G.T. Houlsby, The flow of granular materials - I, Discharge rates from hoppers, Chem. Eng. Sci. 37 (1982) (in press).
2. Tuzun, U., G.T. Houlsby, R.M. Nedderman and S.B. Savage, The flow of granular materials - II, Velocity distributions in slow flow, Chem. Eng. Sci. 37 (1982) (in press).
3. Savage, S.B., R.M. Nedderman, U. Tuzun, and G.T. Houlsby, The flow of granular materials - III, Rapid shear flows, Chem. Eng. Sci. 37 (1982) (in press).
4. Spencer, A.J.M., Deformation of an ideal granular material, in Mechanics of Solids, Rodney Hill 60th Anniv. Vol. (H.G. Hopkins and M.J. Sewell, eds.) Pergamon Press, Oxford, 1981, 607-52.
5. Sayed, M., Theoretical and experimental studies of the flow of cohesionless granular materials, Ph.D. Thesis, McGill University, 1981.
6. Savage, S.B. and M. Sayed, Stresses developed by dry cohesionless granular materials sheared in an annular shear cell, (in preparation).
7. Savage, S.B., Experiments on shear flows of cohesionless granular materials, Proc. of U.S.-Japan Seminar on Continuum-Mechanical and Statistical Approaches in the Mechanics of Granular Materials (S.C. Cowin and M. Satake, eds.) Gakujutsu Bunken Fukyakai, Tokyo 1978, 241-254.
8. Bagnold, R.A., Experiments on a gravity free dispersion of large solid spheres in a Newtonian fluid under shear, Proc. Roy. Soc. Lond. A225 (1954), 49-63.

9. Scott, R.F., Principles of Soil Mechanics, Addison-Wesley, Reading, 1963.
10. Dresher, A. and G. de Josselin de Jong, Photoelastic verification of a mechanical model for the flow of a granular material, J. Mech. Phys. Solids 20 (1972), 337-51.
11. Savage, S.B. and D.J. Jeffrey, The stress tensor in a granular flow at high shear rates, J. Fluid Mech. 110 (1981), 255-272.
12. Jenkins, J.T. and S.B. Savage, A theory for the rapid flow of identical, smooth, nearly elastic, spherical particles, submitted to J. Fluid Mech. (1982).
13. Lun, C., S.B. Savage and D.J. Jeffrey, The stresses developed during the simple shear of a granular material comprised of smooth, uniform, inelastic spherical particles, (in preparation).
14. Ogawa, S., A. Umemura and N. Oshima, On the equations of fully fluidized granular materials, J. Appl. Math. Phys. (ZAMP) 31 (1980), 483-93.
15. Ackermann, N.L. and H. Shen, Stresses in rapidly sheared fluid-solid mixtures, J. Eng. Mec. Div. ASCE 108 (1982), 95-113.
16. Shen, H., Constitutive relationships for fluid-solid mixtures, Ph.D. Thesis, Clarkson College, Potsdam, 1982.
17. Campbell, C.E., Shear flows of granular materials, Ph.D. Thesis, California Institute of Technology, Pasadena, 1982.
18. Hirschfelder, J.O., C.F. Curtus and R.B. Bird, Molecular Theory of Gases and Liquids, J. Wiley, New York, 1954.
19. Carnahan, N.F. and K.E. Starling, Equations of state for non-attracting rigid spheres, J. Chem. Phys. 51, (1969), 635-636.
20. Chapman, S. and T.G. Cowling, The Mathematical Theory of Non-Uniform Gases, 3rd ed., Cambridge University Press, 1970.
21. Ishida, M. and T. Shirai, Velocity distributions in the flow of solid particles in an inclined open channel, J. Chem. Eng. of Japan, 12 (1979) 46-50.
22. Savage, S.B., Gravity flow of cohesionless granular materials in chutes and channels, J. Fluid Mech. 92 (1979) 53-96.

The work reported here was supported by a grant from the Natural Sciences and Engineering Research Council of Canada.

Stuart B. Savage
Department of Civil Engineering and Applied Mathematics
McGill University
Montreal, Quebec H3A 2K6
Canada

Theory and Experiments in the Mechanics of Magnetically Stabilized Fluidized Solids

R. E. Rosensweig, M. Zahn, W. K. Lee, and P. S. Hagan

1. INTRODUCTION.

The magnetically stabilized bed (MSB) represents the bubble-free equilibrium flow of a fluid upward through a quiescent mass of fluidized, magnetizable particles in the presence of an applied magnetic field [1,2]. Our interest in this technology stems from the nearly ideal plug flow countercurrent contacting inherent in the flow of these beds through process vessels [3,4].

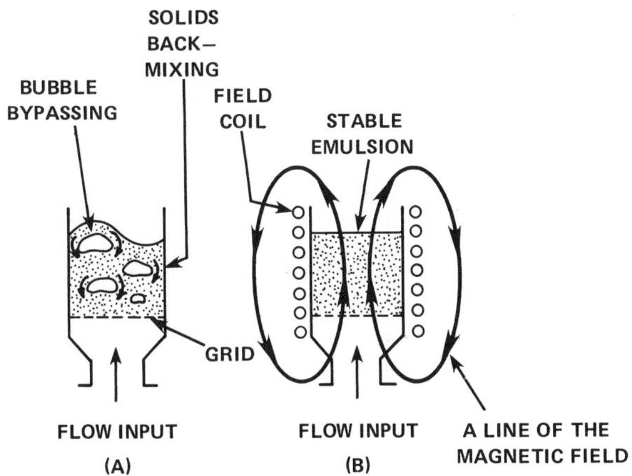

Fig.1
(A) Unstabilized and (B) Magnetically Stabilized Fluid Solids.

ISBN 0-12-493120-0

Figure 1A illustrates conditions in a conventional fluidized bed operated as a batch unit. Beyond the gas input flow rate of incipient fluidization where upward drag just balances the downward gravity force on a bed particle, the excess of gas flowing tends to collect into bubbles. The bubbles are buoyant in the bed and rise up rapidly to escape at the surface. Gas in the bubbles thus tends not to contact the particles, this constituting a source of inefficiency in transferring energy or matter between the phases. Also, flow in the bubble wake stirs the surrounding medium, backmixing the solids which is undesirable in staged processes. A means to eliminate bubble formation completely is illustrated in Figure 1B.

In MSBs, particles of the bed can be either wholly magnetizable or a composite of magnetizable and nonmagnetizable material; the bed of such particles is subjected to a source of applied magnetic field. Figure 2 compares the characteristics of magnetized and unmagnetized modes of operation. In either case pressure drop versus gas flow rate exhibits a break point at the minimum fluidization

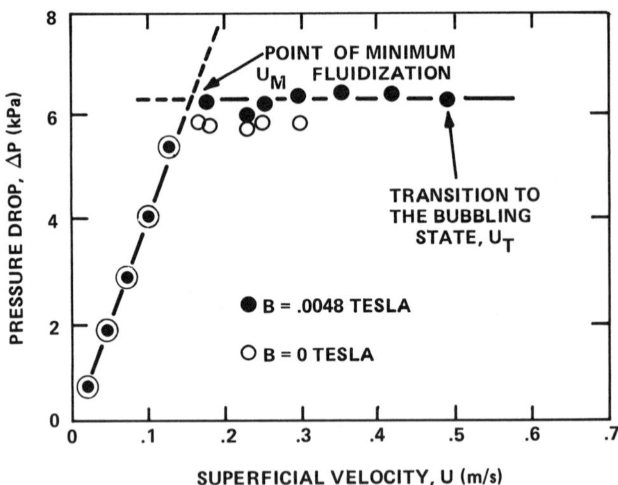

Fig. 2.

Pressure Drop vs. Flowrate is Magnetic Field Independent.

condition, and pressure drop is independent of any further increases in the gas flow rate. In the case of the MSB, the bed accommodates the increased throughput of gas by expanding rather homogeneously, with no bubble formation. Eventually, at a higher flow rate termed the transition velocity the bed suddenly destabilizes producing a magnetized, fluidized, bubbling state.

2. EQUATIONS OF MOTION.

The bed may be regarded as a two phase flow system. At any point within either phase, the motion is determined by expressions of mass and momentum balance. However, we are uninterested in fine details of the microflow field and instead choose to work with averaged equations in which velocity, pressure, and composition of both phases may be regarded as continuous field functions. In this sense, we deal with interpenetrating continua. Research is still in progress to evolve suitable field equations. As our starting point we utilize the general equation set below, which is consistent with the equations of Anderson and Jackson [5] and Drew [6].

Mass Continuity

$$\frac{\partial \varepsilon_f}{\partial t} + \nabla\cdot(\varepsilon_f\, \underline{u}_f) = 0 \tag{1.1}$$

$$\frac{\partial \varepsilon_p}{\partial t} + \nabla\cdot(\varepsilon_p\, \underline{u}_p) = 0 \tag{1.2}$$

Momentum

$$\rho_f \varepsilon_f\left[\frac{\partial \underline{u}_f}{\partial t} + (\underline{u}_f\cdot\nabla)\underline{u}_f\right] = \varepsilon_f \nabla\cdot\underline{\underline{T}}_f - \underline{f} + \varepsilon_f \rho_f \underline{g} \tag{1.3}$$

$$\rho_p \varepsilon_p\left[\frac{\partial \underline{u}_p}{\partial t} + (\underline{u}_p\cdot\nabla)\underline{u}_p\right] = \varepsilon_p \nabla\cdot\underline{\underline{T}}_f + \underline{f} + \varepsilon_p \rho_p \underline{g} + \nabla\cdot\underline{\underline{T}}_p + \nabla\cdot\underline{\underline{T}}_m \tag{1.4}$$

Two Phases

$$\varepsilon_p + \varepsilon_f = 1 \tag{1.5}$$

In the above ε denotes volume fraction, subscript f denotes fluid, and p particle. The fluid-particle interaction force density is denoted $\underline{f}$. $\underline{\underline{T}}_f$ is fluid stress, $\underline{\underline{T}}_p$ particle stress, and $\underline{\underline{T}}_m$ magnetic stress. Constitutive assumptions will be discussed where introduced. With the magnetic coupling it is necessary to include the magnetic field equations.

Magnetostatics

$\nabla \cdot \underline{B} = 0$ Gauss' Law (1.6)

$\nabla \times \underline{H} = 0$ Ampere's Law with no current (1.7)

$\underline{B} = \mu_0(\underline{H} + \underline{M})$ Defining Equation (1.8)

$M = \varepsilon_p M_p$ Mixture Magnetization (1.9)

where the particle magnetization M_p may depend nonlinearly on the particle fraction and the magnetic field intensity.

2. STABILITY OF THE UNBOUNDED BED

Linear stability analysis has been carried out for a magnetized medium of infinite extent and uniform voidage using the following constitutive assumptions.

$\underline{\underline{T}}_m = -\frac{\mu_0}{2} H^2 \underline{\underline{I}} + \underline{H}\,\underline{B}$ Magnetic Stress Tensor (2.1)

$\underline{\underline{T}}_f = -p_f \underline{\underline{I}}$ Inviscid Bulk Fluid (2.2)

$\underline{\underline{T}}_p = 0$ Stress Free Solids (2.3)

$\underline{f} = \varepsilon_f \beta(\varepsilon_f)(\underline{u}_f - \underline{u}_p)$ Low Reynolds Number Drag (2.4)

$\rho_f = 0$ Negligible Gas Density (2.5)

The magnetic stress tensor specifies forces of magnetic polarization and is derived from energy conservation [7,8]. In the form employed, terms of magnetostrictive origin are ignored for simplicity. The fluid phase is treated as inviscid except for mutual interaction with the particle phase. The solids phase is regarded as free of mechanical stress at all times. Mutual interaction is restricted to linear dependence on relative velocity, hence to low Reynolds numbers. The fluid density is assumed negligible compared to that of the solids.

The magnetic force density is computed on the assumption that the material is ferromagnetically soft so that $\underline{M}$ is collinear with $\underline{H}$, or $\underline{M} \times \underline{H} = 0$. Thus, magnetic force density $\underline{F}_m$ is given as

$$\underline{F}_m = \nabla \cdot \underline{\underline{T}}_m = \mu_0 (\underline{M} \cdot \nabla) \underline{H} = \mu_0 \frac{M}{H} (\underline{H} \cdot \nabla \underline{H}) \quad (2.6)$$

Using a vector identity,

$$\underline{F}_m = \mu_0 \frac{M}{H} [\tfrac{1}{2} \nabla (\underline{H} \cdot \underline{H}) - \underline{H} \times (\nabla \times \underline{H})] = \mu_0 M \nabla H \quad (2.7)$$

where M and H are magnitudes and use was made of (1.7).

The constant equilibrium solution in a uniform applied magnetic field corresponding to

$$\varepsilon_f = \varepsilon_0 \quad , \quad \varepsilon_p = 1 - \varepsilon_0 \quad (2.8a)$$

$$\underline{u}_p = 0 \quad , \quad \underline{u}_f = u_0 \underline{e}_x \quad (2.8b)$$

is given by

$$\beta(\varepsilon_0) u_0 = (1 - \varepsilon_0) \rho_p g \quad (2.8c)$$

$$\nabla p_{f,0} = - \beta(\varepsilon_0) u_0 \underline{e}_x \quad (2.8d)$$

Here $\underline{e}_x$ is the unit vector in the vertical direction. In particular, the pressure difference across a bed of height L is $\Delta p = (1-\varepsilon_0)\rho_p g L$.

It should be noted that the magnetics have no influence on the equilibrium solution. This is a direct consequence of (2.7) in that magnetic forces arise only as a result of field gradients while in the equilibrium system the magnetic field is spatially uniform, hence free of magnetic force. This prediction is confirmed by the pressure data of Figure 2.

The system of equations may be linearized around the solution (2.8) assuming the presence of small perturbations [8],

$$\begin{aligned} \varepsilon &= \varepsilon_0 + \varepsilon_1 \\ p_f &= p_{f,0} + p_{f,1} \\ \underline{H} &= \underline{H}_0 + \underline{H}_1 \\ &\text{etc.} \end{aligned} \qquad (2.9a,b,c...)$$

With magnetic field applied collinear to the flow field for linearly magnetizable particles with susceptibility χ the following linear PDE is obtained for axially propagating voidage perturbations.

$$\frac{\partial^2 \varepsilon_1}{\partial t^2} + a \frac{\partial \varepsilon_1}{\partial t} + b \frac{\partial \varepsilon_1}{\partial x} - c \frac{\partial^2 \varepsilon_1}{\partial x^2} = 0 \qquad (2.10)$$

where

$$a = \frac{g}{\bar{u}_0} \qquad (2.11)$$

$$b = \frac{gd}{\varepsilon_0} \qquad (2.12)$$

$$c = \frac{\mu_0 M_p^2}{\rho_p} \left[\frac{(1-\varepsilon_0)}{1 + \chi\,(1-\varepsilon_0)}\right] \qquad (2.13)$$

$$d = [1-2\varepsilon_0 - \varepsilon_0\,(1-\varepsilon_0)\beta'(\varepsilon_0)/\beta(\varepsilon_0)] \qquad (2.14)$$

Here $\bar{u}_0 = u_0\varepsilon_0$ is the superficial gas velocity whose value at minimum fluidization condition is denoted $\bar{u}_m$. Magnetics appears explicitly only in the coefficient c. If ε_1 is considered the (analog) displacement of a longitudinally vibrating bar with damping then the last term in (2.10) represents compressibility influence and $c\rho_p$ is the modulus of elasticity; local expansion or compression of the bed establishes a magnetic force tending to restore the bed to uniformity.

A plane wave disturbance may be considered in the form

$$\varepsilon_1 = \hat{\varepsilon}_1 \, \mathrm{Re}\,[\exp(st)\,\exp(ikx)] \qquad (2.15a,b)$$

$$s = \xi - i\eta$$

where ξ is the growth factor and k is the wave number of the disturbance. An analytical solution is obtained for ξ with the typical behavior shown in Figure 3. Thus when the bed is

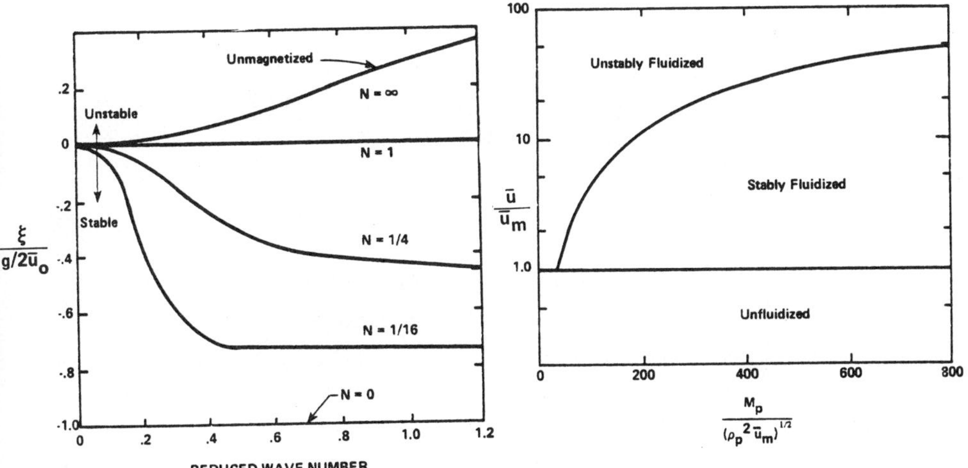

Fig. 3
Growth Factor in Beds of Infinite Extent.

Fig. 4
Phase Diagram of the Infinite Inviscid Bed.

unmagnetized ($M_p = 0$) any disturbance is unstable as $\xi > 0$ for any k. This corresponds to the well known result for ordinary fluidized beds [9]. The growth rate is reduced when magnetization is increased, and neutral stability ($\xi = 0$) is achieved concomitantly for all of the modes. A more detailed analysis shows that this is also true for all wave number orientations when the bed solids are linearly magnetizable. With further increase of magnetization, all modes become stable ($\xi < 0$). The neutral stability condition is given by the criterion

$$N_m N_v = \frac{b^2}{a^2 c} = 1 \tag{2.16}$$

where the dimensionless modulus N_m is defined as

$$N_m = \frac{\rho_p \bar{u}_0^2}{\mu_0 M_p^2} \tag{2.17}$$

and thus N_v is given by

$$N_v = \frac{1+\chi(1-\varepsilon_0)}{\varepsilon_0^2 (1-\varepsilon_0)} d^2 \tag{2.18}$$

More general analysis determines that field oriented transverse to the flow fails to stabilize and that axially oriented field is preferred over any other orientation [1]. Figure 4 is a phase diagram that represents the fixed to fluidized bed transition given by (2.8) and the stable to bubbling bed transition of (2.16) as functions of bed magnetization. A triple point appears where the three states of bed coexist corresponding to the intersection of the transition lines. Data confirming the phenomenon have been published elsewhere [2]-[4].

3. THE SEMI-INFINITE BED

Since real beds are not infinite in extent it is of interest to analyze the stability of a bed which is bounded at the bottom by the flow distribution grid as sketched in Figure 5. It is convenient to employ the same equations and constitutive relationships as in the infinite bed analysis.

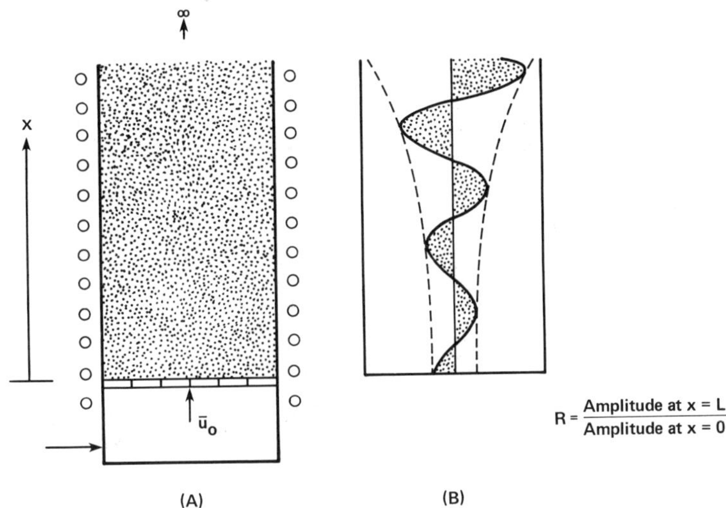

Fig. 5 (A) Semi-infinite Bed and (B) Wave Propagation.

In place of a plane wave disturbance, however, it is assumed that a voidage perturbation is introduced which is fixed in space at the grid plane and which fluctuates periodically with time.

Thus we solve equation (2.10) for $x > 0$ subject to the initial conditions

$$\varepsilon_1(x,0) = 0 \quad , \quad \frac{\partial \varepsilon_1}{\partial t}(x,0) = 0 \quad \text{for all } x > 0 \tag{3.1}$$

and the boundary condition

$$\varepsilon_1(0,t) = \text{Re}[\hat{\varepsilon}_1 e^{i\omega t}] \qquad \text{for all } t > 0 \tag{3.2}$$

The solution of (2.10) with these initial-boundary conditions is

$$\varepsilon_1(x,t) = \text{Re}[\hat{\varepsilon}_1 e^{i\omega t}\exp(x\frac{N\varepsilon_0}{2d}\frac{g}{\bar{u}_0^2}\{1 - [1 - \frac{4}{N}(n^2 - in)]^{1/2}\})] \tag{3.3}$$

plus terms which decay exponentially with time. In equation

(3.3), N is defined to be $N_m N_v$, $n = \omega \bar{u}_0/g$ is the dimensionless frequency, and d is given by (2.14).

Equation (3.3) shows how the amplitude of the long-term sinosoidal response depends on the distance x from the grid, Letting R be the ratio of this amplitude at x = L to the input amplitude at x = 0, we find that

$$\ln R = \left(\frac{N\varepsilon_0}{2d}\right)\left(\frac{gL}{\bar{u}_0^2}\right)\left\{1 - \frac{\sqrt{2}}{2}\left[\left(1 - \frac{4n^2}{N}\right) + \left(\left\{1 - \frac{4n^2}{N}\right\}^2 + 16\,\frac{n^2}{N^2}\right)^{1/2}\right]^{1/2}\right\} \tag{3.4}$$

The behavior of (3.4) reveals that N = 1 again produces neutral stability, i.e., R = 1 regardless of input frequency just as in the infinite bed. N < 1 yields decay while N > 1 produces growth of amplitude with downstream distance. The input modes having the largest rates of growth correspond to infinite frequency, hence these modes control the stability of the process. Allowing n to approach infinity in (3.4) gives the following relationship for R.

$$\ln R = \frac{\sqrt{N}(\sqrt{N}-1)}{\left(\frac{2d}{\varepsilon_0}\right)\left(\frac{\bar{u}_0^2}{gL}\right)} \tag{3.5}$$

In actual beds dissipational effects or finite particle size will limit the highest frequencies which can be found in the medium.

This treatment illustrates that any sufficiently small disturbance introduced at the grid will be swept out of the system without causing a noticeable effect. It seems realistic, however, to assign a large enough value of R that certainly would lead to a bed upset if the inlet disturbance is appreciable. Thus, Figure 6 is a graph of (3.5) with R as parameter and N regarded as normalized velocity. It may be seen that deep beds tend to become marginally unstable for N = 1 independent of the value of the amplitude ratio R that is assigned and that short beds should exhibit greater stability than long beds. In fact, experimental data exhibit just this trend[2].

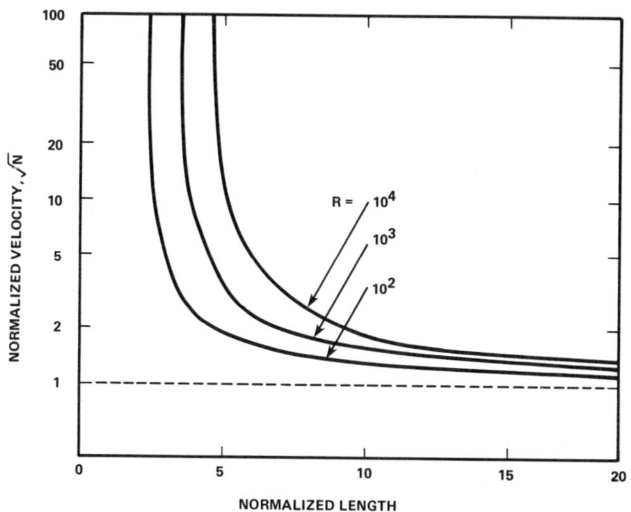

Fig. 6

Theoretical Dependence of Transition Velocity on Bed Length.

4. PARADOXICAL RESULTS OF FINITE LENGTH BEDS

If the bed mechanics are truly understood, then it should be possible to solve the stability problem for a bed of finite length, hence a bed possessing a free surface. The question then immediately arises as to the appropriate boundary condition to be satisfied at the free surface. This point will be returned to later while first a simpler expedient is explored of specifying zero voidage perturbation at the ends[10]. This approach is subject to the criticism that these conditions would be difficult to impose experimentally. The governing equation again is (2.10) and a solution is sought which satisfies the boundary conditions

$$\varepsilon_1(0,t) = \varepsilon_1(\ell,t) = 0 \qquad (4.1)$$

This poses an eigenvalue problem which has an infinite number of solutions, each one referred to as a mode. A superposition of the modes then provides a solution to the general initial value problem $\varepsilon_1(x,0) = h(x)$, $\partial\varepsilon_1(x,0)/\partial t = g(x)$. Equations (2.10) and (4.1) are satisfied by modal functions of the form

$$\varepsilon_1 = e^{\mu x} e^{\sigma t} \sin mx \tag{4.2}$$

where μ is real, σ is complex and $m = mq/\ell$ where q is any integer. Substituting (4.2) into (2.10) and equating coefficients of sin mx and cos mx to zero leads to solutions for μ and σ

$$\mu = \frac{b}{2c} \tag{4.3}$$

and

$$\sigma = \frac{a}{2}[-1 \pm (1 - \frac{b^2}{a^2 c} - \frac{4cm^2}{a^2})^{1/2}] \tag{4.4}$$

Equation (4.4) shows that σ can never have a positive real part and hence all modes decay for any superficial gas velocity or magnetization. This formally stable solution indicates that the use of the pinned boundary conditions is unrealistic. The corresponding phase diagram is shown in Figure 7A.

Notwithstanding the foregoing conclusion, it seems worth noting the role viscosity can play in this problem. Assuming fluid phase dissipation is negligible compared to that which may be attributed to the solids phase, we formally represent the solids phase stress in the Newtonian manner

$$\underline{\underline{T}}_p = - p_p\underline{\underline{I}} + \lambda_p \nabla\cdot\underline{u}_p\underline{\underline{I}} + \eta_p[\nabla\underline{u}_p + \nabla\underline{u}_p^T - \frac{2}{3}\,\underline{\underline{I}}\;\nabla\cdot\underline{u}_p] \tag{4.5}$$

where η_p is the ordinary coefficient of viscosity and λ_p is the bulk coefficient of viscosity. Again, it is assumed $p_p = 0$. The constitutive relationship (4.5) is introduced into (1.4) and linearization carried through as done previously. In place of (2.10) the governing equation for voidage perturbation becomes

$$\frac{\partial^2 \varepsilon_1}{\partial t^2} + a\frac{\partial \varepsilon_1}{\partial t} + b\frac{\partial \varepsilon_1}{\partial x} - c\frac{\partial^2 \varepsilon_1}{\partial x^2} - e\frac{\partial^3 \varepsilon_1}{\partial x^2 \partial t} = 0 \qquad (4.6)$$

where a, b and c are as defined previously and the definition of e is

$$e = \frac{(\lambda_p + \frac{4}{3}\eta_p)}{\varepsilon_p \rho_p} \qquad (4.7)$$

With the boundary conditions of (4.1) the perturbation form (4.2) again satisfies these relationships and now leads to a cubic algebraic equation for σ. The system is formally stable if none of the roots of the cubic have positive real parts. It is not necessary to solve the cubic explicitly to determine stability as one may invoke the Routh-Hurwitz criterion. Qualitative predictions of this treatment are displayed in Figure 7B. Surprisingly, when viscosity is included as an influence the system exhibits ranges of instability even though the inviscid system is formally stable. Thus, although viscosity is ordinarily perceived as a dampening mechanism, in the case of an MSB it may also play a destabilizing role.

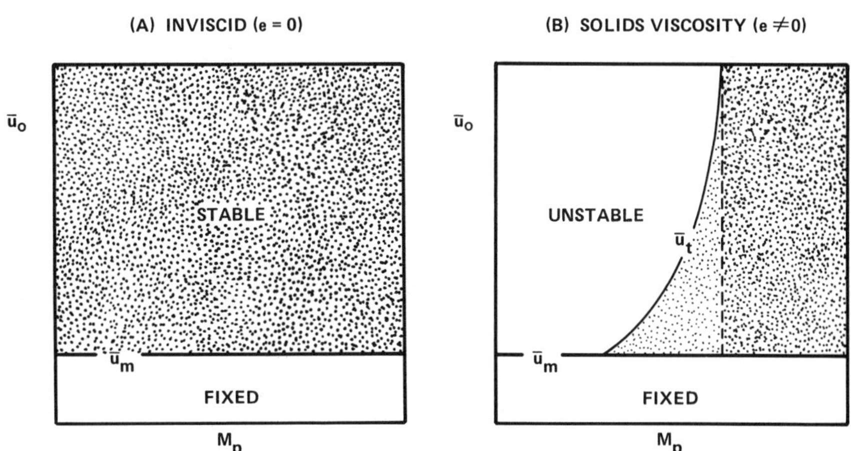

Fig. 7

Phase Diagrams of the MSBs with Pinned Ends. The Severe Constraints Formally Yield Odd Stability Behavior.

Similar instabilities that arise when loss is added to a system occur when an elastic string is convected through a viscous medium at a speed faster than the speed of small signal elastic waves or when in microwave devices an electron beam couples to a resistive wall [11].

Also, the situation is similar to flow in a conduit; the inviscid flow is stable while flow of viscous fluid trips into turbulence beyond a critical flow rate.

5. THE STRESS BOUNDARY CONDITION

The inviscid fluid momentum equation (1.3) may be multiplied and integrated term by term with a volume element of pillbox shape straddling the interface. Only the fluid stress term survives and subject to the constitutive relationship (2.2) it is found that p_f is continuous across the interface

$$[p_f] = 0 \tag{5.1}$$

The solids momentum equation (1.4) may be treated similarly subject to the constitutive relationships (2.1) and (2.2), the condition (5.1) and with $T_p = - p_p \underline{\underline{I}}$. The transformation of the magnetic stress term produces a magnetic surface force density $\underline{F_s}$ that bears special consideration.

$$\underline{F_s} = \int_{[\delta x \to 0]} \nabla \cdot \underline{\underline{T}}_m \, \delta x = \int \underline{n} \cdot \underline{\underline{T}}_m dS = \underline{n}^{(2)} \cdot \underline{\underline{T}}_m^{(2)} + \underline{n}^{(1)} \cdot \underline{\underline{T}}_m^{(1)} \tag{5.2}$$

where superscript 2 denotes the nonmagnetic phase and 1 the magnetic medium. Employing $\underline{\underline{T}}_m$ from (2.1) and redefining $\underline{n} = \underline{n}^{(2)} = - \underline{n}^{(1)}$

$$\underline{n} \cdot \underline{\underline{T}}_m = - \frac{\mu_0}{2} H^2 \underline{n} + B_n \underline{H} \tag{5.3}$$

From (1.6) and (1.7) it follows that the normal induction field and the tangential magnetic field are continuous across the interface.

$$[B_n] = 0 \qquad \text{or} \qquad [H_n] + [M_n] = 0 \tag{5.4}$$

$$[H_t] = 0 \tag{5.5}$$

After some algebraic manipulations, the magnetic surface force density is found from the above as,

$$\underline{F}_s = \underline{n}\,\frac{\mu_0}{2}\,M_n^2 \tag{5.6}$$

The surface force density is directed from the magnetized to the unmagnetized medium. Accordingly, the stress boundary condition of the solids phase, assuming $p_p^{(2)} = 0$, is

$$p_p + \frac{\mu_0}{2}\,M_n^2 = 0 \tag{5.7}$$

where p_p and M_n are both evaluated in the magnetic bed. Since $M_n^2 > 0$, and $p_p = 0$ in a bed that is free of mechanical stress, (5.7) informs us that $M_n = 0$ in contradiction to the assumption of a state of uniform magnetization. This has led us to investigate a radical departure in the analyses of these beds based on the circumstance that bulk magnetization results as the product of particle fraction with particle magnetization, see (1.9). Thus, $M = \varepsilon_p M_p$ and so the condition that $M_n = 0$ at the surface can be fulfilled if

$$\varepsilon_p = 0 \qquad \text{at} \quad x = \ell \tag{5.8}$$

It is noted that the expression for magnetic surface force density of (5.6) figures prominently in the analysis of the magnetic liquid (ferrofluid) normal field instability [7,12] and as the underlying mechanism for the stretching of magnetic fluid droplets in uniform applied magnetic field [13].

6. NON-UNIFORM VOIDAGE DISTRIBUTION IN FINITE LENGTH BEDS

With the boundary condition of (5.8) it is possible to treat the problem of the equilibrium non-uniform voidage distribution in an MSB of finite length in a self-consistent manner. For steady state conditions, $\frac{\partial}{\partial t} = 0$ and $\underline{u}_p = 0$, so the governing equations (1.1), (1.3), and (1.4) for one-dimensional flow with $\rho_f = 0$ become

$$\frac{d}{dx}(\varepsilon_f u_f) = 0 \tag{6.1}$$

$$\frac{dp_f}{dx} = -\beta u_f = -\frac{\beta U}{\varepsilon_f} \tag{6.2}$$

$$\beta u_f - \varepsilon_p \rho_p g + F_{mx} = 0 \tag{6.3}$$

where $U = \varepsilon_f u_f$ is the constant superficial gas velocity and F_{mx} is the vertical component of F_m given in (2.7). From (1.8) and (1.9) with $\underline{B}$ spatially uniform for a magnetically saturated bed,

$$\frac{dH}{dx} = M_p \frac{d\varepsilon_f}{dx} \tag{6.4}$$

Thus the magnetic force density is

$$F_{mx} = \varepsilon_p \mu_0 M_p^2 \frac{d\varepsilon_f}{dx} \tag{6.5}$$

As a model for drag we assume the Carman-Kozeny drag law although studies of particle configurations in the magnetized beds indicate that an anisotropy develops that may range from very small to moderately large [14].

$$\beta = \frac{150\eta_f}{D_p^2} \cdot \frac{\varepsilon_p^2}{\varepsilon_f^2} \tag{6.6}$$

In the above, $\varepsilon_p + \varepsilon_f = 1$ since the flow is two phase. The boundary conditions from (5.1) and (5.8) are

$$[p_f] = 0 \text{ and } \varepsilon_f = 1 \text{ at } x = \ell \tag{6.7a,b}$$

Eliminating u_f, F_{mx} and β from (6.3) using (6.1), (6.5), and (6.6) gives a separable differential equation which is to be integrated with the indicated limits.

$$\int_{\varepsilon_f}^{1} \frac{\varepsilon_f^3 \, d\varepsilon_f}{\varepsilon_f^3 - \alpha(1-\varepsilon_f)} = \frac{1}{\gamma} \int_{x/\ell}^{1} d\left(\frac{x}{\ell}\right) \tag{6.8}$$

The parameters α and γ are defined as

$$\alpha = \frac{150 \eta_f U}{\rho_p g D_p^2} \quad \text{and} \quad \gamma = \frac{\mu_0 M_p^2}{\rho_p g \ell} \tag{6.9a,b}$$

Figure 8 displays a set of voidage profiles computed from (6.8) with $\alpha = 0.1$. When $\gamma << 1$ it may be seen that void fraction is nearly constant over most of the bed length with a rapid change taking place in a surface adjacent boundary layer. In fact, a thickness δ may be assigned to this boundary layer based on the slope of the voidage profile at the surface and $\varepsilon_{p,c}$ the constant particle fraction deep beneath the surface.

$$\delta = \varepsilon_{p,c}\left(-\frac{dx}{d\varepsilon_f}\right)_{x=\ell} \tag{6.10}$$

Differentiation of (6.8) with $\varepsilon_f = 1$ at $x = \ell$ gives

$$\delta = \frac{\mu_0 M_p^2}{\rho_p g} \varepsilon_{p,c} \tag{6.11}$$

The boundary layer thickness is predicted to be independent of bed length or gas velocity.

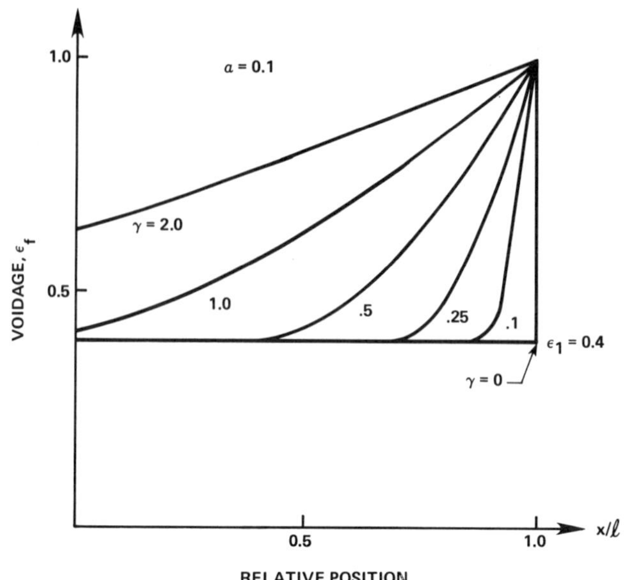

Fig. 8 Theoretical Voidage Profiles in the Inhomogeneous Bed.

Pressure profiles may be computed by integrating (6.2) with the known voidage profile. Figure 9 illustrates the behavior for sets of the parameters α and γ. The linear profile is approximated only for small values of γ. Otherwise the pressure profile is bowed, and approaches the top surface with zero rate of change. The latter is confirmed from inspection of the pressure gradient relationship that results from combination of (6.2), (6.3) and (6.5).

$$\frac{dp_f}{dx} = - \epsilon_p \rho_p g + \epsilon_p \mu_0 M_p^2 \frac{d\epsilon_f}{dx} \tag{6.12}$$

Thus, at the free surface $\epsilon_p = 0$ and (dp_f/dx) vanishes.

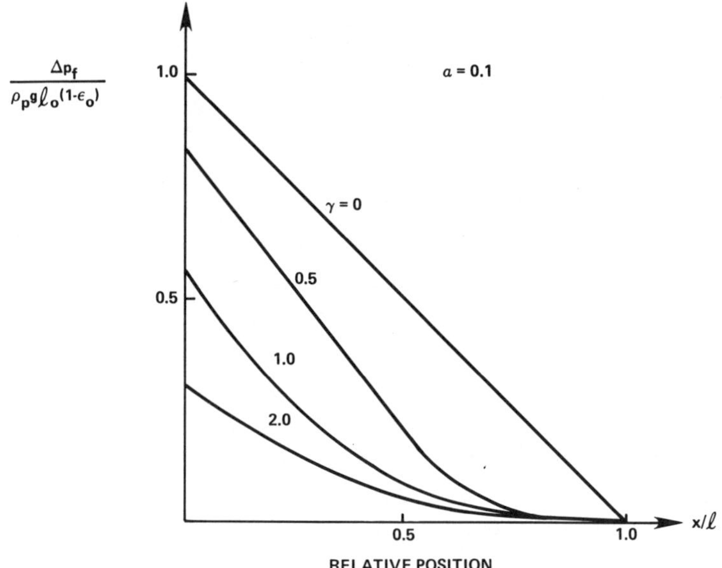

Fig. 9

Theoretical Pressure Profiles in the Inhomogeneous Bed.

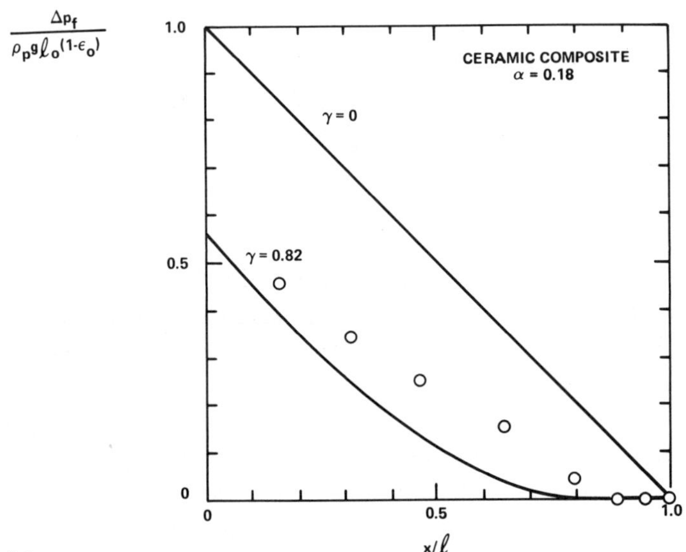

Fig. 10

Experimental Pressure Profile Exhibits the Theoretical Trend.

When pressure profiles are measured experimentally it is found that a hysteresis exists. A freshly dumped bed to which magnetic field is applied, then gas flow admitted, yields the behavior of Figure 2 with a pressure profile in the bed that varies nearly linearly with position in the bed. Maintaining the field after reaching a maximum velocity $\bar{u}_{max}$ the overall pressure drop across the bed does not retrace the horizontal portion of the curve in Fig. 2 but instead decreases nearly linearly as gas flow rate decreases to an operating velocity $\bar{u}_{op}$. The corresponding profile of pressure versus vertical position in the bed is illustrated for a particular case by the test data for which $\bar{u}_m$ = 0.23 m/s, $\bar{u}_{max}$ = 1.28 m/s, and $\bar{u}_{op}$ = 0.495 m/s. Pressure was measured at a series of sidetaps on the vessel wall except for the four data points taken nearest the top of the bed which were detected with a capillary tube inserted into the bed and connected to a manometer.

The bed solid is a composite of ceramic ferrite on an alumina support. The solids magnetization determined with a vibrating sample magnetometer was $\mu_0 M_p$ of 0.0555 T (555 gauss) with applied field B_0 of 0.1 T (1000 gauss). Solids density ρ_p was 1880 kg/m^3. A theoretical curve obtained by numerical integration of (6.12) cast into dimensionless form is plotted in Fig. 10 and labelled γ = 0.82 corresponding to bed length of 0.19 m; the bed had slumped about 0.01 m when turning down the flow, i.e. bed length experiences hysteresis also. It may be seen from Figure 10 that the general features predicted by the theory are supported by the data including the pressure gradient approach to zero at the upper surface. It should be remembered, however, that the data are dependent on the peak value of gas velocity employed in the hysteresis path.

The numerically determined voidage distribution corresponding to the curve for γ = 0.82 of Fig. 10 is shown in Fig. 11. This MSB was operated at high applied field intensity and with high expansion to emphasize the nonuniform aspects of the pressure and voidage distributions; usually, it is expected that the voidage profile will exhibit a much thinner boundary layer.

Theoretically, the voidage is expected to approach unity at the bed top surface whether the gas flow rate is decreased during operation or not. The actual bed appears to satisfy this requirement in the case of no decrease by structuring into fine scale peaks and valleys distributed across the bed top surface; the voidage averaged over a horizontal plane of the flow then varies very rapidly from the bed bulk value to unity in the vicinity of the surface. However, if the particle magnetization is too great, coarser peaks may form and the flow can no longer be regarded as one dimensional.

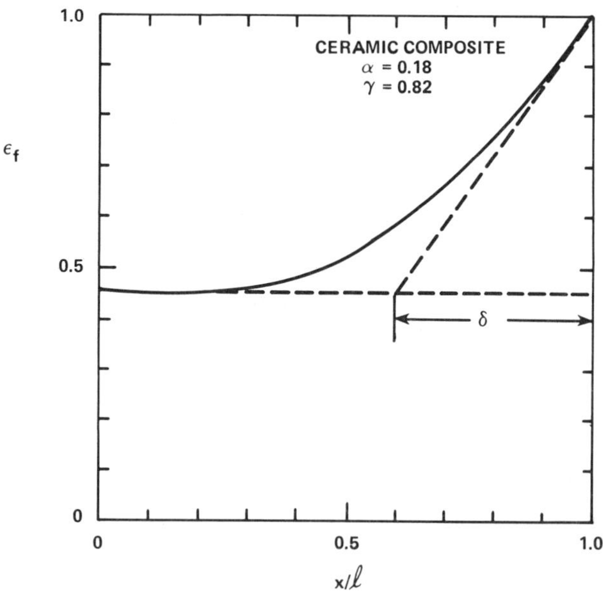

Fig. 11
Voidage Profile Corresponding to the Theoretical Pressure Profile of Figure 10. Thickness δ is Given by Equation 6.11.

7. YIELD STRESS DISTRIBUTION: ANOTHER PARADOX RESOLVED

The magnetic particles of an MSB are in contact with neighboring particles as a consequence of Earnshaw's theorem [15]. The network of contacting particles forms a magnetic gel having a measurable yield stress. Fortunately, a range exists over which magnetization is large enough to produce considerable stabilization but small enough not to hinder the bed flowability to a great degree. Yield stress has been measured in the MSB by noting the force required to withdraw a vertical flat plate having surfaces roughened with glued-on bed particles. Figure 12 illustrates the general trend observed in such measurements. Not unexpectedly, the yield stress increases with applied field intensity as the bed particles become more highly magnetized and attracted to each other. It has been puzzling, however, that the yield stresses approach zero at the bed top surface since heretofore the beds were considered as uniform in voidage.

The deep bed values of yield stress are constant and these asymptotic values have been empirically correlated in a dimensionally sound manner having rather general applicability e.g., to different particle sizes and gas flow rates according to the following relationship.

$$\frac{\tau_y}{\mu_0 M_p^2} = \frac{A}{\varepsilon_f^n} \tag{7.1}$$

where the exponent was found to have the numerical value of n = 12 and A is a constant whose value depends on the nature of the particle (solid ferromagnetic substance or composite) and frequency of the applied field (direct current or alternating current excitation). The value of A is ~1/3 x 10^{-4} for iron spheres.

The voidage distribution given in Figure 11 has been used to compute relative values of yield stress using (7.1) with the resulting distribution shown in Figure 13. The curve in Figure 13 in common with the data of Figure 12 becomes indistinguishable from zero at the bed top surface, i.e., at small relative depth, and approaches a constant value deep into the bed.. The depth interval over which stress varies from its least to its maximum value is greater than the thickness δ because of the strong voidage dependence in (7.1). The data in Figure 12 were obtained in beds not subjected to the hysteresis path. This may account for the flat toe of the curve in Figure 13 whereas none is seen in Figure 12. A further uncertainty is the influence of the flat plate on bed structure during the yield stress measurement procedure.

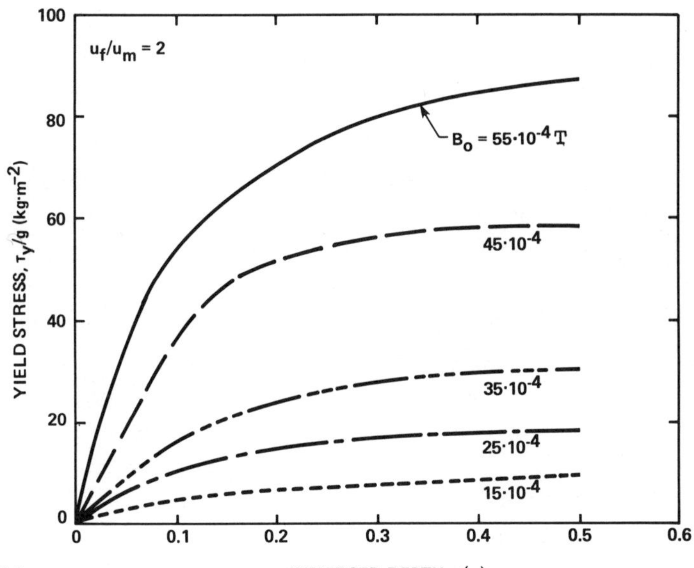

Fig. 12

Experimental Values of Yield Stress vs. Depth in the Stabilized Beds Exhibit Puzzling Variation with Depth.

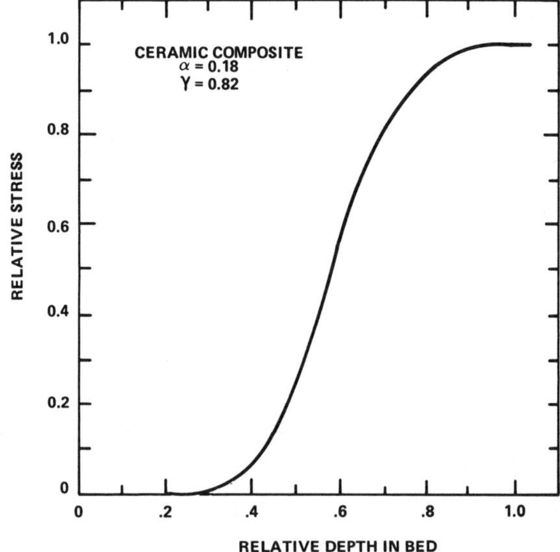

Fig. 13

Yield Stress Distribution Computed From Voidage Profile.

8. CONCLUSION

Mathematical analysis originally predicted the phenomenon of magnetic stabilization of fluidized solids, and now is contributing to the further understanding of bed properties. While the original conception of the beds pictured homogeneity of particle concentration throughout, this picture is beginning to change and, as a result, some anomalous features of the beds can be understood, especially the nature of the surface boundary condition and the variation of yield stress with depth. Perhaps anomalies of transition velocity dependence on particle size and gas viscosity[1] can be resolved in analysis which accounts for inhomogeneity in the beds. It seems also that the stabilized bed offers an opportunity for the incisive development and testing of multiphase hydrodynamic theories in general since the beds are free of turbulence over ranges of flow rates and easily accessible to a variety of measurements against which the theories may be tested.

REFERENCES

1. Rosensweig, R.E., Magnetic Stabilization of the State of Uniform Fluidization, I & EC Fundamentals 18 (1979), 260-269.
2. Rosensweig, R. E., Fluidization: Hydrodynamic Stabilization with a Magnetic Field, Science 204 (1979), 57-60.
3. Lucchesi, P. J., Hatch, W. H., Mayer, F. X., and Rosensweig, R. E., Magnetically Satbilized Beds -- New Gas Solids Contacting Technology, Proc. 10th World Petroleum Congress 4 (1979), SP-4, Heyden & Sons, Phil., Pa.

4. Rosensweig, R. E., Siegell, J. H., Lee, W. K. and Mikus, T., Magnetically Stabilized Fluidized Solids, A.I.Ch.E. Symp. Ser. 77, No. 205 (1981), 8-16.

5. Anderson, T. B. and Jackson, R., A Fluid Mechanical Description of Fluidized Beds - Equation of Motion, I & EC Fundamentals 6 (1967), 527-539.

6. Drew, D. A., Continuum Modeling of Two-Phase Flows, This Proceedings.

7. Cowley, M.D. and Rosensweig, R.E., The Intefacial Stability of a Ferromagnetic Fluid, J. Fluid Mech. 30, Pt. 4 (1967),671-688.

8. Penfield, P. and Haus, H. A., Electrodynamics of Moving Media, M.I.T. Press, Cambridge, Mass., (1967).

9. Anderson, T. B. and Jackson, R., Fluid Mechanical Description of Fluidized Beds - Stability of the State of Uniform Fluidization, I & EC Fundamentals 7 (1968), 12-21.

10. Heard, W. B., Personal Communication (1978).

11. Woodson, H. H. and Melcher, J. R., Electromechanical Dynamics, Pt. 2, John Wiley and Sons, New York, 1968, 608-613.

12. Rosensweig, R. E., Fluid Dynamics and Science of Magnetic Liquids, in Advances in Electronics and Electron Physics (L. Marton, ed.), Academic Press, New York, 1979, 103-199.

13. Arkhipenko, V. I., Barkov, Yu. D., and Bashtovoi, V. G., Shape of a Drop of Magnetized Fluid in a Homogeneous Magnetic Field, Magnetohydrodynamics 14 (1978), 373-375.

14. Rosensweig, R. E., Jerauld, G. R., and Zahn, M., Structure of Magnetically Stabilized Fluidized Solids, in Continuum Models of Discrete Systems 4 (O. Brulin and R. K. T. Hsieh, eds.), North-Holland Publ. Co., Amsterdam, 1981, 137-144.

15. Weinstock, R., On a Fallacious Proof of Earnshaw's Theorem, Am. J. Phys. 44, No. 4 (1976), 392.

This work was done at the Exxon Corporate Research Science Laboratories.

R. E. Rosenweig
W. K. Lee
P. S. Hagan
Exxon Research and Engineering Company
Linden, New Jersey 07036

M. Zahn
Electrical Engineering Dept.
Massachusetts Institute of Technology
Cambridge, Massachusetts

Index